# Environment and Society

## Critical Introductions to Geography

Critical Introductions to Geography is a series of textbooks for undergraduate courses covering the key geographical subdisciplines and providing broad and introductory treatments with a critical edge. They are designed for the North American and international market and take a lively and engaging approach with a distinct geographical voice that distinguishes them from more traditional and out-dated texts.

Prospective authors interested in the series should contact the series editor:
John Paul Jones III
Department of Geography and Regional Development
University of Arizona
jpjones@email.arizona.edu

**Published**
Cultural Geography
*Don Mitchell*

Political Ecology
*Paul Robbins*

Geographies of Globalization
*Andrew Herod*

Geographies of Media and Communication
*Paul C. Adams*

Social Geography
*Vincent J. Del Casino Jr*

Mapping
*Jeremy W. Crampton*

Environment and Society
*Paul Robbins, John Hintz, and Sarah A. Moore*

Research Methods in Geography
*Basil Gomez and John Paul Jones III*

**Forthcoming**
Geographic Thought
*Tim Cresswell*

Cultural Landscape
*Donald Mitchell and Carolyn Breitbach*

# Environment and Society

## A Critical Introduction

Paul Robbins, John Hintz, and Sarah A. Moore

**WILEY-BLACKWELL**

A John Wiley & Sons, Ltd., Publication

Blackwell Publishing was acquired by John Wiley & Sons in February 2007. Blackwell's publishing program has been merged with Wiley's global Scientific, Technical, and Medical business to form Wiley-Blackwell.

*Registered Office*
John Wiley & Sons Ltd, The Atrium, Southern Gate, Chichester, West Sussex, PO19 8SQ, United Kingdom

*Editorial Offices*
350 Main Street, Malden, MA 02148-5020, USA
9600 Garsington Road, Oxford, OX4 2DQ, UK
The Atrium, Southern Gate, Chichester, West Sussex, PO19 8SQ, UK

For details of our global editorial offices, for customer services, and for information about how to apply for permission to reuse the copyright material in this book please see our website at www.wiley.com/wiley-blackwell.

Library of Congress Cataloging-in-Publication Data

Robbins, Paul, 1967–
  Environment and society : a critical introduction / Paul Robbins, John Hintz, and Sarah A. Moore.
       p. cm.
     Includes bibliographical references and index.
     ISBN 978-1-4051-8761-9 (hardcover : alk. paper) – ISBN 978-1-4051-8760-2 (pbk. : alk. paper)
     1. Environmental sciences–Social aspects.   2. Environmental protection–Social aspects.
  3. Human ecology–Social aspects.   I. Hintz, John.   II. Moore, Sarah A.   III. Title.
     GE105.R63 2010
     333.72–dc22
                                          2009039200

A catalogue record for this book is available from the British Library.

Set in 10.5 on 13 pt Minion by Toppan Best-set Premedia Limited
Printed in Singapore by Ho Printing Singapore Pte Ltd

6  2012

*For Elizabeth A. Moore*
*and*
*For Michelle*

# Contents

# Figures

# Tables

# Text Boxes

# Acknowledgments

The book would have been impossible without the impeccably polite prodding of Justin Vaughan at Wiley-Blackwell, an editor whose creative interventions extend beyond editing and were key sparks in imagining the book and setting us writing. He also sprang for dinner that time in Boston.

Paul Robbins and Sarah Moore would like to thank the School of Geography and Development at the University of Arizona for the stimulating environment in which to think and write, and especially John Paul Jones III, Sallie Marston, and Marv Waterstone. They would like to thank the students of their Environment and Society classes for slogging through early performances of some of the material presented here. They owe a debt of gratitude to their current and former graduate students who embody and convey much of the plural thinking on display here. They would also like to thank Marty Robbins, Vicki Robbins, and Mari Jo Joiner. Special thanks to Abby the Great Dane, who was a profound society–environment problem in her own right.

John Hintz would like to thank Bloomsburg University for giving him a course release that provided needed time to work on the book. He would also like to thank the students of his Environmental Issues and Choices classes who are always eager to think about nature and society in new ways. He thanks Dong Constance for generously sharing his files on the tuna–dolphin controversy. Finally, he is deeply indebted to Michelle Hintz for her unwavering patience and support throughout the writing of the book.

# 1

# Introduction
## The View from Clifton Bridge

The Clifton Suspension Bridge, spanning Avon Gorge near Bristol, England. *Credit*: willmetts/Shutterstock.

## Keywords

- Political ecology
- Reconciliation ecology

## Chapter Menu

News headlines from forests, fields, rivers, and oceans suggest we are in a world of trouble. Fresh water is increasingly scarce around the globe, owing not only to heavy water use but also widespread pollution; there is not a single drop of water in the Colorado River in the United States or the Rhone River in France that is not managed through complex dams and distribution systems, or affected by city and industrial waste along their paths to the sea. Agricultural soils are depleted from years of intensive cropping and from the ongoing application of fertilizers and pesticides in the search for ever-sustained increases of food and fiber; in North India, after decades of increasing production, yields of wheat and rice have hit a plateau. Global temperatures are on the rise and, with this increase, whole ecosystems are at risk. Species of plants and animals are vanishing from the earth, never to return. Perhaps most profoundly, the world's oceans – upon which these global systems rest – show signs of impending collapse. The accumulation of these acute problems has led observers to conclude that the environment may be irreversibly lost or that we may have reached "the end of nature" (McKibben 1990).

And yet, looking down from a two-century-old suspension bridge spanning the Avon Gorge in Bristol, England, along the winding waterway towards the lower reaches of the River Avon and the ocean beyond, it is hard not to be swept away in a fantastic environmental romance. Peregrine falcons nest in the walls and can be seen on rare occasions, seeking pigeons and seagulls for prey, flying along the updrafts that rise from the gorge cut by the ancient river, as it makes its way to the sea. Orchids thrive here, as do horseshoe bats. The river is all the more remarkable for its incredible flood tide range, with water rising and falling as much as 15 meters (49 feet) between low and high tide. In the entire world, this dramatic distance is second only to that of the Yangtze River in China. The power of the sea, the river, and the moon is almost palpable as the water passes beneath the bridge out to Avonmouth, adjacent to the vast Severn Estuary, and on to the Atlantic.

But all is not well here. The blessing of a massive tidal flow has made the Severn Estuary a very desirable location for tidal power generation, which captures the energy of the rising and falling water to generate electricity for homes and businesses. Harnessing the tidal power of the area might provide as much as 5 percent of the United Kingdom's power. Because this is a form of power that generates no greenhouse gas emissions, which are the main culprit behind global climate change, this is an environmentally attractive project. The plan, however, does not come without environmental costs and the proposed projects include the construction of vast lagoons, huge underwater fences, and an enormous tidal reef. This could destroy marshland, drive away migrating birds, and harm the fish population in ways that are hard to estimate.

The view from Clifton Bridge makes our global situation easier to understand, though perhaps no simpler to solve. The contradictory proposition – harnessing green power in a way that will paradoxically transform the environment – is a metaphor for the condition of our long-standing relationship to the non-human world. From this view, looking down across the gorge, one can catch a glimpse of heavily quarried sites where the land was mined for lead and iron for more than a thousand years. Yet these quarry sites also provide habitat for wild animals and birds. Everywhere the works of people are in evidence; bridges, tunnels, docks, all formed from brick and iron, span the chasm and snake along the valley floor. Yet these features frequently vanish into the riotous growth of trees and walls of stone. Decisions made at the estuary downstream, therefore, cannot be made solely on the basis that the region is a "natural" one, nor a "social" one. The area is simultaneously neither and both, with environmental features (deep sea inlets, open gorges, and veins of minerals) inspiring invasive human actions (docks, bridges, and mines), which sometimes produce altogether new habitats and environments.

The landscape here has been worked and reworked for centuries by human hands, creating a thoroughly social environment, yet falcons nest here, as do bats and a range of other species, making this an undeniably natural space as well.

If these sorts of decisions are to be made, and the larger complex puzzle of living *within* nature is to be solved, therefore, we need tools with which to view the world in fresh ways and assess possible routes forward. For example, viewed as a problem of ethics, the development of the estuary becomes one of sorting through competing claims and arguments about what is ethically best, on whose behalf one might make such an argument (that of power customers or that of the river itself?), and over which criteria we might use to adjudicate "good" policy. From the point of view of political economy, by contrast, one would be urged to examine what value is created and destroyed in the transformation of the marshlands, which specific technologies are selected for power generation and why, whose pockets are filled in the process, and how decisions are controlled and directed through circuits of corporate power and national authority. Indeed, there is no shortage of ways to view this problem, with population-centered considerations competing with those that stress market logics, and arguments about public risk perception competing with those about the romantic social construction of the river.

## What Is This Book?

This book is designed to explain these varied interpretive tools and perspectives and show them in operation. Our strategy is first to present the dominant modes of thinking about environment–society relations and then to apply them to a few familiar objects of the world around us. By environment, we mean the whole of the aquatic, terrestrial, and atmospheric non-human world, including specific objects in their varying forms, like trees, carbon dioxide, or water, as well as the organic and inorganic systems and processes that link and transform them, like photosynthesis, predator–prey relationships, or soil erosion. Society, conversely, includes the humans of the earth and the larger systems of culture, politics, and economic exchange that govern their interrelationships.

From the outset we must insist that these two categories are interlaced and impossible to separate. Humans are obviously environmental beings subject to organic processes. Equally problematically, environmental processes are also fundamentally social, in the sense that they link people and influence human relationships. Photosynthesis is the basis of agriculture, for example, and so is perhaps the most critical environmental process in the history of civilization. More complex: human transformation of carbon levels in the atmosphere may further alter global photosynthesis in a dramatic way, with implications for human food and social organization. Obviously, it is difficult to tell where the environment leaves off and society begins. On the other hand, there is not universal agreement on exactly of what these relationships and linkages consist. The perspectives summarized in this text present very different views about which parts of society and environment are connected to which, under what conditions these change or can be altered, and what the best courses of action tend to be, with enormous implications for both thinking about our place in the ecosystem and solving very immediate problems like global warming, deforestation, or the decline in the world's fisheries.

In Part I, we lay out some of the dominant ways of interpreting the environment/society relationship. We begin in Chapter 2 with a perspective that is foundational to the history of both the natural and social sciences: population. Here we describe how human population has been viewed as a growing threat to the non-human world, contrasting this with views of population growth as a process that not only consumes, but also potentially *produces*, resources in the world. In Chapter 3 we consider economic ways of thinking about the environment. These views stress the power of markets – a category in which we include systems of economic exchange – to respond to scarcity and drive inventive human responses. This is followed in Chapter 4 by approaches that stress institutions, which we define as the rules and norms governing our interactions with nature and resources. Institutional approaches address environmental problems largely as the product of "common property" problems that are amenable to creative rule-making, incentives, and self-regulation. Chapter 5 examines ethics-based approaches to the environment, with their often radical ways of rethinking the place of humans in a world filled with other living and non-living things. The view of the environment as a problem of risk and hazard is explored in Chapter 6. That approach proposes a series of formal procedures for making the best choices possible, given that environments and environmental problems are inherently uncertain and highly variable. This is followed by a description of political economy approaches in Chapter 7, which are those that view the human relationship with nature as one rooted in the economy, but which insist that the economy is based in, and has fundamental implications for, *power* relationships: who gets what, who works for whom, and who pays. Contrary to market-based approaches, these point to the environmentally corrosive impacts of market economics. Chapter 8 closes this part of the book by describing approaches to environment and society issues that stress social construction, which we define as the tendency for people to understand and interpret environmental issues and processes through language, stories, and images that are often inherited or imposed through systems of media, government, education, or industry. These stories are not harmless, since they can encourage or overlook very real actions, impacts, and behaviors with serious environmental and social consequences.

Within these several ways of seeing are many others, of course. Within political economy, for example, issues of environmental justice are critical to understanding why some people are more heavily exposed to hazards than others. We have nested many of these perspectives within larger categories of thought, though without pretending we can do more than introduce many important concepts. Of particular significance are issues related to gender. These are so important, indeed, that we chose not to set them aside in a separate chapter, but to thread them throughout the book, amidst themes as varied as population and political economy.

Part II presents a set of six critical objects, and examines each of them in turn using a sample of these approaches. Each chapter begins with a "short history" of the object followed by a discussion of ways in which the characteristics of the object present a puzzle or conundrum, and then presents divergent ways of thinking about the object from competing points of view. In Chapter 9, we introduce carbon dioxide, a curious gas with a complicated history on the earth that shows it to vary widely over time, with enormous implications for the forms of life dwelling here. As one of the most important greenhouse gases, moreover, $CO_2$ has become an increasingly contested object, with competing views about its control, regulation, and circulation. In Chapter 10, we discuss trees. These plants have been companions of human civilization since the beginning, though the long relationship has been marked by dramatic ebbs and flows. In this chapter we take the opportunity to introduce varying theories to account for deforestation and reforestation, as well as a startling ethical proposal for trees to legally represent themselves. Chapter 11 is dedicated to wolves, a species with which humans have a current love–hate relationship and whose return throughout North America and parts of Europe and Asia represents a dramatic change in the way humans and animals relate. This chapter stresses diverse cultural understandings of the same animals, and the implications of our ethics and institutions for the many animals that share the landscape with humanity. The tuna takes center stage in Chapter 12, and with it the profound problems faced by the world's oceans. Here, human economics and ethics collide in a consideration of how fish production and consumption are regulated and managed in a complex world. Chapter 13 addresses one of the world's fastest-growing commodities, bottled water. This object has the rare dual role as a solution to problems of water supply in some parts of the world, while being a clear luxury item – with attendant environmental problems – in others. We close with French fries (also called "chips"!) in Chapter 14, a culinary invention that connects the complex centuries-old history of the transatlantic "Columbian Exchange" with the health controversies and industrial food economies of the twenty-first century.

Quite intentionally we have selected *objects* for exploration, rather than *problems*. We do this for two reasons. First, while many objects are obviously linked to problems (trees to deforestation, as we shall see in Chapter 10, for example), *not all human relations with non-humans are problems*. Second, we intend by this structure to invite people to think seriously about how different things in the world (giraffes, cell phones, tapeworms, diamonds, chainsaws …) *have their own unique relationship to people* and present specific sorts of puzzles owing to their specific characteristics (they swim, they melt, they migrate, they are poisonous when eaten …). This is intended as an opportunity to break away from the environment as an undifferentiated generic problem, one universally characterized by a state of immediate and unique crisis. While global climate change is a critical (and sprawling) suite of problems, for example, the long and complex relationship of people to carbon dioxide itself provides a focused entry point, filled with specific challenges and opportunities. We do indeed face enormous environmental problems, but we believe them to be best solved by exploring the specificities and differences, as well as commonalties, of both people and things.

Needless to say, these objects do not encompass the full enormous range of environmental actors with which society is intertwined and, though these chapters do capture a number of key issues and ideas, we do not pretend to have provided an exhaustive list of socio-environmental situations, interactions, and problems. Instead we provide a few key examples to show how objects are tools to think with, and to demonstrate the very different implications of divergent ways of seeing environmental issues.

It is also important to note that this is not an environmental science textbook, though it is a book that takes environmental science seriously. Several key concepts and processes from a range of environmental sciences are described and defined, especially in the latter half of the book, including carbon sequestration, ecological succession, and predator/prey relationships, among many others. These are described in terms detailed enough to explain and understand the way human and social processes impinge upon or relate to non-human ones. Throughout we have drawn on current knowledge from environmental science sources (the report on global climate change from the Intergovernmental Panel on Climate Change, for example), but we intend a book that requires no previous knowledge of such sciences or sources. We believe this book might reasonably accompany more strictly environmental science approaches, or be used in courses that seek to bridge environmental ethics, economics, or policy with issues in ecology, hydrology, and conservation biology, or vice versa.

## The authors' points of view

Finally, we provide many points of view in this volume that directly contradict one another. It is difficult, for example, to simultaneously believe that the source of all environmental problems is the total population of humans on the earth, and to hold the position that population growth leads to greater efficiencies and potentially lower environmental impacts. Even where ideas do not contradict one another (for example, risk perception in Chapter 6 might be seen as a sort of social construction in Chapter 8), they definitely stress different factors or problems and imply different solutions.

With that in mind, it is reasonable to ask what the points of view of the text's authors might be. Which side are we on? This is difficult to answer, not only because there are three of us, each with our own view of the world, but also because, as researchers, we often try to bring different perspectives and theories to bear on the objects of our study, and to foster a kind of pluralism in our thinking.

Nonetheless, we do collectively have a point of view. First, we are each urgently concerned about the state of natural environments around the world. Our own research has focused on diverse environmental topics, including Professor Hintz's work on the status of bears in the western part of the United States, Professor Moore's research on the management of solid and hazardous waste in Mexico, and Professor Robbins' investigation of the conservation of forests in India. From these experiences, we have come to share an approach best described as **political ecology**: an understanding that nature and society are produced *together* in a political economy that includes humans and non-humans. What does this mean? To keep it as straightforward as possible, we understand that relationships among people and between people and the environment are governed by persistent and dominant, albeit diverse and historically changing, *interactions of power* (Robbins 2004). This means that we have some special sympathy for themes from political economy and social construction.

**Political Ecology** An approach to environmental issues that unites issues of ecology with a broadly defined political economy perspective

When Hintz examines the conservation of bears in Yellowstone, for example, he thinks it is critical to examine how bears are *imagined* by people and to know what media, assumptions, and stories influence that imagination, since these prefigure how people do or do not act through policy, regulation, or support for environmental laws. When examining solid waste in Mexico, in another example, Moore thinks the crucial question is who *controls access to and use of* dumps, since this determines, to a large degree, how waste is managed, whether problems are addressed or ignored, and where the flow of hazards and benefits is directed. When examining forests in India, Robbins wants to know how local people and forest officers *coerce one another*, in a system of corruption that determines the rate and flow of forest cutting and environmental transformation. People's power over one another, over the environment, and over how other people think about the environment, in short, is our preferred starting point.

We also share an assumption that persistent systems of power, though they often lead to perverse outcomes, sometimes provide opportunities for progressive environmental action and avenues towards better human–environment relationships. We are stuck in a tangled web, in other words, but this allows us many strands to pull upon and many resources to weave new outcomes.

As a result, we also stress throughout the volume a preference for some form of **reconciliation ecology**. As described by ecologist Michael Rosenzweig (2003), this describes a science of imagining, creating, and sustaining habitats, productive environments, and biodiversity in places used, traveled, and inhabited by human beings.

> **Reconciliation Ecology** A science of imagining, creating, and sustaining habitats, productive environments, and biodiversity in places used, traveled, and inhabited by human beings

This point of view holds that while many of the persistent human actions of the past have stubbornly caused and perpetuated environmental problems, the solution to these problems can never be a world somehow bereft of human activity, work, inventiveness, and craft. Such a point of view does not deny the importance of making special places (conservation areas, for example) for wild animals, sensitive species, or rare ecosystems. But it does stress that the critical work of making a "greener" world will happen in cities, towns, laboratories, factories, and farms, amidst human activity, and not in an imaginary natural world, somewhere "out there."

For all the weight of our own views, however, we strongly believe in the analytical challenges presented by *all* of the approaches described here. It is our intention, therefore, to present the most convincing and compelling arguments of the many and diverse ways of viewing society and environment. We insist that, while it is impossible for us to present a fully unbiased view of the many ways of thinking about nature, it is possible to present fair characterizations of many points of view, characterizations without caricatures. Only the reader can judge our success in this regard.

## Suggested Reading

Robbins, P. (2004). *Political Ecology: A Critical Introduction.* New York: Blackwell.

Rosenzweig, M. L. (2003). *Win–Win Ecology: How the Earth's Species Can Survive in the Midst of Human Enterprise.* Oxford: Oxford University Press.

# Part I

# Approaches and Perspectives

What makes deforestation happen? Why do rivers become polluted? Who is the culprit behind global warming? Walk into any grocery store, barbershop, bar, or internet chat room, and you will hear no end to opinions. There are too many human beings! We live in a consumer society! People have no ethics! It's a government conspiracy! Everyone has some opinion of what causes environmental problems, and many people are not afraid to share theirs. Whether people acknowledge it or not, everybody has an opinion about the environment; everyone clings to a *theory* of what causes environmental problems.

But opinions are *different* from theories in a few critical ways. First, an opinion is rarely put under careful scrutiny to explore the *assumptions* that lie behind it. What ideas and evidence underlie people's ideas? Are they persuasive, plausible, bizarre? Second, opinion is not commonly subject to a serious examination of its implications. If we believe that all environmental problems are caused by overpopulation, for example, what does that mean for preventing deforestation, if anything? Is it of positive use? If we took an opinion seriously, what would the effect be? In a sense then, theories are opinions that have been worked through more rigorously, laid out in ways that their merits can be evaluated, and presented so that they might raise a call to action.

In the first part of this book, we explore the predominant approaches used to address environmental issues, including population, markets, institutions, risk, ethics, political economy, and social construction. Each of these approaches is often an amalgamation of theories, or a cluster of ideas about how things happen. Each of these is also far more than an opinion, insofar as it has been tempered in the fire of debate and usually expressed in reasonably clear terms. All of these theories also inevitably stand on stated assumptions, some of which are sturdy, but many of which (as we shall see) have far more wobbly foundations. Most importantly, all of these theories have implications, some of which are exciting, others of which are highly problematic, or even a little scary. Acceptance of one approach or another will inevitably have important repercussions for what people can and cannot do in the future.

Our selection of theories is by no means comprehensive. There are obviously lots of competing ways of seeing the world (or at least, more than seven!). But experience has shown us that, in one form or another, these are the ones that have returned time and again, and which guide debate in the halls of power, in classroom academic argument, in offices where environmental plans are made, and in any place where environmentally concerned people come together to generate ideas and call for change. These theories also inform assumptions underlying people's opinions around the world; indeed, you are likely to recognize at least some of these ideas in your own thinking.

By becoming conversant with these "big" ideas about what causes environmental change, and their pitfalls and (often dangerous) weaknesses, we hope the reader will come out the other side of the book a little different than when they went in. We do not necessarily anticipate that readers will change their opinions (although that might certainly happen), but we hope that they might have a somewhat greater appreciation for the high stakes of debate.

# 2

# Population and Scarcity

Population control: Chinese billboard advertising a happy future for a couple that has only one child, a daughter. *Credit*: Barry Lewis/Corbis.

## Keywords

- Birth rate
- Carrying capacity
- Death rate
- Demographic transition model
- Ecological footprint
- Exponential growth
- Fertility rate
- Forest transition theory
- Green Revolution
- Induced intensification
- Kuznets curve (environmental)
- Neo-Malthusians
- Shifting cultivation
- Zero population growth

## Chapter Menu

## A Crowded Desert City

A trip into Phoenix, Arizona on almost any day of the week is a journey into a dense haze of exhaust fumes, ozone, and blowing dust. This desert metropolis of four million people is the tangled conurbation of 10 separate cities, together planted squarely in a low desert depression: "The Valley of the Sun."

The city effectively did not exist at the turn of the twentieth century. In a place that receives seven inches of rain a year and where summer temperatures can exceed 120°F for many days in a row, it was a largely overlooked site for settlement during the period of American Westward expansion, though a range of native peoples had adapted and thrived in the area in small numbers in the centuries prior.

Starting in the 1950s, new people began to arrive in the area, bringing with them new demands for land and water. Going from a half million people before 1960 to its present size, the rate of growth in the valley has been on the order of 40 percent per decade. In the 1990s, the population grew by roughly 300 people per day, as a sun belt economy mush-roomed in the region, driven by high tech production, service industries, and retirement communities.

With each new person comes more demand for limited available water, the production of mounds of garbage, and the disturbance of large areas for new home construction. Given weeks on end with temperatures above 100°F (more than 38°C), summer demands for air conditioning are constant. This is made worse by the city's layers of concrete and asphalt that store and radiate the sun's heat all night long through the summer, creating a "heat island effect." The hundreds of thousands of cars in the region (two automobiles per household) each emit roughly their own weight in greenhouse gases per year, contributing seriously to both local air pollution and global climate change. The overall density of people means that newcomers must settle further and further from the city center in search of reasonably priced housing. Such settlement patterns increase the time spent driving to and from work, schools, and stores every day, and the congested traffic load combines with the length of commuting to further exacerbate pollution and fuel demands. The dramatic rate of population growth poses obvious questions about the limits of the land, water, and air to support the city.

Still, there is more than just the *number* of people to concern us here. The average person in Phoenix consumes more than 225 gallons of water per day. This compares poorly even to nearby municipalities like Tucson, where an average individual uses around 160 gallons a day. Where does all this water go? To dishwashers and toilets, among other things, but also to green lawns and vast green landscapes in the desert. Given that by best estimates the minimum amount of water required for human survival is perhaps five gallons a day, it is clear that the affluence and living conditions of people in Phoenix are as important contributors to water use as the overall number of residents.

Either way, the human pressure on resources in the region is tremendous and that impact is by no means limited to water. As new home sites devour land across the valley, the habitat of rare and important desert species (like Gila monsters – rare and fabulous local lizards) declines as well, with implications for global biodiversity. Dense human habitation and activity also encourages the arrival and spread of invasive species and increases the risk of fire hazards across the region. There is a growing human footprint in The Valley of the Sun.

To what degree do the explosive numbers of people in Phoenix represent an environmental crisis? How have affluence and lifestyle influenced this impact? Is there enough water to allow such a city to survive? Many explorations of the relationships between environment and society typically start right here, asking a basic question: Are there simply too many people? Can the world support us all? If not, can human numbers ever be expected to stop growing? How and when?

**Figure 2.1**    The skyline of Phoenix, Arizona, average rainfall seven inches per year, population four million people. How many is too many? *Credit*: Jeff Badger/Shutterstock.

## The Problem of "Geometric" Growth

These questions are by no means new either to the field of ecology or to the examination of society or policy. The concept of overpopulation is indeed ancient, though its most prominent modern adherent lived in the decades spanning the late eighteenth and early nineteenth centuries: Reverend Dr Thomas Robert Malthus. His assertion, in its clearest form, was simply that the capacity of population to grow is greater than the power of the earth to provide resources. Given the procreative capacity of humanity and the inherently finite availability of the earth's resources, in this way of thinking, human population is the single greatest influence on the status of the earth and its resources. Conversely, the earth's resources provide the most definitive and powerful limit for human growth and expansion.

In more careful phrasing, Malthus was clear to describe the mathematical underpinnings of this assertion, stressing that population growth is effectively "geometric" ("**exponential**" in today's terms), since the multiple offspring of a single mating pair of animals or people are each capable of producing multiple offspring themselves. Assuming six children from every mating couple (a family size typical in Malthus' time), for example, means a growth from two to six in the first generation, 18 in the second generation, 54 in the third, and so on. That growth, when graphed, takes the form of a curve, much steeper than a straight line, moving towards an asymptote, that is, a steep increase in a few generations and a large number of individuals, increasing every generation.

> **Exponential Growth**    A condition of growth where the rate is mathematically proportional to the current value, leading to continued, non-linear increase of the quantity; in population, this refers to a state of increasingly accelerated and compounded growth, with ecological implications for scarcity

On the other hand, Malthus argued, the food base for this growing population over time is essentially fixed or, perhaps, amenable to slight alteration through "arithmetic" ("linear" in today's terms) expansion. Food supplies can grow by putting more land under the plow, for example, but not nearly at the rate that population expands. Over time "geometric" growth always outpaces "arithmetic" growth, with obvious implications.

These implications sit at the center of Malthus' key written work, *An Essay on the Principle of Population* (Malthus 1992), which he first published in 1798, but re-edited periodically in the decades after. Here, Malthus first suggested, wars, famine, destitution, and disease are natural limits to growth and act to keep population in check. Second, he maintained that policies promoting the welfare of the poor are counterproductive, because they only encourage unnecessary reproduction and resource waste. Third, he argued that the key to averting periodic and inevitable resource crisis is a moral code of self-restraint.

In terms of natural limits Malthus suggested that famine, starvation, and death were predictable. He insisted, moreover, that the iron laws of scarcity meant that periodic crises and population collapses were practically inevitable, even in a world where some expansion in resources occurred over time. These hypothetical cyclical population-driven crises are sketched in Figure 2.2, which shows a model of the Malthusian dynamic of population versus natural resources.

Malthus freely admitted that the poorest people were the most vulnerable parts of the population. He insisted, however, that efforts to sustain, protect, or subsidize the conditions of the poor were largely pointless, insofar as they bolstered or supported population growth. Malthus, though, was even harsher in his assessment of the poor. He suggested that the poor are reliant on handouts, that they are bad managers of time and money, and that they are given to irrational procreation. Providing them with resources (as the welfare provisions of the time did to a limited degree in the Poor Laws of England) only reproduces these habits, causes population to grow further, and therefore makes resources even scarcer, and misery even more widespread.

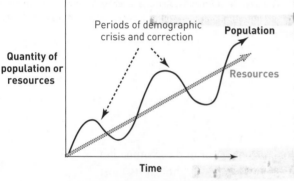

**Figure 2.2** Hypothesized demographic trends in a Malthusian conception. Limits of the environment, though they are amenable to steady increases resulting from growths in resource production, control human population trends with periods of high growth followed by periodic calamities and corrections that bring population back in line with the environment.

> The Poor Laws of England tend to depress the general condition of the poor … They may be said, therefore, to create the poor which they maintain. (Malthus 1992, Book III, Chapter VI, 100)

Rather than provide support for people, Malthus insisted that the best remedy to these crises is the expansion of moral restraint. Specifically he intended the moral restraint of women, whom he held responsible for the maintenance of virtue and, by implication, for population run amuck. He especially focused his criticism on "less civilized" peoples (seen as those from southern Europe at that time) whom he viewed as insufficiently capable of self-control, and so inevitably given to poverty.

> It can scarcely be doubted that, in modern Europe, a much larger proportion of women pass a considerable part of their lives in the exercise of virtue than in past times and among uncivilized nations. (Malthus 1992, Book II, Chapter XIII, 43–4)

> In some of the southern countries where every impulse may be almost immediately indulged, the passion sinks into mere animal desire, is soon weakened and extinguished by excess. (Malthus 1992, Book IV, Chapter I, 212)

The social and political biases of *Essay on the Principle of Population* and the context in which it was written are clear. Malthus developed an explanation for poverty that absolved economic systems, political structures, or the actions of the wealthy or elite from fault. His specific moral vision of women, perhaps even by the standards of his own time, reflects a profoundly biased view of the relationship between women and men.

## Actual population growth

Nevertheless, Malthus' essential formulation, which pitted the dynamics of human population against the semi-flexible conditions of the environment, was crucial in the development of modern population biology and evolutionary theory. In his *On the Origin of Species* (1859), Darwin specifically draws on Malthus' insights for the formulation of the theory of evolution.

Examination of some recent trends also reveals that after two hundred years of demographic history, a few of Malthus' key claims are sustained. To be sure, the exponential nature of human population growth in the past few centuries is quite clear. That growth is roughly shown in Figure 2.3. Where the world at the time of the Roman Empire two thousand years ago contained only 300 million people, today it holds more than 6.5 billion, more than a 20-fold increase, most of which occurred in only the last century.

So even while there are numerous profound limits and problems in this formulation (and more as we will see below), the arguments of Malthus and his present-day followers certainly raise questions about the relationship between society and environment and the nature of resource scarcity, its possible inevitability, and our capacity to overcome it.

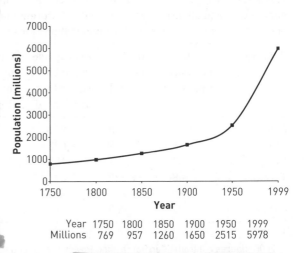

| Year | 1750 | 1800 | 1850 | 1900 | 1950 | 1999 |
|------|------|------|------|------|------|------|
| Millions | 769 | 957 | 1260 | 1650 | 2515 | 5978 |

**Figure 2.3**  World population since 1750. Rapid increases in recent decades reflect exponential growth. *Source*: Following Demeny (1990).

## Population, Development, and Environment Impact

The questions raised by Malthus have been taken up by other scholars interested in relationships between population, economic development, and environmental impacts. One approach, pioneered by Paul Ehrlich and John Holdren (1974), seeks to measure the impact of human beings on the environment, taking seriously not only raw numbers of people but also their overall rate and type of consumption. They proposed that every additional person added an impact on the earth, though the exact rate of that impact was influenced by other factors including the average affluence of a population (a person in Bangladesh uses far less water and energy than one in the United States, for example) and the availability of technology that might lessen human impact (a population using solar power rather than coal power may have far lower carbon emissions, for example, depending on how solar panels are produced and how much energy their owners use). For this relationship they developed a shorthand equation to determine the level of environmental impact (I) as a product of population (P), affluence (A), and technology (T):

$$I = P * A * T$$

Here, environmental impacts are understood broadly as the deterioration of the resource base, the decline of ecosystems, the production of waste, and so on, while population is the number of people in a specific group (usually a country). Affluence, a measure that was not considered in any way by Malthus, is alternatively measured as either 1) the level

**Table 2.1**  Who is overpopulated? Some comparisons of population, per capita gross domestic product, energy use, and other resource demands. Different places have widely divergent levels of population, affluence, and technology, with unclear implications for environmental impact

| Country | Total population (millions)[1] | GDP ($ per capita)[2] | Energy use (kg of oil equivalent per capita)[3] | Annual % total forest cover change (including plantations)[4] | Annual % forest cover change (natural forest only)[4] | Greenhouse gas emissions (tons of $CO_2$ equivalent per capita)[5] |
|---|---|---|---|---|---|---|
| China | 1,294 | 3,936 | 861 | +1.2 | +0.6 | 3.91 |
| USA | 288 | 33,939 | 8,095 | +0.2 | +0.1 | 23.92 |
| Bangladesh | 143 | 1,527 | 133 | +1.3 | −0.8 | 0.38 |
| Turkey | 68 | 6,830 | 1,071 | +0.2 | ... | 4.07 |
| UK | 60 | 23,637 | 3,886 | +0.6 | +1.5 | 11.19 |
| Kenya | 32 | 1,003 | 489 | −0.5 | −0.5 | 0.81 |

1: 2002
2: 2000 (figures controlled for Purchasing Power Parity: equivalence of buying power in local markets)
3: 1999
4: 1990–2000
5: 2005
*Source*: World Resources Institute Data (2005)

of consumption of the population or 2) the per capita gross domestic product. In other words, one considers how many goods per capita (per person) are consumed in that country or area or the total production in the country, divided by the population. Technology, also not considered by Malthus, is the set of methods available to that population to produce the goods that are needed and consumed.

While this formulation certainly makes the relationship between population and environmental degradation more complicated than Malthus did, it has been used by "neo-Malthusians," those more recent adherents to a population-based way of thinking about environmental issues, to argue that population is the paramount factor in this equation. Paul Ehrlich (1974: 1216) explains that population requires the most immediate attention "precisely because population is the most difficult and slowest to yield among the components of environmental deterioration."

There are challenges to this assumption. Critics like Barry Commoner stress that technology has by far the greatest influence on environmental impact, far outweighing the total numbers of people, specifically citing the petrochemical-based economy, pesticides, fossil fuels, and a range of modern developments that increase individual impact enormously. As shown in Table 2.1, environmental impact varies enormously even in current economics. An alternative economy would, by implication, offset population growth (Commoner 1988).

Others have argued that development radically lowers human impact, at a rate far greater than the growth of population. In what some analysts call an **environmental Kuznets curve** (named for economist Simon Kuznets), it is predicted that as development initially occurs, environmental impact increases, with per capita use of resources rising, pollution increasing, and damage to ecosystems like forests rising, and doing so at a rising rate. After a threshold, however, regulation, affluence, and economic transition begin to increase and impacts of humans fall dramatically. Proponents of this argument point out that in many parts of the developing world that have historically experienced high levels of deforestation, urbanization and affluence have left many rural areas abandoned, allowing a **forest transition** back to thick forest cover (Perz 2007, and see Chapter 10).

## Carrying capacity and the ecological footprint

On the other side of the equation, assuming agreement might be reached on how to measure impact per person, the degree to which each such impact is "too much" is also a matter of uncertainty. Just as IPAT and its variations predict future impacts of society on nature, the notion of **carrying capacity** is often invoked to signal the limits beyond which a local area can no longer absorb population. Carrying capacity is the number of people that could theoretically be sustained in one area (or the earth) over an indeterminate amount of time, assuming a particular lifestyle (level of technology and consumption).

**Neo-Malthusians**  Present-day adherents to a position – established by Malthus in the nineteenth century – that population growth outstrips limited natural resources and presents the single greatest driver of environmental degradation and crisis

**Kuznets Curve (Environmental)**  Based in the theory that income inequality will increase during economic development and decrease after reaching a state of overall affluence, this theory predicts that environmental impacts rise during development, only to fall after an economy matures

**Forest Transition Theory**  A model that predicts a period of deforestation in a region during development, when the forest is a resource or land is cleared for agriculture, followed by a return of forest when the economy changes and population outmigrates and/or becomes conservation-oriented

**Carrying Capacity**  The theoretical limit of population (animal, human, or otherwise) that a system can sustain

**Ecological Footprint**  The theoretical spatial extent of the earth's surface required to sustain an individual, group, system, organization; an index of environmental impact

It has been estimated, for example, that if we calculate carrying capacity based on an assumption that all people *lived like people do in the United States*, the earth could sustain only two billion people, or less than one-third of the world's current population (Chambers et al. 2002). If this is taken seriously, we might have to ask how we can possibly decide who should be allowed to live at what standard of living. In other words, is it reasonable to insist that China not develop any further, so that levels of consumption in the United States and Europe can remain the same? Should people in North America demand that India limit its growth so that North Americans (in the United States and Canada) can maintain their own standard of living? For many people, these extremely problematic ethical questions have led to a desire to decrease their own impact on the environment – their **ecological footprint**.

The idea of an ecological footprint analysis is not so much to define the potential of an area to support a particular number of people at a particular standard (although it can do that), but rather to estimate the total area of productive land and water required to produce the resources for and assimilate the waste from a given population. While this can be done at multiple scales – some people use it to analyze the environmental impacts of entire urban areas, or even countries – many people find it useful to estimate how their own daily practices of eating, showering, driving, using the bathroom, washing their clothes, etc. affect the environment. There are many websites available that allow users to enter their own data and receive a number representing their impact on the environment. Of course, you can always view this data with a degree of skepticism, but for many people it is a real eye-opener ("my restaurant dinners add 20 square kilometers to my footprint?!"). So even if human population will never truly "exhaust" the earth, it is reasonable to ask what quality of life might be expected in specific places with large populations (e.g., Phoenix) and to ask what obligations the wealthy have to the earth upon which they tread so heavily.

## The Other Side of the Coin: Population and Innovation

Given the many scenarios of famine, scarcity, and ecological disaster typically laid at the feet of population growth, it might be hard to imagine that there are many thinkers, researchers, and historical observers who actually make the reverse argument: population growth is the root of innovation and civilization. Yet there is a great deal of evidence to support this claim.

In this way of thinking, a growth in human numbers, given a relative scarcity of available resources, induces the search for alternatives and new ways of making more from less. Looking back at agricultural development over thousands of years, there is evidence to suggest that this happens all the time in the provision of food. This is because historically food was produced using extensive production techniques – meaning large areas of land were used for the production of limited amounts of food. Environmental systems (especially soil fertility) impose such limits on production because it is typically the case that the amount of food that can be grown on the same land after a season or two of cropping tends to decrease. The simplest remedy for such soil exhaustion is fallowing of land,

## Box 2.1   The Nine Billionth Person

On October 12, 1999, by all best estimates, the six billionth living person was born (compared to the perhaps hundred billion who have ever lived). She or he entered a world very unlike that into which the five billionth person was born back in 1987. On the one hand, in the intervening period, scarcity of some goods and resources had increased. Many experts believe, for example, that in the interim the world passed its point of peak oil production, after which all petroleum energy would be harder to get and more expensive in perpetuity. On the other hand, many resources actually expanded. Agricultural output per capita increased over the decade, for example, outpacing population. The world of the early twenty-first century also differs from that of the 1980s in terms of where the fastest growth is occurring, as well as where population has leveled off or is in decline. Countries like Russia, Japan, Italy, South Korea, and Ukraine all entered a period of demographic slowing and decline during this period, while India and China continued to grow. A boy from the city of Sarajevo – then just recovering from a terrible war – was symbolically greeted by the United Nations as the world's six billionth person (BBC News 1999), but an abstract composite of the six billionth person based more realistically on probability suggests that they would have more likely been born in Asia, where half of the earth's human population now lives.

The US Census Bureau estimates that the seven billionth person will arrive on the scene in 2012. There are several things to make note of here. First, it is hard to know what the world will look like then. Certainly the overall demands of a global population of this size (some 6 percent larger than today) will be somewhat greater than they are today. On the other hand, even while per capita energy and food demands in China and India are increasing, they are currently extremely far below those of wealthier countries and by 2012 may not be anywhere near current demands in the USA. Second, it is notable that the length of time *between* each increase of a billion people will have begun to increase after decades of decreasing, because global growth rates are slowly declining.

This means that global population will eventually plateau, perhaps around the middle of the century, at roughly nine billion people. If the world of 2050 is like the USA of today, where 86 percent of energy comes from fossil fuels, where every car travels more than 11,000 road miles a year, and each person consumes 1,682 cubic meters of water a year, the arrival of the nine billionth person may be a grim moment indeed. Assuming a world where energy is largely generated through solar power, food production has become highly efficient, and water is used sustainably, on the other hand, this 50 percent increase in the world's population over today's numbers may be an event that goes totally unremarked.

### Reference

BBC News (1999). "World population reaches six billion," October 12. Retrieved March 18, 2009, from news.bbc.co.uk/1/hi/world/471908.stm.

meaning that the field is left to rest for a season or more. During the time that the field is at rest, land is cultivated elsewhere. In traditional **shifting cultivation** of this kind, many fields might be used in a long series of rotations, with people returning to plant in the original fields many years after their first cropping, giving time for the soil to become productive again.

Such a system is perfectly straightforward and feasible for a limited population. But as human numbers grow and demand for food

**Shifting Cultivation**   A form of agriculture that clears and burns forest areas to release nutrients for cropping. Also known as "swidden," this method is highly extensive, typically rotating through areas of forest land for short periods of use, allowing previously used forest land to recover

increases, either the number of fields must be increased (making the production system more extensive – that is, using more land) or the population will have to rotate through their fallow land more quickly, resting the land less often, if at all. If the latter decision is made, then some way of maintaining the fertility of the soil and therefore growing more food on the same amount of land must be devised. The history of agriculture is replete with such innovations, from soil fertilization using manure and more modern inputs to complex systems of intercropping where different crops are used together or in rotation to maintain soil fertility, rather than continuously cropping the same food plant. As explained in Ester Boserup's (1965) now classic analysis, *Conditions of Agricultural Growth*, over long periods of history the amount of food produced on the same amount of land has increased exponentially, because demands for food rise with increases in population. More people mean more food. In some cases, increasing population can lead to improved environmental conditions, moreover, as in the Machakos District of Kenya, where careful management of farmlands and soils and the right combination of development efforts and social action have led to an improved resource base, even and especially as population has increased (Tiffin et al. 1994). This thesis is referred to as **induced intensification** and can be extended to all kinds of other problems and natural resources (see Chapter 3).

**Induced Intensification** A thesis predicting that where agricultural populations grow, demands for food lead to technological innovations resulting in increased food production on the same amount of available land

**Green Revolution** A suite of technological innovations, developed in universities and international research centers, which were applied to agriculture between the 1950s and 1980s and increased agricultural yields dramatically, but with a concomitant rise in chemical inputs (fertilizers and pesticides) as well as increased demands for water and machinery

Global populations have unquestionably increased, and at an increasing rate, over the past 50 years. At the same time, food supplies have also grown faster than they did before. Owing to new cultivation techniques and input-heavy systems of agricultural production, the so-called **Green Revolution**, wheat production in India tripled between 1965 and 1980, for example, far outpacing the rate of population growth. In Indonesia during the 1970s, rice production increased by 37 percent. In the Philippines it increased by more than 40 percent. The period from the middle 1960s to the present in fact has been a period in which more food has been produced than consumed, and in which more people have been moved above the level of starvation than in the century prior.

It should be pointed out that this result does not fully contradict Malthusian thinking, although it makes it more complicated. It does refute the simple assumption of absolute or fixed limits to human population and the necessity of famine to achieve "correct" population levels. Food availability can apparently expand far faster than population growth. Nevertheless, a number of other environmental problems come with the Green Revolution and the expansion of food production, including the loss of unbroken soils across the prairies and rainforests of the earth, and a concomitant loss in biodiversity. Where new land is not used to increase food supplies, increased intensity of production has also meant a massive input of fertilizers and pesticides. These chemicals take a toll on the land and are themselves made from petrochemicals, meaning that they depend upon petroleum extraction and manufacturing, with all of the environmental implications of that system. Energy is necessary, moreover, to produce farm equipment (like tractors, and harvesters), and to power that equipment. All of this energy comes from the ongoing exploitation of increasingly scarce petroleum resources, the production of which is costly and environmentally devastating. Each calorie of food in the years following the Green Revolution has become *far* more ecologically expensive. There appears to be no such thing as a free lunch. In this way, even where population can be a driver of innovation, it may continue to present a chain of related problems for the environment.

These outcomes also raise questions about the problem of *scale* in assessing the impact of population. Agriculture might expand dramatically in a densely populated part of Brazil, for example, to produce soybeans for both consumption and industrial applications, with serious implications for the forest these crops replace. The "population" that drives this demand, however, lives thousands of miles away in Europe and the United States, and thrives in a high-impact lifestyle under conditions of low population density and growth. The circulation of agricultural products makes assessing the impact of local, regional, and global populations extremely difficult.

Nor is any of this to say that unlimited numbers of people lead to unlimited productive capacity and endless plenty, although the view has been described as "cornucopian" (literally "horn of plenty" from the Greek) by its critics. It does raise serious questions, however, about the problems of assuming natural limits. If population growth does not always lead to scarcity, or if it sometimes even results in increased resources, how useful is a Malthusian perspective?

## Limits to Population: An Effect Rather than a Cause?

Beyond this, recent trends in population have made some of the pronouncements and assumptions of population-centered thinking moot. Specifically, the rate of population growth around the world has fallen precipitously in the past few years. Some areas are indeed experiencing negative growth. This should be encouraging for Malthusians. But more profoundly, it raises a basic question about population: Is population a social *driver* of environmental change or is it actually the *product* or outcome of social and environmental circumstances and conditions?

Consider the following: Population growth rates, which measure (as a percentage) the rate of natural increase of the total number of people on the earth, peaked in the period between 1960 and 1970 (Figure 2.4). In the period since, they have only declined, slowly approaching a rate below 1 percent, nearing a state of **zero population growth** (ZPG). Whatever one might think about the danger that population (versus affluence, for example) presents for the environment, one must wonder what causes such a change.

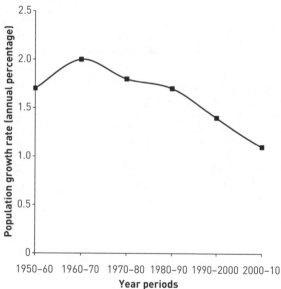

**Figure 2.4** Global population growth rates. Population growth rates peaked in the 1960s and have steadily and continuously declined since then.

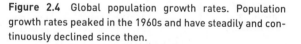

**Zero Population Growth** A condition in a population where the number of births matches the number of deaths and therefore there is no net increase; an idealized condition for those concerned about overpopulation

What makes population growth decline? What are the implications for human–environment relations in a world where population, while continuing to grow, is likely to reach stability in the next 50 years?

## Development and demographic transition

The most obvious precedent for this global shift comes from the demographic history of Europe in the nineteenth and early twentieth centuries. There, population went from a relatively stable state before 1800 to very high levels of growth in the 1800s, followed by a leveling off, and indeed a current state of population decline in many countries. The drivers of this trend can be broken down more carefully. We can examine the specific rate of births and deaths in Europe (or anywhere else) to provide clues as to what happened.

In traditional agrarian societies, the **death rate** (typically measured in numbers of deaths per 1,000 people in a population in a year) and the **birth rate** (births per 1,000 people per year) vary somewhat, but are both relatively high (around 40 or 50 per thousand). They also typically offset each other so that annually no more people are born than die, leading to low or negligible population growth.

In the European case, the initial growth of population was caused by a decline in the overall death rate of the population. The death rate fell largely in response to better medicine and health care, which led to fewer deaths, especially of infants, women in childbirth, and other historically vulnerable populations. Going from a high of around 40 or 50 deaths per thousand to less than 15 deaths per thousand, more people lived, and lived longer, during the nineteenth and twentieth centuries in Europe than ever before. If the death rate falls and the birth rate remains unchanged, population growth occurs, and often at an exponential rate. As long as birth rates are higher than death rates more people enter a population than leave it. Europe grew as a result.

In the years of the late nineteenth and early twentieth centuries, birth rates also began to fall. There are many reasons for this decline, and some remain a matter of debate. Nonetheless, as people moved into cities, the demand for family farm labor fell while the costs of raising and educating children increased. This led families to have fewer children and the birth rate fell, from a high between 40 or 50 births per thousand to a low closer to 10 or 15. Once the birth and death rates came to match, population growth halted. In many countries, like Russia, the death rate is currently higher than the birth rate, leading to a decrease in population.

The **demographic transition model** or DTM (sketched in Figure 2.5) is an abstract representation

**Death Rate**  A measure of mortality in a population, typically expressed as the number of deaths per thousand population per year

**Birth Rate**  A measure of natural growth in a population, typically expressed as the number of births per thousand population per year

**Demographic Transition Model**  A model of population change that predicts a decline in population death rates associated with modernization, followed by a decline in birth rates resulting from industrialization and urbanization; this creates a sigmoidal curve where population growth increases rapidly for a period, then levels off

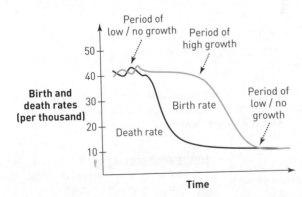

**Figure 2.5** The demographic transition model. In theory, falling death rates lead to population growth, but as birth rates fall thereafter, the rate of growth slows, eventually ceasing when the two reach equilibrium.

of this process that is often used to describe what happened in Europe during this period. Generally it assumes that some form of economic development and change from agricultural to industrial society drive the population towards a period of high population growth, followed by a cessation of growth.

It is tempting to use the DTM to predict what might happen in other parts of the world and in other periods of history. After all, the processes that accompanied demographic transition in Europe are also ongoing in other parts of the globe. Urbanization, decreased proportion of the population in farming, changes in jobs, and the expenses and benefits related to having children are all facts of life in China, India, and much of Africa and Central and South America, although they happen at different times and under different conditions. For demographers, there is a parallel expectation of demographic transition (Newbold 2007). Human pressure on the environment, it might similarly be predicted, will also cease growing (assuming that affluence and consumption do not change for each person!).

## Women's rights, education, autonomy, and fertility behavior

But mounting evidence from around the world demonstrates that population and resource transitions are far more diverse and have far more drivers and explanations than the DTM might suggest. In some countries or regions, population growth has fallen dramatically, for example, in the absence of significant economic growth and change. The state of Kerala in southern India is a paramount example of such an outcome. In the 1950s, this small rural state had the highest population growth rate in India but by the 1990s it was among the lowest, falling from 44 per thousand in 1951 to 18 per thousand in 1991 (Parayil 2000). How was this transition achieved, since – contrary to the DTM – the population remains largely rural and manages with a lower than average gross domestic product per capita, even lower than other states within India?

> **Fertility Rate**  A measure describing the average number of children birthed by an average statistical woman during her reproductive lifetime

The secret may be outside of strictly economic factors. Kerala has a higher than average level of women's education and literacy and a higher than average availability of rural health care – especially for women. This is a universal phenomenon, moreover, and women's education and literacy correlate significantly with lower **fertility rates**, a measure of the number of children the average woman in a place has over her reproductive lifespan (Figure 2.6). Where women's rights are observed and protected, population growth ends.

There are questions about whether correlation here is linked to causality. In countries with lower fertility rates, for example, women are freer to pursue education. Both women's education and lower fertility may themselves be caused by other factors in culture and society. Even so, reduced fertility rates

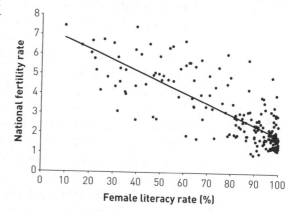

**Figure 2.6** National fertility and female literacy rates around the world: 2006. As female literacy increases, and along with it women's autonomy and employment, fertility rates fall to replacement levels. *Source*: Analysis by authors; data: Population Reference Bureau (2008).

are also associated with the availability of condoms, the availability of women's reproductive health care on demand, and the social/political "autonomy" of women (meaning social ability to make independent decisions) within households and communities. Put simply, it is increasingly clear that the political and economic condition of women in society is the best and most reliable predictor of demographic conditions. To the degree that population is an influence on the environment, the solution to ecological problems therefore lies in women's rights around the world.

## The potential violence and injustice of population-centered thinking

Given that population may be an effect of other economic and political processes, that affluence and consumption have a large hand in determining the environmental impacts of populations of any size, and that some growth in population may lead to increased rather than decreased resources, what are the implications of thinking about environmental problems in strictly demographic terms?

For critics of population control, the implications are serious indeed. For such critics, the history of population politics and efforts at control are fraught with violence and injustice. Consider, for example, the Indian emergency during the period of Indira Gandhi's rule in the 1970s. In 1975 Mrs Gandhi declared martial law and made the end of population growth, through dramatic measures, including mass sterilization camps, the rationing of food and services contingent on family size, and even forced sterilization of some villages and slums, a central part of state policy. Most of these draconian measures were enacted on groups with the least political power, moreover, including marginal caste communities and the urban poor. International neo-Malthusian observers heralded the effort, with some American and European observers calling for logistical support of all kinds in this war on overpopulation (Hartmann 1995).

Yet none of these measures slowed or halted India's population growth, which is only now slowing decades later as a result of complex political and economic factors, including women's rights and access to education. All that these efforts achieved was a horrible violence to those poor and unfortunate enough to be caught up in the panic and a general distrust amongst all Indians for any discussion of population at all. So, the persistence of Malthusian thinking in the international press, in coercive government population policies, and in environmental analyses of various kinds appears to many critical observers to be dangerous indeed, as it distracts attention away from other driving forces (in economy, society, or politics) of environmental degradation. It also tends to unjustly vilify places and people who may have little or nothing to do with ecological change or negative environmental impacts. While there may be a billion people in India, for example, the United States, with one-quarter that population, emits more than five times the amount of carbon dioxide gas – a key driver of global warming (see Chapter 9).

More pointedly, critics maintain that making the politics of the environment a politics of population directs policy action, blame, and social control specifically onto women and their bodies. As Elizabeth Hartmann argues in her critical book *Reproductive Rights and Wrongs*, by seeking to restrict population, neo-Malthusians effectively seek to restrict women, and yet:

the solution to the population problems lies not in the diminution of rights, but in their expansion. This is because the population problem is not really about human numbers, but a lack of basic rights. Too many people have too little access to resources. Too many women have too little control over their own reproduction. Rapid population growth is not the cause of underdevelopment; it is a symptom of the slow pace of social reform. (Hartmann 1995: 39)

## Thinking with Population

In this chapter we have learned that:

- Human population growth holds serious implications for sustainability of environmental systems, especially as growth tends historically to be "geometric" or exponential.
- Environmental impacts of individual people and groups can vary enormously, owing to variations in technology and affluence.
- Population growth has often led to increased carrying capacity, owing to induced intensification and innovation.
- Carrying capacity and ecological footprint analysis can be used as indices to think about impacts of human individuals and populations.
- Malthusian thinking has severe limits for predicting and understanding human–environment relations, since population is an effect of other processes, including development and the rights and education of women.

Even accepting important criticisms, of course, the question of population cannot be fully ignored. Given the lifestyle of the people of Phoenix, Arizona there may well be a limit to the number of people The Valley of the Sun can maintain, and every person over that limit certainly taxes the region's landscapes, with implications for native biodiversity, for open space, and for clean air. Still, as such resources become scarce and more valued, an incentive to provide them may emerge. Environmental groups (like The Nature Conservancy) are increasingly buying lands throughout Arizona, for example, simply to protect them from development, leave them open and unspoiled, and allow them to provide habitat for the flora and fauna of the desert. Advocates of such an approach suggest that the very scarcity of these environmental "services" is the ticket to their conservation. It is to this argument – that markets can provision and produce scarce environmental goods – that we turn next.

## Questions for Review

1. What "crisis" did Malthus predict as inevitable? What was his proposed solution?
2. While Malthus blamed the poor for pending crises, contemporary thinkers like Paul Ehrlich place equal blame on the very wealthy. Why is this the case? (hint: think $I = P*A*T$)

3. Who has a larger ecological footprint, you or a subsistence fisher in coastal Bangladesh? Explain.
4. How can population growth force a transition from extensive agriculture to intensive agriculture? How does this transition often lead to innovation?
5. What factors led to the dramatic decrease in population growth rates in Kerala, India between the 1950s and 1990s? Compare the case of Kerala to India's national population control program put into place in the 1970s.

## Suggested Reading

Boserup, E. (1965). *Conditions of Agricultural Growth: The Economics of Agrarian Change under Population Pressure*. Chicago, IL: Aldine.

Ehrlich, P. R., and A. H. Ehrlich (1991). *The Population Explosion*. New York: Simon and Schuster.

Hartmann, B. (1995). *Reproductive Rights and Wrongs: The Global Politics of Population Control*. Boston, MA: South End Press.

Kates, C. A. (2004). "Reproductive liberty and overpopulation." *Environmental Values* 13(1): 51–79.

Lambin, E. F., B. L. Turner, et al. (2001). "The causes of land-use and land-cover change: Moving beyond the myths." *Global Environmental Change – Human and Policy Dimensions* 11(4): 261–9.

Malthus, T. R. (1992). *An Essay on the Principle of Population (selected and introduced by D. Winch)*. Cambridge: Cambridge University Press.

Mamdani, M. (1972). *The Myth of Population Control: Family, Caste, and Class in an Indian Village*. New York: Monthly Review Press.

Newbold, K. B. (2007). *Six Billion Plus: World Population in the Twenty-First Century*. Oxford: Rowman and Littlefield.

Patel, T. (1994). *Fertility Behavior: Population and Society in a Rajasthani Village*. Bombay: Oxford University Press.

Sayre, N. F. (2008). "The genesis, history, and limits of carrying capacity." *Annals of the Association of American Geographers* 98(1): 120–34.

Warner, S. (2004). "Reproductive liberty and overpopulation: A response." *Environmental Values* 13(3): 393–9.

## Exercise: What Is Your Ecological Footprint?

Go to http://umanitoba.ca/campus/physical_plant/sustainability/funstuff/428.htm and read the page describing ecological footprint analysis. Take the ecological footprint quiz, noting your responses and results. Now take the quiz again, three more times. Each time, change only ONE of your answers from your original answers, as if you had changed one of your own behaviors. Make note of: How does the changed behavior change the resulting final footprint? By how much does the footprint change? And, which particular changes appear to get the most ecological benefit?

What was your original footprint? How did it compare to the national and global averages? In what categories were you higher or lower than this average? How many earths would be required to support a planet in which everyone lived as you do? Why do you think your overall impact is higher/lower/equal to the average? What factors in your life, background, or current situation account for your overall footprint? Will they change in the future? In what direction?

What three changes did you introduce in your proposed behaviors in the three further quizzes? Which of these changes had the greatest impact on your footprint? Which of these is actually the most feasible or likely change (i.e., which would be easiest)? Is the easiest change the one that would make the biggest difference in terms of ecological benefits? Why or why not? What makes some choices more difficult than others? In light of this, to what degree do you feel that global population is a factor in environmental change, relative to consumption, affluence, and lifestyle?

# 3

# Markets and Commodities

*Credit*: Pablo Corral V/Corbis.

## Keywords

- Cap and trade
- Coase theorem
- Externality
- Green certification
- Greenwashing
- Jevons' Paradox
- Market failure
- Market response model
- Monopoly/Monopsony
- Transaction costs

## Chapter Menu

## The Bet

Can using more stuff lead to the availability of more stuff? Can human population growth be *good* for both nature and society? By the end of the 1970s, when population-centered thinking was dominant, such questions were counterintuitive and hard to even ask. At that time, Paul Ehrlich was the most prominent spokesperson for the population crisis and was typically identified as the paramount and persuasive neo-Malthusian of the time (Chapter 2). His book, *The Population Bomb* (1968), was a cornerstone of the thinking and rhetoric of many environmentalists.

It may have seemed surprising, then, for Ehrlich to be challenged to a very public bet in 1980 by a thinker largely unassociated with environmentalism, at least in the public mind. The wager came from economist Julian Simon, who had long maintained that human population growth improved living conditions and environmental quality, because a) more people means more good ideas, and b) more demand for things (including clean air and water) produces an incentive to find, make, and creatively maintain the world. His thinking culminated in a controversial article in the journal *Science* in 1980, entitled "Resources, population, environment: An oversupply of false bad news" (Simon 1980), which maintained that things get progressively better, not worse, with the advent of every birth.

As a journalist for the *New York Times* recorded a decade later (Tierney 1990), the wager consisted of the men betting $1,000 on whether the prices of five metals selected by Ehrlich and his associates – chrome, copper, nickel, tin, and tungsten – would rise or fall in value over the next decade. If Simon was right and the planet was one where the future was always better than the present, the scarcity of these goods would actually decline, owing to increasing human ingenuity and economic growth. If Ehrlich was right and the planet was a finite place bedeviled by rampant consumption, prices should rise considerably. After all, the 1980s was predicted to be a decade of unprecedented growth, with more people born, more quickly, than any time in human history.

Ehrlich lost the bet. The prices of all five commodities fell dramatically as new sources for each were found and new substitutions for each were developed in laboratories and factories around the world (Figure 3.1). For Simon, it appeared to vindicate a view of the world wholly different from that of end-of-the-world Malthusian environmentalism.

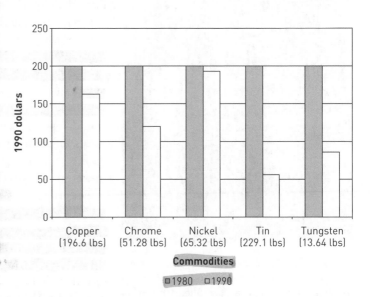

**Figure 3.1** Wagering the future. The prices of all five of the scarce commodities selected by Paul Ehrlich in his bet with Julian Simon fell during the 1980s, a period of unprecedented population growth.

Of course, questions might be asked about the real environmental value of such a bet, and certainly have been in the decades since. To what degree had these two men really "bet the planet"? Had they not really only wagered on the commodity prices of a handful of relatively trivial ecological assets? What if they had bet on whether global temperatures or atmospheric concentrations of greenhouse gases would rise or fall? What would the future hold for these commodities and others as population continued to grow at an expanding rate?

Nevertheless, Simon had, in a very public way, conveyed a view of the relationship between nature and society that was rather different than that of population-centered thinking. Simon's essential economic worldview was reflected in this bet. Beyond its essential optimism, this view held at its heart the *creative potential* of human beings, the importance of *incentives* in producing outcomes, and the utility of *price* for not only measuring but also creating brave new worlds (Field 2005).

## Staining environmental goods: The market response model

Contrary to population-caused visions of scarcity (Malthusian or otherwise), thinking economically suggests that scarcity does not set the limits of the relations between society and environment, but instead operates as the engine of their interaction. Here, scarce resources are made available, or indeed abundant, through the working of supply and demand, which inspires the creative potential of human imagination to be unleashed by economic incentives.

In this way of thinking, available resources, because they are valuable, are exploited and used for social good. Oil is burned to propel cars that take people to work, landfills are made to dispose of waste far away from populations, and vegetables are consumed to maintain people's health. Exploitation of environmental goods, even renewable resources like forests or fisheries, does tend to decrease their supply (just as any good Malthusian would insist!). But when things become scarce, their price in a market tends to rise; consider: people pay more for gold than lead. This increase in prices, rather than spelling immediate or imminent ecological disaster, presents producers and consumers with new and interesting choices (Figure 3.2).

For producers, the increase in prices may open up innovative opportunities for finding new sources of resources or developing new technologies to extract, produce, or synthesize environmental goods, including those techniques previously too expensive, relative to the price of the resource, to consider. In the southwestern United States, for example, where it had dominated for a century, copper mining came to a sudden halt in the 1960s and 1970s. Mineral stocks had become too diffuse to merit the expensive effort of extracting them from the ground, where only traces remained. But when

**Figure 3.2** Environmental scarcity drives markets. Shell gas station operator Steve Grossi's gasoline price board at his Shell station in Huntington Beach, California, May 21, 2006. *Source*: Reuters/Courtesy Brook Grossi/Handout (Wed May 24, 2006 9:24am ET171). *Credit*: Robert Galbraith/Reuters.

the price of copper rises, as it has in the last years of the first decade of the twenty-first century, applying more expensive extraction techniques to long-dormant surface mines begins to become profitable, leading to renewed mining and increased supplies.

Similarly, as the price of one good rises relative to the price of a less frequently used alternative, people might turn to this alternative as a cheaper substitute. History is full of such substitutions, from whale oil being replaced by carbon mineral oils to copper pipes giving way to polymer plastic ones. Innovations driven by such responses to scarcity create new economies in themselves, employing technicians, workers, and designers in new, previously unimagined production systems.

Producers and suppliers are not the only creative actors in such markets for environmental goods. Those who consume these goods, either firms that need them for production or individual people, also respond to prices. For consumers, an increase in the price of a good typically leads people to use that good less – decreasing demand. They may reuse or recycle the goods they have already used, come up with new substitutes of their own, or increase their efficiency of use. If the price of water rises to a point where it is too expensive to water the garden, consumers might abandon outdoor plants, substitute less water-demanding species, or reuse the "greywater" from their washing machine and sink to care for their landscape. All of these efforts at conservation, it should be noted, are driven not by altruism or green sensibilities, but rather by a simple response to market forces.

This process, where scarcity is relieved by laws of supply and demand, which together govern and sustain the relationship of people to nature, is what resource economists and geographers call the "**market response model**" (Figure 3.3). Here, price signals are

> **Market Response Model**   A model that predicts economic responses to scarcity of a resource will lead to increases in prices that will result either in decreased demand for that resource or increased supply, or both

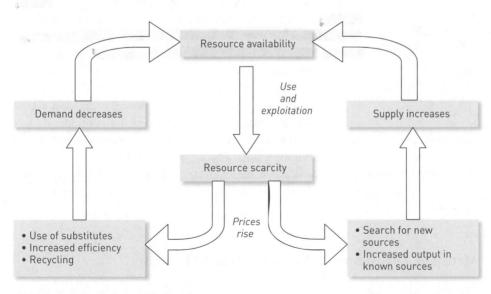

**Figure 3.3**   The market response model. In theory, scarcity of environmental goods and services sets into motion a series of adaptations to rising prices, actually resulting in *increasing* resource availability. *Source*: Adapted from Rees (1990), p. 39.

translated into adaptations by rational and creative people in the market, providing abundance under conditions of scarcity. It is this logic that apparently allowed Simon to prevail over Ehrlich in their 1980 bet.

## Jevons' Paradox

The model holds one further counterintuitive surprise for traditional demographic environmentalism. If it is true that consuming goods drives scarcity, which leads to innovation in a market context, what does imposed or even voluntary conservation lead to? In theory, this apparently desirable and frugal behavior may lead to perverse outcomes in the market. Conservation and conservation technology, if carried out at a wide scale, theoretically lead to a market response where prices for a conserved good fall, which may lead to increased consumption of the good by other consumers, who might not have otherwise utilized the good at all.

Environmentally, this has strange implications. For example, if I live in a quickly growing city like Atlanta, Georgia where a series of droughts has curtailed water availability in the region, it may make sense to conserve water, to not leave my tap running, and to restrict my planting of outdoor lawns. Indeed, it may seem sensible to impose such restrictions by regulation throughout the city. But if all my neighbors practice the same methods, or if we regulate our behavior as a group, this only serves to increase the supply of water available. The result would be a decline in water prices, which actually encourages the growth of new subdivisions, which might not have appeared under conditions of higher water prices. These new subdivisions make demands not only on water, but also on land and new services. In this case, water conservation has theoretically led to more water use and environmental stress.

**Jevons' Paradox**  The somewhat counterintuitive observation, rooted in modern economic theory, that a technology that increases the efficiency of resource use actually increases, rather than decreases, the rate of consumption of that resource

This apparent contradiction is typically referred to as **Jevons' Paradox**, named for the early economist William Stanley Jevons who observed that coal demand actually increased when more energy-efficient steam engines were developed in the 1800s. By analogy, do hybrid sport utility vehicles (SUVs) reduce the use of petroleum, or instead artificially dampen otherwise rising gas prices, leading to increased road mileage in the population overall? If the latter is true, is personal or community conservation a good idea or does it make more sense to simply tax gas across the board (see below)? This also raises some profound questions about counting on innovation (solar energy, water purification, etc.) to reduce consumption of resources. If anything, Jevons demonstrates that, at least under capitalism, innovation may accelerate resource decline, degradation, and other environmental problems.

## Managing Environmental Bads: The Coase Theorem

Environmental problems and issues are not always about the scarcity of discrete, individual things, like copper or oil. How can market thinking apply to all environmental objects and conditions (e.g., swimmable streams, diverse rainforests, clean beaches), not just traditional commodities (e.g., milk, tungsten, codfish)? In theory, if human activity makes clean air or water scarce, and I value clean air or water, I should be willing to pay for remediation of the situation. If *someone else* is polluting that air or water, however, and I am experiencing the effects of that deterioration downstream or further away, how could *my* willingness to pay be realized?

Many economists have addressed this problem, but the most prominent solution came from a Nobel laureate in economics, Ronald H. Coase. In 1960, Coase laid out one of the founding propositions of contemporary economics (often called the "**Coase theorem**") which held that in cases of competing interests, the most efficient outcomes will occur through bargaining between property owners (Coase 1960).

> **Coase Theorem** A thesis based in neoclassical economics, holding that externalities (e.g., pollution) can be most efficiently controlled through contracts and bargaining between parties, assuming the transaction costs of reaching a bargain are not excessive

What Coase proposes is that many environmental problems can be solved most effectively through contracts. Consider, for example, a family living in a beautiful valley spot in Montana next to a cattle ranch. As it turns out, the cattle ranch is both noisy and somewhat smelly. Such an effect, where one person's economic activity comes at the expense of another, is called an externality, and as we will see below and in Chapter 4, it is common to environmental problems. Such situations also cause lots of nasty arguments in Montana, moreover.

How can such a situation be regulated? One solution might be for the county government to ban ranching in the area, in order to protect homeowners, causing the ranch to shut down. Another would be to make it clear that cattle are more important than housing in rural areas, thus shutting down housing developments in the area. Alternatively, complex rules could be designed, telling developers how to build smell- and sound-proof housing and cattle ranchers how to move and keep their cattle more quietly. None of these solutions is necessarily socially fair or just to one side or the other. The last solution, while perhaps fairer, may be economically inefficient and far more expensive than the other alternatives. Who pays for these redesigning efforts and on what terms? What approach would produce an optimal outcome?

In Coase's way of thinking, it would be better to let the two parties sort the problem out themselves through contracts and, in the process, discover the *real* costs and values of ranching, mountain views, and cattle smells. By coming to terms with one another, *whatever they decide* is always the most efficient outcome, no matter what the initial rights are in the situation.

If the family moved in after the ranch had long been in business, for example, or if the family had no legal right to limit the rancher's actions, they would simply have to tolerate it, or absorb the cost of the smell by paying some higher price for property elsewhere, and moving far away from the ranch. The difference in the price of their current house and the new one reflects how much the people are willing to pay to not live around cattle. Perhaps, however, the cost of the new home is far higher than the cost they might be able to pay the rancher for simply moving her cattle shed to the other side of the property. By paying the rancher a lower sum, the family maintains its pretty view and reduces the smell while offsetting the rancher's costs for doing so. In that case, the price of the environmental nuisance is discovered through the negotiation between the two parties (rather than by relying on a regulator), and everyone experiences an optimal outcome.

If, on the other hand, the family came first or it held some kind of county-given right to a non-smelly environment, other options emerge. In that case, the cattle owner might cease ranching, at whatever cost, and sell to a housing developer. It may be even cheaper for the rancher to directly pay the homeowners, essentially bribing them to put up with it. On the other hand, if it turns out to be even less expensive to move the cattle shed than it is to pay off the neighbors, the rancher can always choose this option.

Notably, it does not matter what the configuration of legal rights is at the outset of the scenario – ranchers might hold the right to ranch, families might have the right to be free of cattle smells, or neither may pertain. In any case, if contracts can be worked out and enforced between the two parties, they always reach a decision that is economically most efficient. Such a determination is fully in line with the market response model, but extends its crystalline logics to the complex world of environmental **externalities**.

> **Externality**  The spillover of a cost or benefit, as where industrial activity at a plant leads to pollution off-site that must be paid for by someone else

Coase stipulated two key assumptions, however, that were required to be true for such smooth efficiencies to prevail: property rights have to be exclusive and the transfer and protection of contracted rights has to be free. This means, for the efficiencies of Coase to be realized in the above example, 1) both the rancher and the homeowner must have the full ability to control their land and the decisions made on it and, more importantly, 2) their negotiations and contracts must not cost time or money to negotiate, write, and enforce. Put in other terms, a free market system is efficient to the degree that actually sorting out agreements, coming to understandings, and designing fair rules and restrictions are socially and economically free or cheap. Similarly, it depends on enforcement (policing, monitoring, and punishing violations) of contracts and rights having no costs.

And in reality, of course, this is totally untrue. For our rancher and homeowner, the time it takes to negotiate the contract, the possible cost in lawyers, and the hidden cost of maintaining county court houses and civil servants to process and administer the contract are indeed quite high. Defining property rights for more intangible goods and services (e.g., biodiversity) is even more daunting. Enforcing contracts over incredibly complex systems (e.g., global climate) appears all the more impossible. The problem comes in the practical difficulty of assigning private rights to "fugitive," mobile, intangible things – like air – clearly contracting the relations between owners, and enforcing the results. For an "air market" to function, the air must be owned by someone who paid for it, who can get value from it, who has an individual interest in keeping it clean, and who can legally challenge someone else who dirties it or violates a contractual agreement over its condition.

While in some ways such a market seems inconceivable (because it is difficult to enclose, see Chapter 4), recent evidence suggests that it may be possible. Rather than assigning rights to clean air, recent efforts in the United States and elsewhere have worked to give *rights to pollute*. In a specific example, in the early 1990s, the United States Environmental Protection Agency (EPA) set limits on industries for the emission of sulfur dioxide, a primary cause of acid rain. But rather than setting a limit on each factory, the total allowable level of pollution was divided into units and distributed to producers, in the form of credits that could be sold. Should a company find a cheap way to reduce their sulfur dioxide production below the level for which they held credits, they could sell the spare credits for a profit. More radically, if environmental groups believe that the limit on total emissions provided in the credit system is too high, and they are *willing to pay* to reduce it, they have the right to buy credits on the market, like anyone else, and simply take them out of circulation. This also has the effect of raising the scarcity and cost of pollution credits, creating incentives for industry to become even more efficient. Such a market has been in operation for 15 years and is only one of many such efforts (see more on "cap and trade" below).

Whatever the flaws in such a system, it demonstrates that markets can function for all sorts of environmental goods and services, but that – as Coase suggests – to make them work, private property rights to nature must be clearly assigned to corporations or people. This is a logical and practical prerequisite to any market solution, but certainly one with serious social, environmental, and political implications as we shall see.

## Market Failure

There are several ways in which these market and contract-governed ways of living in nature might fail. Such **market failures** emerge from a mismatch between the assumptions of the market model and the real world. Chief challenges to market assumptions include the facts that 1) transactions are not by any means free (as per Coase's assumptions), 2) contracts and property rights have to be defined and enforced often at great legal and regulatory expense, and 3) not all parties to negotiations have perfect and equal information. This is especially the case for environmental goods and services that are spread across a large population of individuals.

> **Market Failure**  A situation or condition where the production or exchange of a good or service is NOT efficient; this refers to a range of perverse economic outcomes stemming from market problems like monopoly or uncontrolled externalities
>
> **Transaction Costs**  In economics, the cost associated with making an exchange, including, for example, drawing a contract, traveling to market, or negotiating a price; while most economic models assume low transaction costs, in reality these costs can be quite high, especially for systems with high externalities

Consider, for example, the problem of the ranch and residence provided above, but now imagine it with thousands of scattered homeowners and hundreds of ranches. Under such circumstances, the complexity of working out discrete contracted negotiations becomes enormous, as does the problem of monitoring and enforcing the rights of different ranchers each operating with different rules with differing property owners.

There is also always a temptation for some people to wait to accrue benefits from other people's negotiations without taking the time or energy to negotiate for themselves. Such a "free-rider" problem is typical of common property environmental problems (see Chapter 4). In such a case, the **transaction costs** of getting the problem sorted out contractually are simply much higher than the cost of the problem. Typically such cases lend themselves to regulatory and treaty-based, rather than market-based, solutions. For example, the system of tradable pollution credits that made acid rain reduction a success in the United States had to be created and enforced by the Environmental Protection Agency of that country, an entity with police powers paid for through federal taxation. Consider too, that reduction of sulfur emissions in Europe was managed through the creation of the "Helsinki Protocol," a treaty agreement to achieve 30 percent reductions by 21

**Monopoly**  A market condition where there is one seller for many buyers, leading to perverted and artificially inflated pricing of goods or services

**Monopsony**  A market condition where there is one buyer for many sellers, leading to perverted and artificially deflated pricing of goods or services

nations signing the document. Efficient markets require public investments. Free markets are rarely free.

Other asymmetries also plague markets. One of the most serious is that of **monopoly**, where many buyers face one service provider or owner, or **monopsony**, where many sellers face a single buyer. In either case, the individual or firm is in a position to set prices and buy and sell goods or services free from competition and with no incentive to be efficient. Neither are such cases rare; the histories of the American and European capitalist economies are filled with cases where monopolies and monopsonies emerged through the concentration of wealth (in railroads, meat packing, and communication, among many). For environmental goods and services, the record has been equally spotty. Most municipal water provision in the United States, for example, was developed by private companies in the 1800s. The failure of these utility monopolies to efficiently manage and price water, however, led to the transition of most such utilities to state control.

A further problem is raised if we consider that many of the potential parties in a contractual arrangement or in a market have not yet been born. People may negotiate with one another over the relative value of cutting down a forest or enjoying its timber for construction, but what about people a hundred years from now? Do they have a place in such a market? A strict adherent to market logics would make no provision for such people. Getting environmental economics right is difficult enough without considering future generations, after all! Alternatively, it could be argued that both economic development and conservation in the present, in whatever market-negotiated combination, are always in the interest of future generations, who benefit from better economic and environmental conditions.

## Market-Based Solutions to Environmental Problems

We have seen that markets can fail and even vociferous supporters of market environmentalism admit the necessity of some form of regulatory guidance of the economy. Nevertheless, a range of market-based policy solutions have been introduced in recent years to solve countless environmental problems. These each, in some way, for better or for worse, use the concepts of incentives, ownership, pricing, and trading to address environmental problems (Table 3.1, p. 38).

### Green taxes

One of the most direct ways to harness the market, and therefore to influence environmental decision-making by people and firms, is through artificially altering prices. According to the market response model, after all, it is increasing prices that drive providers to search for new sources, innovators to substitute, consumers to conserve, and alternatives to emerge. Taxing certain goods or services, and so increasing prices, should result in either decreased use of these resources or creative innovation of new sources or options.

## Box 3.1  Peak Oil

Petroleum is a fossil fuel, formed from biotic materials (like ancient fern forests) in the deep past through geological pressure beneath the earth's surface. This means that all the petroleum there will ever be, over human lifespans, already exists. Each drop of petroleum withdrawn from the reserves of the earth is one drop closer to the inevitable limit of supply. If this is the case, it is reasonable to assume that there comes a point in time when oil production reaches its highest level, after which scarcity makes extracting the remaining supplies harder. At this point we will begin to draw upon the remaining "second half" of the earth's supply, tapping a dwindling, scattered, and increasingly difficult-to-access supply. That theoretical moment, sometimes called "peak oil," is more formally termed "Hubbert's Peak" for M. King Hubbert, a geoscientist who famously predicted in 1956 that US oil production would peak around 1970, which it did.

Following on from Hubbert's work on individual oil fields, and aggregating over the entire geosphere, debate concerning peak oil rages on, largely over when it will arrive, if it hasn't already. The intuitive nature of the concept makes it extremely compelling and many peak oil advocates suggest that the point of decline cannot be more than a decade away, or has more likely passed already, perhaps in the 1970s.

The potential impact of declines in oil supplies after a peak is also somewhat frightening. Once peak "production" is past (production here is taken to include mining/drilling the raw petroleum, as well as processing and distributing it), it can be assumed that prices will rise steeply and fluctuate far more wildly, leading to uncertainty, speculation, and exponential increases in costs for transportation, home heating and cooling, and industrial production, all of which depend heavily (and often exclusively) on petroleum worldwide. Many suggest that the economic disorder of a rapid breakdown in the petroleum economy could easily ricochet into political and social disorder, spinning scenarios of disaster.

Critics of peak oil present two central arguments. First, many argue that the reserves of oil are simply larger than others acknowledge and that improving exploration technology continues to find new reserves all the time. Those maintaining a market-oriented approach, secondly, point out that any immediate decrease in available supply of oil is communicated through the market response model to decreased demand and more capital availability to find new sources or substitutes. Consider that during the dramatic price run-ups of 2006–7, when gas prices in the USA went from $2 per gallon to almost $5 per gallon, more fuel-efficient vehicles went on the market and Americans actually *decreased* their driving by more than 4 percent over one year, the largest such dip ever recorded. At the time of writing, a general economic slowdown has led to less demand and lower fuel prices in all sectors.

On the other hand, oil exploration technology is an advanced science, and the balance of evidence suggests we are likely well past the peak. Moreover, substitution has not shown a propensity for sustainable technology, with advocates of coal-burning, nuclear power, and biomass fuels continuing to win most arguments over either more sustainable production technologies or – more compellingly – a curtailment of consumption. Rates of petroleum use continue to rise throughout the world on a planet where petroleum is ultimately limited. There will come a time after oil, and it is likely close at hand. Just what that transition is going to look like, whether a shift to a green economy or a spiral into an unprecedented global economic catastrophe, is another question.

The money raised through the tax can be used directly by the government either to provision services or to search for alternatives.

Many examples of such "green taxes" exist. Facing landfill costs, labor expenses, and related costs in the provision of garbage disposal, for example, some municipalities have required households to dispose of all waste in special trash bags, purchased by consumers

**Table 3.1** Market-based solutions. An overview of some dominant environmental regulatory mechanisms that involve market components and are based in part on market logics. Note that in all cases the state remains an important player in making markets work and achieving environmental goals

| Regulatory mechanism | Concept | Market component | Role of the state |
|---|---|---|---|
| Green taxes | Individuals or firms participate in "greener" behavior by avoiding more costly "brown" alternatives | Incentivized behavior | Sets and collects taxes |
| Cap and trade | Total amount of pollutant or other "bad" is limited and tradable rights to pollute are distributed to polluters | Rewarding efficiency | Sets limits and enforces contracts |
| Green consumption | Individual consumers choose goods or services based on their certified environmental impacts, typically paying more for more benign commodities | Willingness to pay | Oversees and authenticates claims of producers and sellers |

themselves, and often costing a dollar or more each. The results have been greatly increased recycling and more careful attention by consumers to packaging and waste. By internalizing the costs of trash to consumers, there has been an observed decrease in the flow of garbage from households.

More radically, such taxes have been proposed for the control of greenhouse gases that drive global warming. Sweden enacted a carbon tax in 1991, followed by other countries, including the Netherlands, Finland, and Norway. This tax is leveled against oil, natural gas, coal, and a range of fuels. Such taxes have also been considered in the United States and the European Union, although they face significant political opposition.

## Trading and banking environmental "bads"

**Cap and Trade** A market-based system to manage environmental pollutants where a total limit is placed on all emissions in a jurisdiction (state, country, worldwide, etc.), and individual people or firms possess transferable shares of that total, theoretically leading to the most efficient overall system to maintain and reduce pollution levels overall

Also prominent among these market approaches are policy efforts that draw upon Coase's insights to reduce environmental problems as efficiently as possible, using contractual exchange. Such mechanisms usually take the form of "**cap and trade**." Here, the state determines a regulatory maximum for emissions of an environmental hazard and allows firms to meet the goal themselves or to pay other firms, who are able to reduce outputs more efficiently, to do

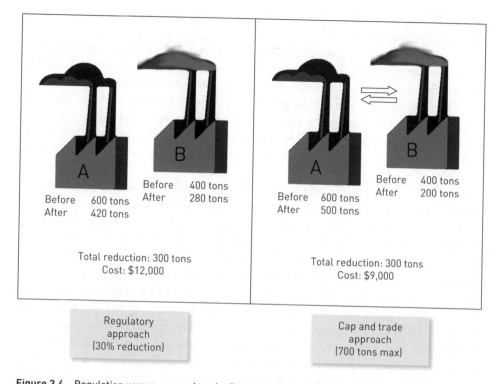

**Figure 3.4**  Regulation versus cap and trade. Both approaches result in net desired pollution reduction, but the cap and trade approach is theoretically cheaper overall.

so for them. This achieves the same results as traditional regulation, but does so at lower overall cost (and so economists describe this outcome as "more efficient").

As shown in Figure 3.4, regulation can be used to reduce emissions in an industrial setting by demanding a 30 percent cut in tons of pollutants for both factories A and B, resulting in a total removal of 300 tons of pollution from the atmosphere, leaving 700 tons overall. Significantly, however, Factory B, owing to its technology and system of production, can eliminate pollution at a cost of $25 per ton, while the older Factory A requires $50 per ton invested to do the same. Rather than spending an overall total of $12,000 to reduce pollution, therefore, it would be more efficient to simply cap the total amount of pollution at 700 total tons and allow the two firms to trade permits if they desire. In that case, Factory A might make some reductions of its own, but purchase the remainder of pollution credits from the more efficient Factory B, whose aggressive reductions result in meeting the net target. Both systems then result in the same amount of pollution reduction, but cap and trade does so more efficiently.

The basic idea is that firms who are able to reduce emissions more cheaply (because of available technology, know-how, or experience) can do so *for* other firms who are less able, and then be paid for the trouble. The US $SO_2$ "Acid Rain" trading system began in 1995 and reports significant emissions reduction. Proponents claim that the trading system yielded 30 percent more reductions than non-flexible methods, where every factory had to meet the exact same requirements (see the bold claims of the global shipping industry, for example, at www.seaat.org/).

Such systems are not without problems and limitations, obviously. No allowance here has been made for geography, for example. If individual factories are allowed to pollute on site, offsetting their emissions with reductions at faraway locations, the environmental effects might be locally disastrous, even if the overall reduction meets a target. The problems of actually setting the cap, determining the limits, and monitoring and implementing the reduction all remain as well. The use of the market tool also does nothing to depoliticize pollution control, since determining *how much* pollution is allowable requires difficult tradeoffs and regulatory experience.

In a variation upon this approach, some regulations allow markets to be extended to "banking" and "withdrawing" environmental services through third-party providers. Most notably, wetland protection in the United States has moved to a banking system. Here, a cap is set on the total amount of wetlands, enforcing no *net* loss of wetlands overall, but allowing that any local loss of wetlands can be offset by the creation of a wetland elsewhere, so long as it is approved by the EPA in terms of its delivery of similar "services." A new big-box retail chain, for example, which destroys a wetland in the construction of a new store, is required under law to create a similar one somewhere else. Since building wetlands is a specialized practice, however, probably out of the technical and ecological capacity of the person or firm who destroyed the original ecosystem, a third party might be called upon to create and manage the new wetland. More radically, a savvy developer might build an enormous wetland or set of wetlands – a kind of bank – speculating that new constructions will need to purchase portions or shares in the future to meet their obligations.

Such a system presents any number of practical problems, which raise questions about the ecological efficacy of any such effort. In the case of wetlands, it is necessary, for example, to assure that newly built environments actually deliver the ecosystem services lost in the destruction of the original landscape. Such monitoring and oversight demand extensive regulatory efforts, many ecological experts on site examining the system, and careful scrutiny over time. Indeed, a market-based approach, often vaunted for its ability to cut out government intervention, may demand an *extension* of state regulations, with increasing numbers of state scientists and monitors, paid at the public expense, assuring the legitimacy of transactions in the market (Robertson 2006). Cap and trade mechanisms have been roundly criticized, especially for greenhouse gases (see Chapter 9), but even its proponents concede the inevitability of a strong role for governmental regulation in fostering and maintaining a market in the first place.

### Green consumption

Market-based solutions to environmental problems also tend to stress the power of consumer demand for changing environmental conditions. Pointing to an overall social shift towards green values, market advocates suggest that the most powerful way to change production systems is to allow consumers to "vote with their money," and select and purchase green products, often at a premium. The success of organic foods, which are now very much in the mainstream of consumer culture, provides an example of consumers paying extra and creating incentives for more producers to change their methods and technologies.

Such approaches have drastic limitations. How does a consumer know, after all, the specific environmental impacts of products that carry handsome green labels? How do they know such products and companies have not been merely "**greenwashed**," presented and advertised heavily as environmentally sound or benign with little substantive change in actual production practices, packaging, or disposal? Indeed, for many firms far more time and money are invested in green advertising than green practices (TerraChoice Environmental Marketing Inc. 2007).

> **Greenwashing**  The exaggerated or false marketing of a product, good, or service as environmentally friendly
>
> **Green Certification**  Programs to certify commodities for the purposes of assuring their ecological credentials, such as organically grown vegetables or sustainably harvested wood products

One way of confirming truth in green advertising is through the process of **green certification**, in which a third party monitors production of a range of products and provides a confirmatory "seal of approval" for products meeting specific standards. A number of governmental and nongovernmental green certification systems exist, including those for timber, organic foods, and energy-efficient appliances, among many others. As green certification systems continue to proliferate, however, their reliability and consistency become more questionable. Some certifications, for example, are established by companies themselves rather than third-party observers. Moreover, many countries have adopted their own standards. For example, the country of Malaysia has instituted its own independent certification for sustainable timber, which competes directly with international standards. This makes global trade in eco-friendly goods a confusing smorgasbord.

## Beyond Market Failure: Gaps between Nature and Economy

Leaving aside problems in implementing market-based environmental policy, larger questions loom. Adopting an economic logic for nature presents more basic problems: it makes it difficult to maintain ecocentric values (see Chapter 5), because there is a mismatch in the behavior of money and ecosystems, and because basic economic inequity presents a barrier for reconciling markets and the environment in a socially just way.

### Non-market values

At bottom, the market response model can only be deemed to be operating successfully if success is measured in strictly economic terms. When sperm whale oil became scarce and was replaced with fossil fuels, as actually happened in the nineteenth century, the outcome for people was negligible or beneficial. New resources, propelled by the scarcity of old sources, were made available through the magic of the marketplace. But what is the lesson of this story for the sperm whale itself? Long before market responses "kicked in" to send humans in search of new oil sources, the sperm whale was driven to the brink of extinction, a result that was only forestalled by the creation of international bans on whaling in the twentieth century. Indeed, global green markets have proven slow to adjust to the decline of rainforests, the plunge in biodiversity, and the potential catastrophic implications of global warming, raising questions about the capacity of trade to capture the value of these things, at least within the urgent timeframe in which they will need to be addressed.

The problem is therefore not simply that the market may fail on its own terms (and indeed it does by the admission of many of its most enthusiastic supporters), but that its success can only be judged in economic and therefore anthropocentric terms. If there are values that cannot be captured in a market, like the evolutionary, aesthetic, or moral "value" of a species, what difference does it make that its depletion eventually allows its *substitution* by something else? While economic valuation allows us to identify what people might be willing to pay for abstract and intangible goods and services (i.e., the presence of sperm whales somewhere on the planet), it does not necessarily point a way for them to be valued, in monetary terms.

## Money and nature

Beyond this, the valuing of ecological conditions through the market, and specifically in money terms, is fraught with other very basic problems. First, the history of capitalism has shown that markets are highly volatile, and given to bubbles and busts. This is not necessarily a bad thing for capital, which travels from crisis to crisis, moving from investments in forestry to plastics to biofuels. But these rapid fluctuations of money values of different natural objects, typically driven by speculation, may be out of step with both cycles of environmental systems and changes in social values. Without recourse to some other system of valuation, however, these eruptive metrics are the sole measure in a market. As geographer David Harvey describes, this becomes "a tautology in which achieved prices become the only indicators we have of the money value of assets whose independent values we are seeking to determine. Rapid shifts in market prices imply equally rapid shifts in asset values" (Harvey 1996: 152). But the crashes and crescendos of historical markets continuously show that market volatility may not reflect the social values they are understood to measure. Can we trust a turbulent commodity exchange market to reflect the slow and steady pace of changing environmental values?

A commodity approach to the environment also tends to stress the exchange of discrete and specific items or services. The complex ecosystem of a river is most effectively managed in a market through the discrete components that can be valued by differing parties. For example, the river in one condition or another may be able to provide wetlands or offer flood control, or to maximize trout habitat, or to facilitate transport or recreation. In theory, to the degree to which these services are mutually exclusive, a market can best adjudicate what is most desired and provide it from the river. In reality, however, these functions are connected and interdependent in many ways, precisely because the ecology of streams is complex. Discrete markets are "anti-ecological" in that many of the river's unvalued parts may be valuable to other components, mutually produced, and interdependent. By separating the system into marketable services, these interrelations are severed, displaced, and divided. Can the functioning of whole ecosystems be assured in markets that capture only the value of discrete goods and services?

## The crisis of equity: Turning economic injustice into environmental injustice?

Applying the logics of the market to the environment raises basic questions about equity and rights. This is because, to the degree that the environment is "marketized," the ability of individuals and groups to participate in environmental action and remediation, or indeed even to have access to basic environmental services (e.g., clean air or wilderness), is limited by their available capital. This holds implications for democracy. By "democratic" here, and elsewhere in this book, we refer to people's ability within a society to have an equal voice in political decision-making and outcomes. Market environmentalism is democratic, therefore, only to the degree that the financial resources available are equally distributed throughout the population.

Nothing, of course, could be further from the truth. Turning decisions over nature into decisions within a market can be considered undemocratic because money is almost never evenly distributed within a polity. In the United States, an enormously wealthy country by global standards, the richest fifth of the population received 49 percent of the nation's income in 1999, while the poorest fifth received less than 4 percent. In terms of overall wealth, in 1998 the top 5 percent of the population owned 60 percent of the country's wealth. Globally, the statistics are more striking; the richest fifth of the world's population earn 83 percent of all income and the richest 10 percent of adults control 85 percent of the world's total assets. The specific concentration of wealth and income in the hands of corporate entities, rather than people, is also notable. So, too, is the unevenness over control of money and finances within households around the world, where women may be excluded from access to and control of money, even though their labor and effort provide household income. Given this reality, making politically charged environmental decisions dependent solely on economic willingness to pay may represent a profound subversion of democracy, in a world where ability to pay is so widely uneven.

As Sharon Beder has argued, moreover, the appearance of economic and scientific neutrality that market-based solutions possess may in part further disguise their fundamentally political characteristics:

> The portrayal of economic instruments as neutral tools removes them from public scrutiny and gives them into the hands of economists and regulators … A market system gives power to those most able to pay. (Beder 1996: 61)

Market advocates respond that individuals with even paltry means exercise significant power through market actions, especially when aggregated into large bodies of consumers. So, too, they argue that most environmental values are universally desired, and "we" lack only the means to efficiently achieve these global goals. Nevertheless, it is reasonable to ask whether current fundamental inequalities in income and capital should be the ground on which to construct environmental governance. Can markets produce not only efficient but also democratic relationships between people and the environment?

Finally, many observers insist that depending on markets to solve environmental problems is a problematic place to start considering the overall, and apparently unstoppable, growth of the global capitalist economy (see Chapter 6). As global trade continues to devour, mobilize, and dump resources, objects, and fuels at an accelerating pace, it becomes difficult to imagine how such energies could ever be harnessed and simultaneously controlled. And yet the language of "markets," "free trade," and "ecological economics" must be admitted to be the most dominant, widespread, and uncritically accepted ways of thinking about the environment in the early twenty-first century. Faith has abated slightly in the wake of recent financial shocks, but proponents continue to insist on the ability of the economy to solve problems that the market itself has had a strong hand in creating.

## Thinking with Markets

In this chapter we have learned that:

- A dominant school of thought holds that, as long as environmental goods and services can be sold or traded, scarcity will be diminished by economic forces through the market response model.
- The market response model alleviates scarcity by creating incentives that either increase the supply of environmental goods and services or reduce demand for them.
- Environmental externalities can be mediated, in this theory, through private contracts more efficiently than through regulation.
- Many market-based mechanisms therefore may exist for solving environmental problems, including green taxes, markets for pollution, and green consumer choices.
- Markets, however, can fail, raising questions about holding faith in them for consistently solving environmental problems.
- Other problems face market-based environmentalism, including the fact that some environmental goods are difficult to value, that markets can be volatile and fickle, and that economic solutions are not necessarily democratic ones.

## Questions for Review

1. Compare/contrast Julian Simon's and Paul Ehrlich's views on the general effects of human population growth on environmental conditions (include the term "scarcity" in your answer).
2. Provide an example of the "law" of supply and demand influencing the ways humans use or exploit a particular natural resource.
3. Using the logic of Jevons' Paradox, explain how putting more fuel-efficient cars on the road *could* result in increased petroleum use worldwide.
4. Which of the following environmental problems is better suited to solutions derived from the Coase Theorem?: a) land use disputes on adjacent parcels of private property; or b) reducing water pollution across a region (explain).
5. How does the ecologically complex nature of a river (or any similar "piece" of nature, for that matter) make it difficult, if not impossible, to value in monetary terms?

## Suggested Reading

Crook, C., and R. A. Clapp (1998). "Is market-oriented forest conservation a contradiction in terms?" *Environmental Conservation* 25(2): 131–45.

Field, B. C. (2001). *Natural Resource Economics: An Introduction*. Long Grove, IL: Waveland Press.

Godal, O., Y. Ermoliev, et al. (2003). "Carbon trading with imperfectly observable emissions." *Environmental and Resource Economics* 25(2): 151–69.

Johnson, E., and R. Heinen (2004). "Carbon trading: Time for industry involvement." *Environment International* 30(2): 279–88.

Randall, A. (1983). "The problem of market failure." *Natural Resources Journal* 23: 131–48.

Rees, J. (1990). *Natural Resources: Allocation, Economics, and Policy*. New York: Routledge.

Robertson, M. M. (2006). "Emerging ecosystem service markets: Trends in a decade of entrepreneurial wetland banking." *Frontiers in Ecology and the Environment* 4(6): 297–302.

Sen, A. K. (2001). *Development as Freedom*. Oxford: Oxford University Press.

Taylor, P. L., D. L. Murray, et al. (2005). "Keeping trade fair: Governance challenges in the fair trade coffee initiative." *Sustainable Development* 13(3): 199–208.

TerraChoice Environmental Marketing Inc. (2007). *The Six Sins of Greenwashing: A Study of Environmental Claims in North American Consumer Markets*. Reading, PA: Author.

## Exercise: The Price of Green Consumption

Go to a grocery store or supermarket near you. Select four or five different types of products (for example: fruits, vegetables, packaged goods, meats, paper products, cleaners, etc.). Find a conventional version of this product as well as a "green" alternative. This may include an "organically" grown fruit or vegetable, a "free range" meat, "locally grown" produce, "green" or "eco-friendly" products, or products made from "recycled" or "recovered" materials, for example. What is the price difference (per unit where appropriate) between the "green" and conventional versions of each product?

What is the average percentage increase in cost of the groceries if your "green" products are selected instead of conventional ones? The average American family of four spends $8,500 per year on groceries (the average British family spends approximately $6,300). Assuming your percentage increase is typical and that all conventional groceries have "green" alternatives, how much more would the average family have to pay for only "green" groceries? Who can afford to pay such extra costs for groceries?

What is the benefit from this extra cost? Why are "green" alternatives more expensive to produce? Where does the extra money spent on each product go? How would you know? Where would you go to find out?

# 4

# Institutions and "The Commons"

Credit: Eric Vidal/EPA/Corbis.

## Keywords

- Common property
- Game theory
- Institutions
- Prisoner's Dilemma
- Social Darwinism

## Chapter Menu

## Controlling Carbon?

There is a growing consensus on the science of climate change and the urgency of the problem. According to the Intergovernmental Panel on Climate Change, whose job it is to vet the science surrounding global warming, global temperatures are rising, sea-level rise is expected, water availability may be impeded, and extreme weather events are likely to become more common and more severe. These changes will be felt unevenly but no country will be immune to such effects.

*Given the widespread and severe nature of the problem, why are global carbon emissions so hard to control?* One might think of many good reasons for this to be true. People tend to ignore problems they cannot see, governments are under the influence of oil companies, and so on.

One compelling argument holds that the root of the problem is that carbon quite simply *does not stay put.* With every combustion event (driving a car, burning a log, firing a coal plant for electricity, etc.) the carbon that is released quickly finds its way into the atmosphere. Carbon released in one country is – *instantly* – a burden shared by all countries. From the point of view of any individual country, moreover, carbon reductions are not "free," since they require creating new rules, potentially reducing or redirecting economic production. A "carbon-reduced" product – whether a car, computer, or vegetable – may be more difficult to produce than its "status quo" counterpart. If such a product is more expensive as a result, it may not be competitive on a global market, especially if it must compete against products produced where carbon reductions have not been imposed. Many governments, including that of the United States, therefore express fear that if they make sacrifices in this direction while others do not, they will no longer be competitive. The benefits of carbon reduction, in a parallel way, would be experienced by all countries, but must be paid for by individual countries.

For many environmental problems, costs are often borne collectively, while benefits accrue to individuals; on the other hand, individual costs may lead to collective benefits. Nations must cooperate where there is no "super-government" to impose environmental laws and where there are few or no punitive measures for not cooperating. Such mismatches mean that hoping for spontaneous global action on climate change might be unrealistic. Arguably, some kind of rules and trust must be established to cooperate when incentives constantly lure people and states to follow their own self interest … towards collective planetary ruin. In this chapter, we address this persistent and vexing problem: *How, if at all, can rules and norms of global behavior be fashioned to encourage shared costs and collective benefits? At what scale is cooperation possible?*

## The Prisoner's Dilemma

Such questions are of universal concern since problems of this sort are ubiquitous. There are countless examples from our own daily ecologies as well as those around the globe.

Neighbors, for example, need to pull weeds from their lawns to keep them from spreading across property lines. By failing to do so, any one non-cooperating neighbor enjoys some benefits from their neighbors' sweaty work (fewer dandelions in their yard), but simultaneously provides an ongoing seed source for dandelions that might spread to the lawns of others.

It is tempting for someone to spill the small quantity of paint left over from renovating a bathroom into a storm drain rather than take the time to dispose of it legally at an approved dump site. Should a single household be the only one to behave in this way, it enjoys a lower labor burden *and* benefits from the healthy environment resulting from the majority of hard-working cooperative neighbors who do not dump paint into public waterways. But since such temptation exists to all, the incidence of paint dumping is far higher than anyone individually desires.

Nor is the situation of a homeowner in urban America or Europe entirely different from that of a villager in rural India. There, the benefits from a community forest include the ability to graze livestock in the shade of trees and the availability of seeds and pods that fall from trees, among other resources. These collectively enjoyed benefits are only available if the trees of the forest are not cut down for fuel or construction timber. It is inviting for any individual household to cut down a tree or two for their own purposes, making only an incremental change in the density and health of the village forest. As long as others refrain from doing the same, these minimal costs are spread across the village and the benefits to the household, which may need to go great distances to collect fuelwood otherwise, are obvious. But again, since the incentive to break the rules is shared by all, and in anticipation of the fact that others may well "cheat" and cut trees, householders might well be tempted to cut trees as soon and as quickly as they can, in an effort to gain resources before their neighbors do.

In all of these cases, cooperation is necessary for the best outcome, but an incentive exists to take a "free ride" and let others invest time or money or restrain their behavior while we do not. Since the incentive exists for everyone, the possibility of total failure always looms.

A popular metaphor for situations such as these is the story of the **Prisoner's Dilemma.** In this tale, so often replayed on contemporary television crime dramas like *Law and Order* or *The Closer*, you are asked to imagine two people charged with a crime, perhaps something like burglary, in which the best evidence that the police have (though not the only evidence) would be the testimony of one of the burglars against the other. Both suspects are taken into custody by the police and interrogated individually. Each is separately told that if they testify against the other person, they will go free or have a greatly reduced sentence. The logical decision in a perfect world is for both of the suspects to keep their mouths shut. With no one talking, they might each serve a short jail term, if any term at all. The problem stems from the fact that each knows that the other might well "rat" him out. If one chooses not to talk while the other "rats," the reticent one will do hard time while the other one goes free. Since no one wants to be a "sucker" and suffer at the hands of the other, the predictable outcome is that both testify against the other, leading to the worst possible mutual outcome, one in

**Prisoner's Dilemma** An allegorical description of a game-theoretical situation in which multiple individuals making decisions in pursuit of their own interests tend to create collective outcomes that are non-optimal for everyone

which both do hard time. Trying to anticipate and avoid the punitive "defection" of the other person, the two partners act in service of the police. Such a situation not only invites each person to "rat" out the other, it actually makes it *rational* to do so, even though the outcome is inevitably bad for everyone, something no rational person would choose if they could control the behaviors of *both* parties.

These sorts of fascinating dilemmas are the province of **game theory**, adopted by thinkers who employ a form of mathematical analysis of decision-making, examining those sorts of situations that might be expressed in game terms. For game theorists, certain games provide models for how people think and behave. Of most interest are games where the crux of the problem is anticipating what the other player might choose to do, where bluffing and second-guessing are paramount. Such is the case of our two prisoners, who individually reach what is ultimately a bad mutual decision because they are trying to *anticipate* what the other prisoner might decide to do. Along these lines, and as established by the refugee scientist John von Neumann after World War II, game theory understands a "game" to be "a conflict situation where one must make a choice knowing that others are making choices too, and the outcome of the conflict will be determined in a prescribed way by all the choices made" (Poundstone 1992: 6).

> **Game Theory** A form of applied mathematics used to model and predict people's behavior in strategic situations where people's choices are predicated on predicting the behavior of others

In game theory, as in the Prisoner's Dilemma, the best outcomes are typically achieved by cooperating while individual incentives tend to lead to a reason not to cooperate. The problem of our two suspects might be expressed in abstract terms by game theorists as in Figure 4.1. These games can be made ever more complicated by introducing quantitative cash rewards and punishments and making the decisions more complex, involving, for example, investment in a public good versus a private bank, and so on. Complex mathematics can be used to predict behaviors, including retaliations, mutual learning, and system collapse.

|  | Prisoner B "clams up" | Prisoner B "rats" |
|---|---|---|
| Prisoner A "clams up" | Both prisoners do light sentences | Prisoner B walks free, Prisoner A goes to jail |
| Prisoner A "rats" | Prisoner A walks free, Prisoner B goes to jail | Both prisoners do hard time |

**Figure 4.1** The Prisoner's Dilemma in game theoretical terms. The best outcome, in the upper left, is also the least likely, since each player prefers to avoid the worst individual outcomes, in the upper right and lower left corners, leading to the worst collective outcome, in the lower right.

## The Tragedy of the Commons

The first applications of game theory mathematics were directed to the Cold War logic of mutual nuclear annihilation. Funded by the RAND Corporation and the US Pentagon in the 1950s, game theorists asked the unthinkable: Is it rational to strike first with nuclear weapons, not knowing whether and when the enemy might do so? Is it rational to drop the bomb?!

But what might any of this have to do with environment and society? Applying this kind of thinking to our interactions with the natural world leads to some potentially grim and tragic conclusions. For while there is a possibility of cooperation around environmental conservation, there is a potentially overriding incentive to "defect," in the language of game theory, leading to a general inability to manage or control our consumption and use of the environment, and so to environmental destruction.

Thinking along these lines, Garrett Hardin presented one of the most compelling, persistent, and in some ways problematic arguments linking environment to society through the commons. In his article "The Tragedy of the Commons," published in *Science* in 1968 (where von Neumann is prominently cited), he directed this logic to the problem of overpopulation. He argued that while the advantages for any individual or family of reproducing freely are immediate, their costs are diffused across the planet, increasing incrementally the burden of humanity upon the earth. This is a Prisoner's Dilemma, insofar as some people may choose to forgo more children in the interests of the planet, but others will inevitably "defect" or cop a "free ride." The worst outcome is much more likely (now sometimes called the "Nash Equilibrium" for its mathematical discoverer John Forbes Nash, made famous in the film *A Beautiful Mind*), at least without some form of coercive restraint on people's behavior. Overpopulation, as this logic goes, is inevitable without some form of enforcement mechanism (Chapter 2).

The article was made more compelling by its use of an agricultural metaphor for the problem. Rather than think directly about people's reproduction, Hardin asked us to "picture a pasture open to all …" in which numerous herdsmen managed their individual herds. Following precisely the logic of the Prisoner's Dilemma, it is in the interest of each herder to increase the size of their own herd, Hardin argued, since each new animal costs him nothing but gains him much. But since all herdsmen enjoy the same incentive, the inevitable result is a destroyed pasture. Because it belongs to everyone, the resource belongs to no one, and will inevitably be grazed into destruction. In language typical of the article he explains:

> Ruin is the destination toward which all men rush, each pursuing his own best interest in a society that believes in the freedom of the commons. Freedom in a commons brings ruin to all. (Hardin 1968: 1243)

Conscience and goodwill, Hardin further asserted, were useless in the face of compelling, internal, adaptive, evolutionary logics. Real solutions, however distasteful, must inevitably take some other form. People of the earth must choose either coercion ("mutual coercion mutually agreed upon": p. 1247), to tyrannize ourselves into control, or turn to strict forms of private property and inheritance so that all impacts of poor decision-making will be visited only upon the owner of that property. The former approach was rejected by Hardin, as he concluded that the problem with tyranny is that there is always a possibility that a system of governance will come under the undue sway of one of the users of the commons and cannot itself be controlled: "*Quis custodiet ipsos custodes* – 'Who shall watch the watchers themselves?'" (pp. 1245–6). The latter approach – privatization – was preferred and defended by Hardin since, no matter how unjust it might be (not all rich people are smart people, he pointed out: "an idiot can inherent millions," p. 1247), such a solution was the best one available. This argument for private control of common goods parallels that proposed by those with faith in markets for solving environmental problems (see

Chapter 3). In either case, whether state or private control, some form of *enclosure* was deemed as essential, where an "open access" resource is bounded and given over to control either by individual owners or by a strong state management body.

The power and influence of this argument were enormous, and remain so to this day. It continues to be perhaps the most cited academic article in the social sciences, it provided foundational arguments for fields as widespread as evolutionary biology and economics, and it is typically invoked in debates over environmental scarcity. Because Hardin convincingly used an environmental crisis as his metaphoric example, his essay on population quickly became the key defining metaphor for many people (managers and scholars) in guiding their thinking about *all* environmental problems more generally. Nature in all its forms (fisheries, oil fields, climate systems) might be seen as commons, those difficult-to-enclose systems that invite free riding and defection.

Viewed this way, the solutions to environmental problems do appear to take the form Hardin suggests: either some form of environmental super-police state, or private property rights over all environmental systems or objects. Environmental commons in this way of thinking lead inevitably to tragedy and so must be made into non-commons through the power of law and property. There is close match of this way of thinking with the market logics (i.e., internalizing externalities) reviewed in Chapter 3. This congruence of the two ways of thinking may have made the "tragedy of the commons" easier to swallow in places like the United States and the United Kingdom, where trust in markets (whether rational or irrational) is a long-established part of the culture.

## The Evidence and Logic of Collective Action

Exactly at the same time that the logic of the Prisoner's Dilemma and the Tragedy of the Commons was becoming widely accepted, confusing and incongruent evidence was beginning to mount. As anthropologists, sociologists, historians, and geographers observed resource management around the world, they continued to report behaviors that in no way fit Hardin's predictions. Specifically, they found countless examples of complex systems for management of difficult-to-enclose resources – ranging from fish, trees, and pasture to computer processing time – that relied neither upon some form of tyrannical enforcement authority nor upon the assignment of exclusive private property rights to the resources in question. Some other form of management appeared not only possible but actually predominant in natural commons around the world.

Lobster fisheries in Maine, for example, were observed to be supported and maintained less through centralized state law than through the behavior of lobster fishers themselves, who self-limited the number of boats allowed in the fishery and their distribution of traps. Village irrigation systems in southern India were shown to be carefully self-managed systems where irrigators followed careful rules concerning the opening and closing of flood gates to water their fields while keeping the array functioning for the flow of water to downstream users. Tree tenure traditions in East Africa were organized to allow the use of forest products by households who did not hold title to the land on which the trees stood, while limiting their access to sustain the harvest. Across the world, countless, highly vari-able systems of institutions appeared to violate the iron laws of the Prisoner's Dilemma through some form of local organization.

What all of these cases appear to have in common is the presence of some form of institutions, understood here as systems of recog-nized constraints on individual behavior, including formal laws, but also unofficial rules or even strong social norms that guide people's expectations of one another's behaviors, leading to orderly and con-strained use of natural resources. Such institutions, even in their most informal manifestations, can be quickly recognized all around us. People typically queue at the box office of a movie theater rather than shove one another in a pack at the ticket window, for example. For more complex problems, sustaining a fishery for example, the rules may be quite complex, and the mechanisms of self-governance and enforcement may be deeply rooted in traditional social systems. Nevertheless, the root principle applies; collective good and environ-mentally sustainable outcomes were achieved through cooperation.

**Institutions**    Rules and norms governing collective action, especially referring to rules governing common-property environmental resources, like rivers, oceans, or the atmosphere

**Common Property**    A good or resource (e.g., bandwidth, pasture, oceans) whose characteristics make it difficult to fully enclose and partition, making it possible for non-owners to enjoy resource benefits and owners to sustain costs from the actions of others, typically necessitating some form of creative institutional management

Observers of these successfully cooperative systems needed to assemble new stories, metaphors, and theories, however, to better explain what they saw. All of these revisions of "Tragedy" thinking shared one thing: a need to explain how rules and norms were able to constrain behavior and achieve cooperative outcomes.

Chief among these reconsiderations is an effort to try to define the difference between these observed systems and the property imagined in the "Tragedy of the Commons." Acknowledging that the total absence of rules would lead to tragic outcomes, it remains necessary to also acknowledge the existence of forms of property that function through customs, rules, and/or regulations, but not in the form of exclusive private ownership rights. "Common property" is a descriptor that includes all of these diverse forms. Different from wholly unowned resources (Latin: *res nullius*), common property (Latin: *res com-munes*) includes some form of group ownership, so that it is neither open to everyone in the world nor necessarily held exclusively by an individual (Ciriacy-Wantrup and Bishop 1975). Such a group or community of owners may take many forms, of course, to include people who know one another face to face in a village fishery or those more far-flung across a city, who together hold cooperative rights to urban garden plots.

But even where a right can be held by a community, this does not explain how and why members of that group might be able to achieve mutual understandings, agreements, and most importantly constraints. Rather, there must be an underlying logic that allows people to overcome, through collective action, the worst environmental case outcomes of the "Tragedy of the Commons." Returning to the game-theoretical logic of the Prisoner's Dilemma, a generation of "neo-institutionalist" scholars, led by political scientist Elinor Ostrom, began to critically analyze the assumptions of game theory. Specifically, they asked: What if our two hypothetical prisoners had been allowed to speak with one another prior to their interrogation and "get their stories straight"? Would that have changed the outcome? Logic holds that it might. Subsequent economic experiments – in which people played complex cooperation games for money incentives – bore this out as well. When the players of a game can collude together or negotiate, they are far more likely to cooperate (Ostrom 1990). And where the "payouts" or ecological benefits of cooperation increase, the allure of sacrificing immediate individual gain for larger group benefits improves. Certain conditions seem to make communal management of natural resources quite possible, and indeed likely. In this way, the "neo-institutionalists" (adherents to a school of economics that stresses rules and social organization) do not deny the logical incentives that underpin Hardin's tragedy, but do point to conditions where commons are not "free" but instead governed by rules that encourage cooperation.

Nor was this "discovery" of cooperation really all that new. Indeed, in 1888, geographer, explorer, and anarchist philosopher Peter Kropotkin published his classic book *Mutual Aid*, an account of the evolutionary advantages of within-species cooperation, documenting collective action amongst wild animals and insects and simultaneously tracking it through the history of civilization. Accepting that competition between species is an essential part of Darwinian evolution, he argued from a range of evidence that cooperation was equally essential. Rejecting social Darwinism, which holds that competition between people is both natural and advantageous, Kropotkin's book painstakingly documents how cooperation in tribes, cities, factories, and farms has led to technological innovation, increased productivity, and the rise of effective forms of self-governance around the world. In typically emphatic language, he rejects social Darwinism as pseudo-scientific, on the evidence of the countless institutions he observed around him, just as Ostrom and other institutionalists would again later:

> **Social Darwinism** The use of Darwinian evolutionary theory to explain social phenomena; as individuals are viewed as naturally and inherently competitive and selfish beings, social Darwinism typically rationalizes war, poverty, and hierarchically stratified social systems

> the nucleus of mutual support institutions, habits, and customs remains alive with the millions; it keeps them together; and they prefer to cling to their customs, beliefs, and traditions rather than accept the teachings of a war of each against all, which are offered to them under the title of science, but are no science at all. (Kropotkin 1888: 261)

Later, the economist John Commons reiterated this conclusion, suggesting that all economics, and so all relationships between people and nature, were guided by institutions, which he defined as: "collective action in control of individual action." Rather than countless individuals acting out their individual personal *preferences*, economics was instead directed by laws, rules, and norms. An efficient economy, and a sustainable one by implication, is one that "overcomes scarcity by cooperation" (Commons 1934: 1–6). Collective action, it has long been maintained by institutional thinkers, is the rule and not the exception.

## Crafting Sustainable Environmental Institutions

A century later, however, it is clear that not all forms of communal resource management succeed. Environmental "tragedies" do indeed occur, most commonly in one or a combination of the following situations: 1) when laws poorly match the ecological systems they

## Box 4.1    The Montreal Protocol

Chlorofluorocarbons (or CFCs) are artificial chemical compounds that have a range of industrial uses, including applications in refrigeration and fire fighting and uses ranging from solvents to propellants in spray cans. Part of the chemical revolution of the 1920s and 1930s, CFCs were viewed as a flexible, cheap, and efficient set of chemicals and they were widely used throughout the twentieth century. The most exciting thing about these chemicals, from an economic point of view, was that they are highly non-reactive; they break down only very, very slowly.

This blessing proved to be a curse, as soon as it was discovered that the chlorine in CFCs, when exposed to the atmosphere, broke down ozone, leaving behind more trace chlorine for hundreds more such reactions. Such ongoing reactions can lead to whole gaps in the ozone layer shrouding the earth: an ozone hole. That hole in the stratosphere was observed by scientists in the 1970s and became a source of grave concern. Because atmospheric ozone keeps deadly radiation from reaching the earth, and because trace amounts of CFCs can destroy lots of ozone, and since CFCs are otherwise so non-reactive that they can survive in the atmosphere for more than a century before becoming inert, it was determined by the late 1980s that the world had an extremely serious problem on its hands. Worse still, by that time, CFCs were employed in an unimagi-nable range of economic activities, with many industries depending heavily on the chemicals; chemical companies in particular defended their product.

If this was not bad enough, the ozone crisis represented a classic common property problem and a prisoner's dilemma. The cost for any firm or country to switch to alternatives was extremely high, so unless everyone stopped using CFCs at once, the opportunities and incentives to "free-ride" – by continuing to use CFCs and underselling other countries and companies with cheaper goods and services – were extremely high. While some individual countries and states (Oregon, for example) did create their own bans on CFC use, the problem could not be solved without collective action.

Remarkably, the world community overcame the cost and difficulty of coordinating their actions and sat down for an international meeting in Montreal in 1987. They signed a treaty in that year that resulted in a global reduction (indeed a near-total worldwide ban) on most CFCs. This Montreal Protocol was strengthened over the years with follow-up meetings and decisions, and the treaty represents perhaps the single most effective global environmental agreement in history. Many look to the successes of the Montreal Protocol as a model for future global agreements, especially surrounding the very different problem of climate change and green-house gas emissions.

are intended to govern, 2) when rules are not supported by sufficient social and political authority to be respected by the people who are expected to obey them, and 3) when norms fail to materialize at all because environmental conditions change too quickly.

So to understand how "the commons" actually works, institutional thinkers have stressed certain rules or principles that tend to lead to sustainable outcomes. In any real-world commons, the central challenge of managing the resource sustainably focuses on a number of discrete grounded problems, each of which poses difficult questions. Consider, for example, the problems of managing a fishery. Here is a resource that is largely invisible, highly mobile, depletable (if overfished), and impossible to enclose. For institutionalists, the central challenge becomes: *How do fishers avoid a "free-for-all" where each mounts increasing efforts to compete for dwindling resources, removing fish faster than the rate at which the fish population can reproduce?* The broad challenge now defined, we are immediately confronted with a slew of other questions:

- How do fishers keep the number of fishing boats to a reasonable level?
- How do fishers compensate individuals for time or effort expended in managing the fishery?
- How does the group reach decisions about what rules are fair?
- How do they know if rules are being followed, given that fish populations are hard to track and count?
- What do they do to rule breakers who over-harvest fish at times the group have decided to be restricted?
- How do they solve conflicts over rights?
- What keeps any locally crafted system from being nullified by a higher authority from the central state or "federal" level?

As most commons management challenges share similar sets of issues, general design principles for management of such resources have been developed. Following Ostrom (1992), successful commons management must include the following.

## Boundaries

The resource and the user group should have clearly defined boundaries. That means that the fishery in our example should be a specific territory or population of fish rather than a nebulous area. Equally important, the fishers who have rights to use the fishery by necessity must be specified; it cannot be open to anyone with a boat that simply sails into the area from elsewhere.

## Proportionality

Costs accrued in managing should be in line with benefits. People who bear the costs of organizing or monitoring our hypothetical fishery should enjoy equal or higher access rights than those who do not. There should be some form of compensation for any investments in equipment or labor that members of the commons commit to the group.

## Collective choice

Arrangements need to be in place so that the specific rules for managing the resource are made by the resource users themselves and/or can be modified through some kind of deliberative group forum. Fishers, in our example, should be able to set the limits for fishing together.

## Monitoring

Some system of monitoring needs to exist so that people's behaviors and uses are known to the group and so that the status of the resource itself is checked in order to allow for adjustments. This means that some resources must be dedicated to keeping an eye on what vessels are coming and going from the fishery and to taking a reliable sampled census of fish stocks. In line with previously noted principles, the costs of such activities need to be borne fairly throughout the group, and the system for implementation should be decided collectively.

### Sanctions

Sanctions must be imposed on violators, but these should be graduated, meaning that the system should encourage voluntary compliance with rules, have low punishments for first offenders, and only turn to coercion as a last resort. For our fishery, this means that fishers should expect to monitor one another and comply with rules voluntarily. Should a fisher be found in violation of a limit on the amount of fish they take or some other provision (through the monitoring system established above), they can be encouraged to return to compliance without undue or disproportionate duress or expulsion.

### Conflict resolution

Social mechanisms must be developed to resolve conflicts between users. There are many possibilities for mutual complaints in a common property system, and in the case of our fishery, a robust management system would allow a low-cost way to work out mutual grievances without turning towards expensive litigation or calling in higher-order authorities. These mechanisms might be a small council of respected citizens, a mediation system using an outside third party, or any number of other socially appropriate systems.

### Autonomy

For a common property management system to work, it is essential that it is allowed at least some measure of autonomy from higher or non-local authorities. Imagine dedicating several years of careful work in developing a community management system of fishers described above, only to have a government official from a distant municipality arrive, review the rules, and begin meddling with their specifics. Where this can be expected to happen, it is unlikely fishers would take the time and effort to craft such a system in the first place.

Given the apparent complexity of making common property systems work, it would seem they would be very rare indeed! Nothing could be further from the truth. The world is brimming with "commons," once a neo-institutional eye is brought to bear on the question. Indeed, it is a sad commentary on our prevailing collective wisdom that cooperation is treated as an oddity or an exception, when indeed it is quite often the rule.

## Ingenious flowing commons: Irrigation

Most of the world's crops, including almost all of the vegetables and many of the grains you probably consume, come from irrigated fields. Getting water to food plants might be the oldest problem in civilization. Indeed, the difficulties in maintaining and managing irrigation have occupied environmental managers for thousands of years. One of the great-

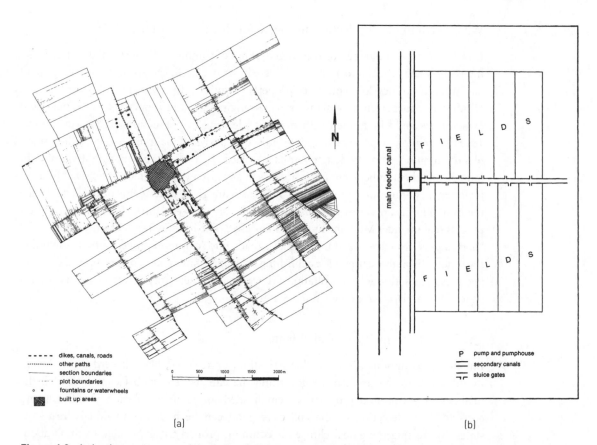

**Figure 4.2**  Irrigation systems are labyrinths of sluices, canals, and gates, which test the limits of people's ability to cooperate in managing environmental goods. This example, from the village of Musha in the Nile Valley, shows the inter-relationships of one private field to the next, tied together by the mutual need for water held communally. *Source*: Turner and Brush (1987), pp. 227, 235.

est challenges in managing irrigation is that, typically, many water users are connected through the same complex system of ditches and walls, through which water flows from the highest point (the system "head") to its lowest fields (its "tail"). Fields may be held privately, but the irrigation water must be managed collectively. Such systems are labyrinths of sluices, canals, and gates, requiring that each user follows a careful set of rules that allow them their share of the water for a time, but also maintenance of the supply so other users, especially downstream, get their fair share as well (Figure 4.2). The opportunities for system failure are obvious. If one person at the head of the system fails to open or close a gate after they use their supply of water, the downstream user will receive none. If all parties do not work together to flush the system as a whole, the water can become salty, leading to the loss of the crops of all farmers. These difficulties notwithstanding, the world is filled with local irrigation systems where multiple users cooperate in the actions, make decisions together, monitor the infrastructure, and achieve equitable outcomes with little waste or loss of precious water.

### Wildlife commons: Collective management through hunting

Even the world's wildlife can be considered a kind of commons. In the United States, where herds of elk and other important species were in serious decline a century ago, management has worked to develop common property solutions to problems of over-hunting. Historically, since such animals were wide-ranging and the property of no single land owner, they could be hunted with impunity, leading to population declines in the late 1800s. Current systems of management in states like Montana utilize many of the principles of common property design. Limits are placed by the government on the number of hunters and the number of hunting licenses in any given year based on extensive monitoring of game populations. Preferences for licenses are given to residents of the state. While the overall limits are set by officials, all the rules are overseen through a collective review process that includes Montana hunters themselves. The result is a system where a potential "open access" resource (free-ranging elk) is made into "common property" by 1) excluding some potential outside users, 2) establishing rules and limits, and 3) reviewing and overseeing these rules through consultation with resource users themselves. Versions of this system are in place across the United States.

### The biggest commons: Global climate

But not all commons are local, like irrigation, or regional, like elk herds. This brings us full circle to the problem of governing the climate. The global climate has all the qualities of a common property system headed for failure: exclusion is difficult and costs to defer depletion of the collective good can be high to individuals, firms, or states. By treating the global climate as common property, it is possible to think about it in a new way, however. As a commons, we can imagine climate as a shared good, and that people polluting it might constrain their behavior through some kind of collective agreement.

Clearly the possibilities for collective action exist, and many new systems to manage the problem have emerged in the last decade, including the Kyoto Protocol (see Chapter 9). That agreement essentially imposes a mutual set of restrictions that countries must follow on their emissions, with mechanisms for crafting rules and making decisions even in the absence of any kind of higher authority; there is no real "world government" to enforce global agreements, after all.

The problems facing the common property of a fishery are largely similar to global climate, of course. It is hard to monitor who is doing what. Sanctions are difficult to impose on free-riders who do not comply with the rules or on users of the resource (polluters) who are not part of the agreement (e.g., the United States). The presence of collective choice systems for setting and revising the rules is also unclear, beyond the fact that signatories participate in negotiation rounds to work out provisions. For these reasons, an institutional analysis of the climate problem sheds light on the prospects for success in controlling climate change by identifying areas where creativity will be necessary to solve it as a common property problem.

# Are All Commoners Equal? Does Scale Matter?

As attractive and effective as an institutional perspective appears to be, it is not without detractors. At their core, these criticisms revolve around some of the assumptions that common property theory inherits from game theory. Specifically, these theories, as embodied in the Prisoner's Dilemma and the Tragedy of the Commons, assume individual free agents of relatively similar power in both determining the outcome of the game and making choices to cooperate. In a world of complex political power and inequality, such assumptions are unrealistic and potentially dangerous.

Consider the ways that access to resources and the rules that govern them have been heavily influenced by social and political conditions surrounding, for example, gender. In many places ownership or inheritance of property is traditionally patriarchal and patrilineal, meaning women cannot own, or at least not inherit, property. At the same time, women's labor may be essential for the maintenance or use of commons, including the drawing of water, the management of irrigation infrastructure, the tending of community gardens or forests, among many others. What this means for common property management is important, since matching rules with users and developing systems of collective choice becomes extremely difficult in places where not all users have equal access or responsibilities.

For these reasons, developing common property management regimes in village India, for example, can be difficult. Here, traditional decision-making bodies (called *panchayats*) tend to be dominated by men, and usually men from elite families or castes. If these decision-making bodies are the ones that make the rules about access to village pastures for grazing, they may be viewed as totally illegitimate by the people who use the resource, including women and more marginal groups who tend cattle. If, on the other hand, new collective decision groups are formed to make new rules, and these groups include women and more marginal groups, they might still be dominated by historically powerful communities, again leading to a lack of legitimacy. In either case, a weak authority may lead to indiscriminate use of the grazing lands by people who feel excluded or, worse, view the establishment of rules as effectively representing a taking-away of their traditional rights. Forming clear common property systems may be desirable in complex power-laden contexts, but it may not be easy by any means (Figure 4.3).

**Figure 4.3**  A woman tending her herd in India. In many parts of the world the responsibility for using common property and the control over the rules of the commons are often split along lines of gender, class, or race, portending problems for management.

This insight (from the field of feminist political economy: see Chapter 7) can be extended to all commons, whether users are differentially empowered because of income, race, or a host of other discriminatory factors. Under such conditions, the dynamics of the commons become dramatically more complex and sometimes untenable. Social difference and social power make collective decision-making far more difficult, since parties may not trust one another and because the self-imposed group rules determined from common property management may suit the interests of only a small proportion of commons users. Bear in mind that the first and foremost rule of common property resource management is the bounding and exclusion of some potential user populations (Ostrom 2002). Where such exclusions are founded on basic inequalities, many people will be less inclined to cooperate and failure becomes inevitable. So while "getting the rules right" is the favored approach for an institutional solution to environmental crises, some problems, especially problems of power, come far prior.

A final problem of institutional analysis emerges from disagreement over the degree to which the lessons of the countless local common property regimes around the world can easily transfer to larger problems, spread over more diverse populations, at scales beyond the daily experience of people. Some have argued that collective action can only occur in small groups, where face-to-face interaction builds trust, as in an irrigator community. Other evidence suggests, however, that distant people can come together in cooperative regimes, because they are united by the common property they share, as in the case of the atmosphere and climate system.

Can we "scale up" from the local commons to the global commons? It is possible that the future of the planet rests on the resolution of this very question.

## Thinking with Institutions

In this chapter, we have learned that:

- Many environmental problems appear intractable because they are prone to problems of collective action.
- Coordination around such problems fails owing to the "Prisoner's Dilemma," a metaphor describing the tendency of individuals to rationally seek their immediate gain at the expense of greater gains that might have been made through cooperation.
- Such failure to cooperate around environmental problems typically leads to a "Tragedy of the Commons" where collective goods (e.g., air, water, biodiversity) are degraded.
- Evidence exists from around the world, however, that people succeed at cooperating to preserve common property.
- Theories of common property have therefore emerged, which stress that Hardin's tragedy might occur where there are absolutely no owners or responsible parties, but most commons are owned and controlled by groups as common property.
- By crafting and evolving social institutions that direct cooperative behavior on common property, communities can overcome commons tragedies.

• Barriers to institutional formation and collective action emerge from social, political, and economic inequities, which make cooperation difficult or impossible.

## Questions for Review

1. What is the inevitable and "tragic" behavior exhibited by the herdsmen in Hardin's hypothetical grazing commons?
2. For Hardin, what are the only two options for averting "tragedies of the commons"? Which option does Hardin prefer, and why?
3. Describe a real-world example of a common property resource that is not (tragically) overexploited (use the term "institution/s" in your answer).
4. In a valuable ocean fishery, how can well-crafted boundaries help create a manageable *res communes* (while keeping it from degenerating into an unmanageable *res nullius*)?
5. Why might the atmosphere stand as the most difficult to manage of all imaginable common property resources?

## Suggested Reading

Benjaminsen, T. A., and E. Sjaastad (2008). "Where to draw the line: Mapping of land rights in a South African commons." *Political Geography* 27(3): 263–79.

Commons, J. R. (1934). *Institutional Economics*. New York: Macmillan.

Community Economies Project (2005). "Community Economies." Retrieved May 30, 2005, from www.communityeconomies.org/.

Hardin, G. (1968). "The Tragedy of the Commons." *Science* 162: 1243–8.

Kropotkin, P. (1888). *Mutual Aid: A Factor in Evolution*. Boston, MA: Porter Sargent.

Mansfield, B. (2004). "Neoliberalism in the oceans: 'Rationalization,' property rights, and the commons question." *Geoforum* 35: 313–26.

Ostrom, E. (1990). *Governing the Commons: The Evolution of Institutions for Collective Action*. Cambridge: Cambridge University Press.

Ostrom, E. (2005). *Understanding Institutional Diversity*. Princeton, NJ: Princeton University Press.

St. Martin, K. (2001). "Making space for community resource management in fisheries." *Annals of the Association of American Geographers* 91(1): 122–42.

Tucker, C. M., J. C. Randolph, et al. (2007). "Institutions, biophysical factors and history: An integrative analysis of private and common property forests in Guatemala and Honduras." *Human Ecology* 35: 259–74.

## Exercise: A Commons Nearby

Name a common property resource you have access to. This might be an environmental resource (a local park) or a shared good (a computer network). With whom is access to this resource shared? What risks are there to depletion, overuse, problems, or degradation from failures of cooperation or collective action? Is there a tacit set of norms or rules governing the use of the resource? Could these rules or systems of management be improved by application of the principles of institutional analysis? What barriers are there to improving the way the resource is used or shared?

# 5

# Environmental Ethics

Credit: © Richard Levine/Alamy.

## Chapter Menu

## Keywords

- Animal liberation
- Anthropocentrism
- Conservation
- Deep ecology
- Dominion thesis
- Ecocentrism
- Ecology
- Environmental justice
- Ethics
- Factory farms
- Holism
- Intrinsic value
- Moral extensionism
- Naturalistic fallacy
- Pragmatism
- Preservation
- Scientism
- Social ecology
- Stewardship
- Utilitarian
- Wilderness

## The Price of Cheap Meat

**Factory Farms**  Intensive animal-raising agricultural operations; factory farms attempt to maximize production by raising as many animals in as little space as possible, often resulting in significant air and water pollution

Being a pig just isn't what it used to be. Gone are the days when hogs had the run of the barnyard, served as garbage disposals for kitchen scraps, and, of course, wallowed in that delectable mud on steamy hot days. Gone, at least, for the roughly 80 percent of all hogs in the United States that are being raised on industrial-scale "**factory farms**."

Take a "breeder sow," for example (a female pig raised solely for the purpose of birthing and nursing future pork). At about 8 months old, a sow is (artificially) impregnated for the first time. Immediately after being impregnated, the sow is moved into a "gestation crate." A box. To live in. The crate is usually about two by seven feet and not even large enough for the sow to turn around, nor can she stand up or lie down without rubbing up against the metal bars that form the walls. Just before giving birth, she is moved to a "farrowing crate," which is basically a modified gestation crate (Figure 5.1), with pockets for the piglets to nurse without being crushed (though many still wiggle out of the pockets and are soon crushed to death). Almost as soon as the piglets are weaned, the sow is impregnated again, denied the normal

**Figure 5.1**  A sow's "farrowing crate," where she will remain for the 35 (or so) days her young are nursing. Crating is an economically efficient method of raising hogs, but *is it wrong*? *Credit*: © Animal Rights Advocates Inc. 2004. Image Courtesy of Against Animal Cruelty, Tasmania, www. AACT.org.au.

period of five to six months "off" between breeding new little ones (Marcus 2005).

So why are these animals penned in such cramped conditions? Why are they not allowed to rest and recover after each period of pregnancy and nursing? For the sake of industrial efficiency. These factory farms are, after all, run more like factories than farms. Smaller crates mean more sows in the same amount of space. Confining the sows allows their gestation periods and projected birthing dates to be tracked with incredible precision. A non-pregnant, non-nursing breeding sow is, by the logic of business, wasted capital.

Many people have strong opinions – some for and some against – on the increasingly industrialized nature of hog farming. One proponent might like the fact that more animals can be raised on less land, freeing up space for other economic development activities. Another might cite the fact that – courtesy of factory farming (and government subsidies and other often-overlooked variables) – the price of meat in the United States has dropped considerably over the past 30 years. Families living near the officially designated poverty line can, for perhaps the first time in history, afford meat every day, if not multiple times per day. Granted, these arguments are most vociferously (and somewhat cynically) espoused by representatives of the industry itself, but they do represent a set of grounded claims – reasons to justify this action. Utility, most notably, is at the root of this set of arguments, with the case for more economic growth and more available consumer goods taken to be the "good" reasons for factory farming.

Simultaneously, opponents of factory farms also have varied arguments. Some speak out against the de-peopling of rural communities, a process concomitant with industrial intensification, as small farms go belly-up in the face of competition across the Corn Belt and the Southeast. Others speak out against the (very considerable problem of) air and water pollution produced by these mammoth-scale operations – some holding tens or even hundreds of thousands of hogs on a single production unit, with problems for other residents and for the costs related to cleaning periodic disastrous spills. These are compelling arguments, located in a (mostly) human-centered view of the problem, and the "bad" reasons against factory farming.

Many, if not most, vocal opponents of factory farming, however, oppose it for what appears on the surface to be the simplest and most straightforward argument of all: Animals are not "capital." They are living, sentient, beings. For them, it is just plain *wrong* to treat animals in this manner. Factory farming, for those critics, is an *ethically indefensible* practice. This is again an ethical argument, like the ones above, but it is one grounded in a somewhat different place: the experience of the hog.

In each case, ethics is at work: the study of right and wrong. What *ought* people do in certain situations and why? Throughout history, codes of ethics have provided guiding principles for civilizations, including a number of key philosophies and religions. There are countless such ethical traditions in the world, from the precepts of Buddhism to the ethical laws of scientific humanism. As you can see from our opening argument, however, for the purposes of this chapter we are especially interested in the ethical considerations and ways of viewing environmental problems that question human-centered logics and propose an environmentally centered alternative.

Unlike in Buddhism and many other traditional ethical systems, in the specific ethical traditions of "Western" civilization, such reconsiderations (from human-centered to environmentally centered views of right and wrong) are remarkably recent. Throughout much of "Western" philosophical tradition, questions of right and wrong action have largely centered on actions *of people towards other people*. Indeed, many important environmental movements, regulations, and ideas are themselves founded on such ethical concerns of humans for other humans. Consider, for example, the concerns of **environmental justice** (see Chapter 7 for more detailed discussion), where environmental risks are criticized because of higher exposures of poor and minority populations. These are fundamentally ethical arguments. Indeed, all views of the environment (from population to markets to institutions) contain their own ethical dimensions, especially as humans relate to one another.

Only recently, however, have ethicists expanded inquiry and consideration to include *environmental ethics*. That is, what are right – and wrong – ways of acting in nature and

**Ethics/Ethical**  The branch of philosophy dealing with morality, or, questions of right and wrong human action in the world

**Environmental Justice**  A principle, as well as a body of thought and research, stressing the need for equitable distribution of environmental goods (parks, clean air, healthful working conditions) and environmental bads (pollution, hazards, waste) between people, no matter their race, ethnicity, or gender. Conversely, environmental injustice describes a condition where unhealthful or dangerous conditions are disproportionately proximate to minority communities

toward the *non-human* world? For example, one might ask, after reading the story that opens this chapter, does a sow need to be able to turn around to live a full life? Is the full life of a sow important? *Is it wrong to treat sows this way?* But before we can address this and other contemporary questions from an environmental ethics perspective, let us backtrack a few centuries, to some influential documents that laid out the foundations of right action toward the non-human world.

## Improving Nature: From Biblical Tradition to John Locke

Many environmental thinkers cite two common beliefs that have governed the relationship of Western civilization to the natural world, in one way or another, until the last century. The first of these is the idea that humans are separate from, and superior to, nature. The second is that nature is only as valuable as it is useful to humans. Where do these ideas come from? In the influential essay boldly titled "The Historical Roots of Our Ecologic Crisis," Lynn White (1968) argued that the **dominion thesis** from the Old Testament established humans as the sole earthly creatures made in God's image (therefore, logically separate from and superior to the rest):

> Be fruitful and multiply, and replenish the Earth, and subdue it; and dominion over the fish of the Sea, and over the fowl of the air, and over every living thing that moves upon the Earth. (Genesis, 1:28)

This is, to be sure, no small order. There are neither places nor creatures – *anywhere on Earth*, it would seem – that should not be "subdued" by humans. If people have indeed been commanded by an all-powerful deity to do this, then quite clearly, not only is subduing and dominating all of Earth a permitted goal to pursue, it is, moreover, the *right* thing to do, and part of a larger plan. As such, according to White (and many others), biblical tradition provided humans with an environmental ethic, albeit one that is at best silent about, or at worst hostile toward, those animals, plants, and places not useful to humans, or under their direct control. With such a literal reading of the Old Testament, is it any wonder that the puritans of colonial New England so feared and despised the wilderness? Is it any wonder that the first governmental environmental action in North America was not the creation of a park, or the establishment of regulatory limits on the cutting of trees, but rather a publicly subsidized wolf eradication program for the Massachusetts colony (in the sixteenth century, no less!)?

On the other hand, the inherent hostility or indifference of the Bible to the non-human world can be strongly debated. For example, many find throughout the Bible a strong call for human **stewardship** of nature, the moral responsibility to care for and protect the natural world. While humanity might still be fundamentally separate from nature, as stewards the mandate to subdue is subsumed under an ethic to care. Such a religious ethic of stewardship can be seen in human activities and movements around the world; witness the Kibbutz movement in Israel, the Amish in the US Northeast, the

**Dominion Thesis**   Arising from the Book of Genesis, the dominion thesis states that humans are the pinnacle of creation; as such, humans are granted ethical free rein to use nature in any way deemed beneficial

**Stewardship**   Taking responsibility for the property or fate of others; stewardship of land and natural resources is often used in a religious context, such as "caring for creation"

anti-corporate French "peasant" farmers' movement, and the recent energetic movement of evangelical churches in the United States to halt global warming. But from whichever angle – dominion, steward-ship, or somewhere in between – Western civilizations, especially those of the modern era, do appear to demonstrate an **anthropocentric** (literally, human-centered) ethic that, for many environmental advocates, leaves much to be desired.

These ethics, however, may be less rooted in the Bible than they are in political philosophy. Fast-forwarding from the Old Testament to seventeenth-century Britain, we find the philosopher John Locke penning a key document that further articulated a Western, human-centered environmental ethic.

> **Anthropocentrism**  An ethical standpoint that views humans as the central factor in considerations of right and wrong action in and toward nature (compare to ecocentrism)
>
> **Utilitarian**  An ethical theory that posits that the value of a good should be judged solely (or at least primarily) by its usefulness to society; following the eighteenth–nineteenth-century philosopher Jeremy Bentham, usefulness is equated with maximizing pleasure or happiness and minimizing pain and suffering

With his *Second Treatise on Government*, written in 1690, Locke wrote what would become the definitive statement of libertarian philosophy. In this unabashedly political work, Locke's "chief concern … was to establish the legitimacy of a popularly elected democratic government, and thereby to oppose the prevalent view of the divine right of the monarchy" (Katz 1997: 225). Not surprisingly, Locke was widely read and his anti-monarchist views were widely popular in eighteenth-century America. In the newly independent United States, it was Locke's theory of *property* in the *Second Treatise* that would, perhaps more than anything else ever written, lay the foundation for a new Enlightenment-era rationale for the human domination of nature.

Political libertarians believe in government, but always and only in a very limited role. For Locke, and most libertarians who have followed, government should function only as is necessary to protect the "natural" freedoms of individual citizens. Freedom, in this context, is the right and ability to acquire, possess, and maintain property. Without property, government would be unnecessary, because with no possessions, there is nothing to defend.

What, then, is property? In Locke's view, an individual's property begins with their physical body. Beyond the body, an individual's physical labor is also part of his property. A free individual, therefore, has the sole right to the use of their body and labor. It is worth noting here that Locke was referring to *free men* of the time, which, for the eighteenth-century reader, would have excluded the bodies of many, especially women and non-whites, from their own control. In any case, acquiring (or "appropriating," for Locke) property from nature merely requires that an individual mixes labor with some or another "chunk" of nature. By shooting a deer, picking an apple, cutting down a tree to make sawlogs, or clearing a patch of forest to plant a field, an individual has turned external nature into personal property. They have, moreover, done the *right* thing.

*Using* nature, for Locke, meant very specifically the transformation of external nature (which has little or no value), through individual human labor, into *property* (improved nature, now *valuable*). So not only is the environmental ethic established by Locke clearly anthropocentric, it is also purely **utilitarian**. This means that the *value* of any part of nature is determined solely by its *usefulness* to humans. The implications, for the newly independent United States and beyond, cannot be overstated. A domesticated animal (*property*, the product of human labor) is, by fiat, more valuable than any wild animal.

An agricultural field or a forest cut for timber is more valuable than any native prairie or wild forest. The US West – the frontier – was, for Locke, a literal wasteland. Even as it may have been a storehouse of *potential* value, the "Wild West" (its indigenous human communities included) was valueless insofar as it was unused.

Locke's theory of the ethics of appropriating nature did come with two (arguably rather minor) constraints on what qualifies as just uses of nature. An individual should not, first of all, take more than they can use before it would spoil or go to waste. Further, no individual should appropriate more land than he can work. As such, Locke's theory was a perfect match for early American ideals, justifying Westward imperial expansion, but at the same time advancing an ideal that honors the sanctity of the individual citizen and egalitarian opportunity. Nearing the end of the nineteenth century, this anthropocentric, utilitarian environmental ethic would find an articulate and powerful proponent in Gifford Pinchot, good friend of Theodore Roosevelt and first chief of the US Forest Service. Pinchot's strong advocacy for the efficient use of natural resources would not, however, go unchallenged.

## Gifford Pinchot vs. John Muir in Yosemite, California

Gifford Pinchot is a justifiably famous – some might even say legendary – figure in the history of environmental conservation. After graduating with a bachelor's degree from Yale, Pinchot received postgraduate training in scientific forestry in France (there was, at the time, no college of forestry in North America). Just a few years later, President Grover Cleveland appointed Pinchot to the federal Division of Forestry, and during the subsequent administration of Theodore Roosevelt, Pinchot became the first chief of the US Forest Service. Pinchot was a lifelong advocate of what he called "**conservation**" – the efficient and sustainable use of natural resources, always for "the greatest good for the greatest number." This phrase, Pinchot's most famous aphorism, epitomizes his staunchly utilitarian environmental ethic.

**Conservation**   The management of a resource or system to sustain its productivity over time, typically associated with scientific management of collective goods like fisheries or forests (compare to preservation)

At the time, it should be noted, a vast proportion of the United States was held by the Federal government, especially in the western part of the country (indeed, almost 30 percent of land in the United States remains under Federal control today). These lands comprised forests, deserts, mountains, rivers, wetlands, and gorges, which were rich with resources, including timber, minerals, and hydropower. While a lot of land had been given over to private interests, including farming and the railroads, enormous areas remained. How these lands should be managed, and for what purposes, put Pinchot at the center of an ethical debate that remains at the heart of environmental politics today.

Pinchot viewed himself as a centrist. Opponents on one side of him were lumbermen, mining interests, and other large corporate interests and firms who wanted (legal and ethical) free reign to chop and dig over the privately and publicly owned lands of the country as quickly and profitably as they could. These "timber barons" and their allies had had a free hand in using public resources largely as they pleased throughout the nineteenth century. But with the arrival of President Roosevelt and other trust-busting political populists at the turn of the twentieth century, Pinchot and his vision of state intervention to rationalize resource use could take center stage.

On Pinchot's other side, however, were proponents of **preservation** – an environmental ethic that would rival conservation. Preservationists argued against the human exploitation of nature, efficient or otherwise. Nature, particularly unspoiled or scenic tracts of "**wilderness**," should be left alone, preserved *from* human use and abuse. This tradition of preservation itself developed over the nineteenth century, with its national roots in philosophical American Transcendentalism, advocated by Henry David Thoreau and his contemporaries, which held that human potential lay in intuition and a close relationship to the natural world. This philosophy, embodied in the National Parks movement, stressed protection of nature for its own sake.

> **Preservation**  The management of a resource or environment for protection and preservation, typically for its own sake, as in wilderness preservation (compare to conservation)
>
> **Wilderness**  A natural parcel of land, more or less unaffected by human forces; increasingly, wilderness is viewed as a social construction

At the turn of the twentieth century, the most articulate and famous spokesperson for preservation was the wiry Scottish immigrant John Muir. Muir came to the United States at age 11, and grew up on a farm in Wisconsin. After three years of college at the University of Wisconsin, he left the Midwest and traveled south to Florida, across the Gulf of Mexico to Central America, then by boat to Alaska and California. It was there, in the Sierra Nevada mountains of California, that Muir eventually settled. In 1892 he founded the Sierra Club, and it was largely through the lobbying of Muir and the Sierra Club that Yosemite National Park was expanded to its current boundaries.

It was in his beloved Yosemite that, a couple of decades later, Muir would face off with his former environmental ally Pinchot in what would become one of the defining environmental debates in American history. The debate was over the fate of Hetch Hetchy Valley. Hetch Hetchy was a deep, scenic glaciated valley in the remote northeast corner of Yosemite, similar to but smaller and less famous than Yosemite Valley. The city of San Francisco wanted to dam Hetch Hetchy Valley to create a large reservoir and permanent municipal water source for the growing municipality (Figure 5.2).

For Pinchot, the decision was an easy one. Which choice, reservoir or wilderness, was the *right* one? It was simple. As Pinchot testified before Congress, he explained that the reservoir provided a greater benefit for more people than preserving the valley in its natural state. Through this simple calculus, damming the valley was argued to be the right thing to do. Muir, on the other hand, saw it differently. As he had argued years before, when he successfully lobbied to get permanent protection for Yosemite Valley and the Mariposa Grove of giant sequoias, this park was a unique natural treasure. It was a storehouse of scenic wonder, a place for weary, modern humans to witness the spiritual grandeur of the natural world, a place *not* dedicated to human progress.

He was arguing for preservation for its own sake and the protection of scenic landscapes with little to no human imprint: *wilderness*. For nearly a century, the preservation movement argued for wilderness along these intangible, non-utilitarian lines. The right thing to do, according to these advocates, was to leave alone what little pristine nature was left. As Yosemite, Yellowstone, Glacier, and other national parks had already been established, lands were in the process of being set aside from direct, obvious human exploitation. A large part of the reason these lands were "set aside" in the first place, however, was because

**Figure 5.2**   Hetch Hetchy Valley, undammed before 1914 (left), and today (right), now dammed to provide the city of San Francisco with a clean and reliable source of fresh water. The fate of this valley was debated between leaving it "wild" versus appropriating it for its direct human usefulness. *Credit*: (i) F.E. Matthes/United States Geological Survey Photographic Archive; (ii) Anthony Dunn/Alamy.

they were viewed as having no "better" use than being more or less left alone, as parks. What marked Hetch Hetchy as such a watershed in the history of environmental politics was that it was the first highly publicized debate over the fate of a particular place directly drawn along utilitarian versus non-utilitarian lines. Previous allies in the fight against the unchecked abuse of nature, conservationists and preservationists parted ways over Hetch Hetchy. This divide – conservation vs. preservation or wise resource use vs. wilderness preservation – would persist within North American environmentalism through the twentieth and into the twenty-first centuries. By mid-twentieth century, however, the burgeoning science of ecology would provide an impetus for a new environmental ethic, a "third way," broader in scope than the narrow confines of either anthropocentric conservation or wilderness preservation.

## Aldo Leopold and "The Land Ethic"

**Ecology**   The scientific study of interactions amongst organisms and between organisms and the habitat or ecosystem in which they live

Ecology – the scientific study of the interconnections between organisms and their (living and non-living) environment – is a relatively young field of inquiry. Indeed, the term "ecosystem," central to ecology and now common parlance, was not coined until the 1930s. One of the leading early ecologists in North America, and one who, more than any other, would weave the insights of ecology into an environmental ethic, was Aldo Leopold. Trained as a forester at the Yale School of Forestry (founded by Pinchot, and the first school of scientific natural resource management in North America), Leopold was one of the most influential and prolific environmental writers in the world. His clearest articulation of an ecologically informed environmental ethic is his most famous essay, written late in his life, "The Land Ethic," from his 1949 book *A Sand County Almanac* (Leopold 1987).

## Box 5.1   Endangered Species Act

In 1973, both houses of the US Congress passed the Endangered Species Act (ESA) with overwhelming bipartisan support (votes were 90–0 in the Senate and 390–12 in the House). The ESA provides regulatory protection for imperiled animal and plant species and the ecosystems on which they depend. "Species" is broadly defined to include any species or subspecies of plant or animal. For vertebrate animals, "distinct population segments" can also be listed. The ESA provides two classifications of protected status, endangered and threatened. An endangered species is "any species which is in danger of extinction throughout all or a significant portion of its range" while a threatened species is "any species which is likely to become an endangered species."

Once a species is listed, the US Fish and Wildlife Service (the agency in charge of listing and managing endangered species) must design and implement a recovery plan for the species. The stated (and logical) goal of listing is timely recovery of the species and removal from the ESA. If a species is designated as recovered, it is "delisted" and the management of the species reverts to the states.

There are two regulatory provisions that give the ESA its teeth. First, the ESA prohibits the federal government from taking any action that would further jeopardize listed species. This requirement has been rigorously upheld by federal courts since the *snail darter* controversy in 1978 that temporarily halted construction of a large dam on the Tellico River for protection of that relatively small fish. Second, the ESA prohibits the killing or harming of listed species on all lands, whether publicly or privately owned. This clause is perhaps the single most controversial component of the ESA, as it severely limits what private landowners can

do with their property, if endangered species are present. Perhaps not surprisingly, an anti-ESA countermovement known as the "property rights movement" argues that federal regulations should not dictate what landowners do on their own land, as long as it does not directly harm others.

Currently there are over 1,200 animals and 700 plants listed. Most listed species are not well known, like the Ozark big-eared bat or the San Joaquin kit fox. A few, however, are high-profile conservation icons, like the grizzly bear and the California condor. Some critics of the ESA note that it has a low "success rate," meaning that few species recover to the point of delisting. While that may be true, it could also be noted that only a very few species have actually gone extinct after listing (the Eskimo curlew, a medium-sized shorebird now thought to be extinct, would be one of only a few examples). The ESA does have its share of notable successes as well. The bald eagle and brown pelican, for example, once decimated due to the effects of the insecticide DDT, are now quite common and have been delisted throughout most of their range. In addition to the snail darter mentioned above, there have been several endangered species whose protection invoked firestorms of controversy. Perhaps the most notable recent example would be the northern spotted owl, whose listing resulted in a great reduction of logging in the old-growth Douglas fir forests of the Pacific Northwest.

### Reference

US FWS (2009). *Endangered Species Act of 1973.* Retrieved September 17, 2009, from www.fws.gov/endangered/pdfs/esaall.pdf.

"The Land Ethic" stands as an important essay for several reasons. First, it was groundbreaking in integrating the new insights of ecology into an ethical framework. It was also notable for the clarity and simplicity of its argument. Additionally, perhaps the most extraordinary thing about this essay has been its persistence. It remains one of the most widely read and influential statements of environmental ethics ever written.

Though philosophers dispute the exact accuracy of Leopold's account of the historical development of ethics, his system is clear and easy to appreciate nonetheless: there are two existing and problematic forms of ethics and a missing but needed third form. According to Leopold, the earliest formal ethics were statements of right and wrong action between human individuals, as with, for example, the Ten Commandments. Not long after, ethical relations between individuals and society (for example, the Golden Rule) and between society and individual (for example, democratic governance) were established. Missing, according to Leopold, was an "ethic dealing with man's relation to land and to the animals and plants which grow upon it. Land ... is still property. The land-relation is still strictly economic, entailing privileges but not obligations." This missing, third form of ethics is an "ecological" necessity. Now more enlightened by scientific ecology, and with our knowledge of biological evolution, humans should realize that not only are they, as a species, part of the natural order, but that they are also wholly dependent on their environment – the lands, the plants, the animals, and the healthy functioning of our ecosystems.

The notion of dependence is apparent in the institutions of human *community*. People typically realize that they must cooperate, giving up some of their liberties and freedoms, to be an "interdependent part" of a human community, so they do so willingly. With this understanding of ecological connections – our *interdependence* with our environment – it is only logical and reasonable to extend outward the boundaries of the community to "include soils, waters, plants, and animals, or collectively: the land." Since one cannot at once be a conqueror and member of a community, we must change our relationship with nature from "conqueror ... to plain member and citizen" (Leopold 1949: 204).

So how to judge right versus wrong action? The most famous and pithy maxim from "The Land Ethic" proposes a litmus test: "A thing is right when it tends to preserve the integrity, stability, and beauty of the biotic community. It is wrong when it tends otherwise" (Leopold 1949: 224–5). The land ethic, as such, parts ways with the simpler ethics of conservation and preservation, where *human use* is more or less right (à la Pinchot) or wrong (à la Muir). Use of the land will – *must* – continue (indeed, Leopold wrote considerably on agricultural land use and conservation; he was not just an advocate for wild nature). Land use must now, however, be guided by our knowledge of ecological connections and ecosystem health. Indeed, the land ethic stands as the key **ecocentric** (ecology-centered) environmental ethic, though the term "ecocentric" itself would not be coined until decades later. Not surprisingly, many contemporary environmentalists – from biodiversity conservationists to proponents of "ecological sustainability" – claim Leopold as their own.

Beyond the recognition of ecosystem health as fundamental to any sufficient environmental ethic, a fundamental contribution to be sure, this essay marked another key idea in environmental ethics. "The Land Ethic" was one of the first widely read pieces to argue for **moral extensionism**, that is, extending our sphere of moral consideration beyond the human realm (now, to "include soils, waters, plants, and animals, or collectively: the land"). A couple of decades later, the specific idea of granting animals (including domestic

**Ecocentrism**  An environmental ethical stance that argues that ecological concerns should, over and above human priorities, be central to decisions about right and wrong action (compare to anthropocentrism)

**Moral Extensionism**  An ethical principle stating that humans should extend their sphere of moral concern beyond the human realm; most commonly, it is argued that intelligent or sentient animals are worthy ethical subjects

animals) moral consideration would rise to prominence, spark considerable debate, and inspire a popular social activist movement.

## Liberation for Animals!

When Aldo Leopold argued that animals, plants, and soil should receive ethical consideration, he was thinking as a scientific ecologist. The animals Leopold had in mind, therefore, were not the individual critters themselves but rather species, like the gray wolf, the grizzly bear, the peregrine falcon, or even a less "charismatic" species like the now-extinct dusky seaside sparrow. In the 1970s, new social movements dedicated to extending our range of moral consideration arose. The "**animal liberation**" and "animal rights" movements argued that individual animals – wild or domestic – were worthy of our moral consideration.

**Animal Liberation**    Named after Peter Singer's groundbreaking 1975 book, a radical social movement that aims to free all animals from use by humans, whether those uses are for food, medical testing, industry, personal adornment, entertainment or anything else

Peter Singer's book *Animal Liberation* (1975) is notable in environmental ethics writing because it is perhaps the only such book to become a popular bestseller. Not surprisingly, as such, it also inspired the rise of a vocal – and sometimes quite effective – social movement, the animal liberation movement.

Singer's argument is straightforward enough, and for many people, resonates strongly as a reasonable call to *right action*. He asks: How is it that we have (quite *rightly*) broadened our moral horizon to those populations who have historically had few or no rights (including women, minorities, the infirm, and the handicapped) without similarly extending beyond the human species to include animals as well? Distinctions drawn, Singer argues – whether drawn along lines of race, class, or gender, or along the human/non-human species divide – are arbitrary. How did people once rationalize denying moral standing to members of other races? Only through now discredited *racist* theories. In extending our sphere of moral standing, our *motivation* – that others have interests that deserve equal consideration – was correct. Moreover, our *direction* – increasingly inclusive, outward from an unjustifiably narrow conception of who counts – was correct and laudable as well.

The shortcoming of these previous successes, for an animal rights advocate, is simply that they did not go far enough. The reason? According to Singer, it is because they failed to use a sufficient measuring-stick to test the notion of who (or what) deserves moral standing. Following the nineteenth-century philosopher Jeremy Bentham, Singer argued that ethics should strive for maximizing the "good" that society can produce. A prerequisite for maximizing good (as pleasure and happiness) is to eliminate, as much as possible, suffering. As such, all beings who can suffer (Singer calls them "sentient beings") have interests, and warrant equal consideration in matters of ethics.

Contrary to what some detractors might claim, Singer does not advocate exactly *equal treatment* of all animals. He does not, for example, claim that the life of a rat and the life of a human are equal, and should be treated with equal weight in questions of ethics. Rather, he claims that all sentient beings deserve *equal consideration*, that minimizing or even eliminating their suffering should be part of any ethical decision. Through this calculus, Singer concludes that almost any use of animals for human purposes is unethical.

He staunchly opposes the use of animal-based products in cosmetics and fashion. He also opposes almost all use of animals in medical research. Not surprisingly, he also advocates the elimination of animals from agriculture. Through vivid and often-harrowing accounts of these practices, Singer has convinced thousands of activists worldwide to fight for the elimination of animal suffering at the hands of humans. Not all animal rights advocates are as absolutist (e.g., *no* animals in agriculture, *no* animals in research, etc.) as Singer, but even the most moderate supporter of this view would be an advocate for major changes to many aspects of modern life. Stated another way, to put into action what Singer advocates – *animal liberation* – would indeed require a radical restructuring of our current culture. Another self-styled radical environmental ethic that inspires activism, if of a different type, is deep ecology.

## From shallow to deep ecology

**Deep Ecology**  A philosophy of environmental ethics that distances itself from "shallow" or mainstream environmentalism by arguing for a "deeper" and supposedly more truly ecologically-informed view of the world

**Intrinsic Value**  The value of a natural object (e.g., an owl or a stream) in and for itself, as an end rather than a means

Norwegian philosopher Arne Naess (Naess 1973) coined the term "**deep ecology**" in the short essay "The Shallow and the Deep, Long-Range Ecology Movement: A Summary." As the title of the paper suggests, this was at once a positive formulation of a new, "deep ecology" and a critique of what he termed "shallow ecology." These divergent "ecologies" were not divisions within scientific ecology but rather branches of the environmental movement. Consumed with the search for piecemeal solutions to particular *issues* (e.g., pollution or resource depletion), shallow ecology failed to ask "deeper questions" about the *causes* of ecological problems and therefore could never hope to solve the ecological crisis itself. Deep ecology, on the other hand, offered a wholesale critique of human society and particularly the human relationship with non-human nature.

The bookends of Naess's philosophy of deep ecology are "self-realization" and ecocentrism. "*Self-realization*," for Naess, is the logical conclusion of any truly deep ecological questioning. When "we" (individuals) realize the interconnectedness of all things, it becomes evident that any concept of the self must expand beyond the individual to include all things. *Ecocentrism*, the second key component of Naess's deep ecology, is a logical derivation of self-realization. Once an individual realizes that he or she is not a narrow enclosed self and properly identifies with all of nature, anthropocentric (human-centered) thought or action becomes – simply – *illogical*.

Deep ecology was relatively unheard of in North America until 1985 with the publication of Devall's and Session's *Deep Ecology: Living as if Nature Mattered*, which presents a "platform" for a "deep ecology movement." The platform was based on the fundamental tenet that nature has "***intrinsic value*** ... independent of the usefulness of the nonhuman world for human purposes." The platform also called for a reduction in human population, a decrease in human interference in the natural world, a change in environmental policies, and a personal commitment to environmental activism. Simple, direct, and resonating with familiar themes (e.g., Malthusian overpopulation, see Chapter 2), deep ecology has made its mark on environmental activism, perhaps most indelibly on the radical wing of the North American wilderness preservation movement. Like all of the strands of environmental ethics reviewed in this chapter, deep ecology is certainly not without its critics, and in the following concluding section of the chapter, we review some criticisms of deep ecology and other strands of environmental ethics.

## Holism, Scientism, and Pragmatism? Oh My!

Is protecting the environment *right*? Is overexploiting nature – forests, animals, ecosystem processes – *wrong*? Posing environmental dilemmas in this manner can, at least for some, make reworking our environmental ethics seem like a logical first step toward reorienting society in a needed direction. There are critics, though, who fear that some popular environmental ethics would, if taken to a logical end, lead to a reactionary, authoritarian politics.

Leopold's land ethic and the environmental activism it has inspired are perhaps the most common targets of the "*eco-authoritarian*" critique. This critique rests on misgivings about the land ethic's **holism**, that is, its insistence that protection of *wholes* (e.g., ecosystems, species, natural processes) takes precedence over the protection – or rights – of *parts* (e.g., individual animals, humans included). It is hard not to read Leopold's famous "a thing is right when it tends to preserve the … integrity of the biotic community" maxim as advocating such a preference. This could mean that, if an action bolstered the good of the whole (for example, sacrificing individuals to decrease population pressures), it might well be deemed not just permissible, but *right*. Critics on all ends of the political spectrum fear that, in the wrong hands, ecocentric ethics could justify the subversion of basic individual liberties. Defenders of the land ethic disagree, arguing that the authoritarian charge relies on comically inflexible reading of the theory. The land ethic argues for integrating a missing, needed criterion into our moral consideration, not using ecology to override all existing criteria.

**Holism**  Any theory that holds that a whole system (e.g., an "ecosystem" or the earth) is more than the sum of its parts

**Naturalistic Fallacy**  A philosophically invalid derivation of an ethical "ought" from a natural "is"

**Scientism**  Usually deployed as a term of derision; refers to an uncritical reliance on the natural sciences as the basis for social decision-making and ethical judgments

**Social Ecology**  A school of thought and set of social movements, associated with the thinker Murray Bookchin, asserting that environmental problems and crises are rooted in typical social structures and relationships, since these tend to be hierarchical, state-controlled, and predicated on domination of both people and nature

Another charge against the land ethic, deep ecology, and their offshoots is that they are guilty of what philosophers call the "**naturalistic fallacy**." By using the findings of ecological science to figure out what is good and therefore what is right, an "ought" has been derived from an "is," or a value from a fact, a practice now almost universally viewed as philosophically indefensible (all kinds of things exist that are not right, after all). Recently, critics have ramped up this charge even more, accusing some deep ecologists of "**scientism**," meaning raising (supposedly objective and value-free) science up to the level of *ultimate authority*. Authority of this form of knowledge over all other sources of value and judgment, for example, religion, philosophy, and humanistic notions of justice, in this sense, means unchecked expert power flying in the face of democratic ideals. Advocates of ecocentric ethics might, therefore, be wary of the philosophical and political pitfalls of staking too much of their claim on ecology.

The most sustained and challenging critique along these lines has come from proponents of **social ecology**, particularly through four decades of writing by the late Murray

Bookchin. Informed by a form of radical political economy (see Chapter 7), rather than natural science or philosophical ethics, social ecologists argue that deep ecologists and land ethics proponents are aiming for the wrong target. Society's root problems are not matters of misguided ethics, bogus ideas, or ecological naiveté. Rather, our ecological ills are social in nature. This tradition of thought, related to social anarchism and rooted in the research and thinking of nineteenth-century researchers like Peter Kropotkin, suggests that domination and exploitation of nature is inevitable in a society rooted in social hierarchies, domination and exploitation of other humans. Until an egalitarian, *just* society is established, people have no hope of ceasing environmental destruction, no matter what ethical lens people use to view nature.

Animal rights advocates, as well, are far from immune to political critique. Practically speaking, what does it mean to say that an animal has interests equal to a human? Do we differentiate between human and non-human animal interests? If so, how? If, from an ethical standpoint, the line between the species is arbitrary, then is it right and reasonable (or even possible) to make decisions along the human/animal line when, in specific instances, the welfares of humans and animals are at odds? All but the most radical (or misanthropic) animal rights activists would likely respond that questions such as these are straw-men (overstated and unfair caricatures of positions that make easy targets for critics) thrown up to intentionally make the ideas presented sound inhumane and unworkable. Fair enough, but that does not solve the problem for activists who ultimately must (or at least should) be able to address difficult, actual conflicts with defensible solutions.

So far we have reviewed *external critiques* of environmental ethics, meaning that these are criticisms drawn mostly from outside the circle of self-identified environmentalists. Environmental ethics, a field comprising both academic scholars and environmental advocates (and many scholar-advocates), also includes considerable internal debates as well. Perhaps the most well developed of these intramural debates is between (some) animal rights advocates and (some) land ethics advocates. To put it simply, the former care about individual organisms while the latter care about protecting species, ecosystems, evolutionary processes, and the like. These two priorities can and often do clash, for example, along the issue of hunting. Many, if not most, animal rights advocates disdain hunting in almost all its forms (it does, after all, cause sentient beings to suffer). But just as many environmentalists see hunting as an ecologically useful – even *necessary* – activity. Which can lead one to ask: Must people really choose between extending ethics outward (to non-human animals) or upward (to ecology and evolution)? Many on both sides of the debate answer this with an unapologetic "yes," seeing the other "side" as an obstacle in their path to right action. Others argue that this is an unnecessary impasse, a false wall erected by people more interested in maintaining ideological purity than achieving workable solutions.

Some reconciliation may be on the horizon. **Pragmatism**, a branch of philosophy explicitly concerned with the *practical consequences* of theory, has inspired a relatively new and robust critique of environmental ethics. *Environmental pragmatists* argue that too many environmentalists have created unnecessary and counterproductive theoretical divisions. Deep ecology, for example, forges a rift between its adherents and all those (by default) "shallow" ecologists who do not

**Pragmatism**  A branch of philosophy that arose in late-nineteenth-century North America, pragmatism considers real-world consequences and effects to be constituent components of truth and reality

openly fly the flag of deep ecology. Similar divisions are drawn by those who advocate for an environmentalism drawn along a strictly ecocentric ethics. Are all those who might argue along anthropocentric lines for environmental protection somehow inadequate environmentalists (even when they may agree with ecocentrists on specific questions of policy)? Environmental pragmatists argue that since a larger, more inclusive environmentalism is needed to push through necessary reforms, such inherently divisive rhetoric should be avoided.

Does anthropocentric thinking necessarily lead to an anti-ecological ethics? Are there fundamental problems with the human relationship to non-human animals? Environmental ethics poses these and other fundamentally challenging questions to society. If, as many environmental ethicists argue, the answer to these questions is "yes," then fundamental, ideological changes are truly needed if people are to cure their planetary ills. Even so, holding to a personal or group ideal, ideology, or vision of right behavior may be only a small part of a complex puzzle, where economic drivers, political forces, and institutional impasses all provide barriers to individual action. While the theoretical and practical challenges to environmental ethics are daunting on their own, the deeply entrenched political and economic conditions of the current world deserve scrutiny in and of themselves.

## Thinking with Ethics

In this chapter we have learned that:

- Competing ethical systems have influenced human treatment of non-human nature for thousands of years, including especially systems that are anthropocentric versus those that are ecocentric.
- In the contemporary era, priorities of conservation have competed with those favoring preservation.
- More recently, the land ethic has been forwarded as a way to value nature from an ecological standpoint without eliminating a role for humanity.
- More radically, deep ecology represents an ethic that staunchly asserts the value of nature in, of, and for itself.
- Other critical and pragmatic traditions have raised questions about the possibility of absolute, clear-cut, and universal ethical positions to govern human actions towards the environment, though not without stressing the importance of underlying ethics to guide behavior and decisions.

## Questions for Review

1.  Summarize the "dominion thesis." Is it an "ethical theory"? Explain.
2.  How does John Locke's theory of property put forward a particular environmental ethic? What, for example, would it suggest is the proper use for a tract of uncut "old-growth" forest?

3.  The famous debate between Pinchot and Muir over Hetch Hetchy Valley marked a split between "conservationists" and "preservationists." Summarize the debate around an explanation of these two terms.
4.  Peter Singer makes a case for "animal liberation." From what does he argue animals should be liberated, and how should we go about making this happen?
5.  Explain the "deep" in "deep ecology." What does it mean to say that deep ecology espouses an *ecocentric* ethics?

## Suggested Reading

Abbey, E. (2000). *The Monkey Wrench Gang*. New York: Harper Perennial Modern Classics.

Des Jardins, J. R. (2006). *Environmental Ethics: An Introduction to Environmental Philosophy*. Belmont, CA: Thomson Wadsworth.

Light, A., and H. Rolston (eds.) (2003). *Environmental Ethics: An Anthology*. Malden, MA: Blackwell.

Locke, J. (1690). *The Second Treatise on Government*. Publicly available in several locations on the World Wide Web.

Norton, B. G. (2003). *Searching for Sustainability: Interdisciplinary Essays in the Philosophy of Conservation Biology*. New York: Cambridge University Press.

Smith, M. (2001). *An Ethics of Place: Radical Ecology, Postmodernity, and Social Theory*. Albany, NY: State University of New York Press.

## Exercise: Pass the Bacon (or don't)

Considering the theories (as well as their critics) surveyed here, let us reexamine our opening story and ask the question: *Is factory farming a defensible practice?* Environmental ethics can be helpful in answering this question, but the answers that arise are far from simple or self-evident. Using books, magazines or the internet, find at least three arguments each from opponents and proponents of factory hog farming. Read the material and then answer the following questions: *How much of their arguments are based in ethics? Of the ethical arguments, are they based on ecological or animal rights grounds (or both)? What other factors enter into their support or opposition? Having read from both sides, what is your opinion?* Complete the exercise by writing a two-paragraph statement of opposition to or support of this practice.

# 6

# Risks and Hazards

*Credit*: James Dawson/Image Farm Inc/Alamy.

## Keywords

- Affect
- Cultural theory
- Externality
- Hazard
- Risk
- Risk perception
- Uncertainty

## Chapter Menu

## The Great Flood of 1993

In mid-June of 1993, all across the US Midwest, it started to rain. It rained in Iowa. It rained in Illinois. It rained in Missouri. It rained in Minnesota, Nebraska, Wisconsin, Kansas, and Indiana, day and night for weeks, and did not stop until August. As the rain continued to fall steadily, the water behind thousands of dams in the Mississippi River Basin rose and millions of people in small towns and big cities across the region all nervously watched the water inch towards a point of crisis. At last, around 1,000 of the 1,500 federal, state, and municipal levees (earthen embankments) on the river, which hold the water back from populated areas, broke or were overtopped. Water poured across cities and farms. The "Great Flood of 1993" was an event that killed 48 people and cost US $20 billion in damages.

It was the largest and most significant flood event in US history and the costliest flood to that date. Though the events of Hurricane Katrina of 2005, where thousands died and countless others were left homeless, and other disasters would later dwarf its impact, the consequences for people in the region and the nation are difficult to overstate. Fifteen million acres of farmland were inundated. River sediments, containing a range of toxins and even DDT (a persistent pesticide deposited there decades before and long since banned), were dumped across some of the country's most valuable growing areas, rendering them useless for years. Seventy-five towns were engulfed and 10,000 homes totally destroyed.

The event was certainly unusual. The record rainfall of that summer was 200–300 percent higher than average and achieved flood levels seen only every 100–300 years in some parts of the region (Johnson et al. 2004).

On the other hand, flooding of the Mississippi River is by no means unprecedented. As early as 1885, the Mississippi River Commission was charged with controlling the vast and unpredictable river, and chose a "levees only" policy rather than maintaining wetland floodplains or encouraging river outlets to absorb catastrophic flows. Seeing the folly of such an approach, Mark Twain wrote in his 1896 book *Life on the Mississippi*: "ten thousand River Commissions … cannot tame that lawless stream, cannot curb it or confine it, cannot say to it, Go here, or Go there, and make it obey; cannot save a shore which it has sentenced; cannot bar its path with an obstruction which it will not tear down, dance over, and laugh at" (Twain 1981: 138). His words were prophetic; in 1927 the river overflowed its banks, broke its levees, and flooded 27,000 square miles, only a year after the Commission and the US Army Corps of engineers predicted that the levees would hold forever.

Given what we know about the Mississippi River, was the "Great Flood of 1993" unpredictable? Was it caused by underinvestment in levees and other infrastructure? Or did over-engineering the river exacerbate the crisis? Or was the event turned from a relatively benign weather event into a national disaster by poor floodplain decisions that put settlers and property into the inevitable path of periodic high water, letting people build and live in dangerous places? How do you plan for disasters that you cannot predict? Who should pay for the restoration of the land? Those who live there and took "irrational" risks? Those of us who ate cheap food off that land for a century, enjoying the benefits of floodplain development without exposing ourselves to a hazard? The government that subsidized growth in the Midwest while investing little into mitigating possible disaster? How does society adjust to a chaotic, more-than-human world in a way that is socially just and environmentally sustainable? These are the questions that come to our attention when we choose to think about the environment as a hazard and to consider social relationships to nature as management of the constant risks that the non-human world inevitably presents.

## Environments as Hazard

For those who examine the environment in this way, a set of clarifying concepts help sort out problems in the world: hazard, risk, and uncertainty. A **hazard** is a thing (e.g., UV radiation in sunlight), a condition (e.g., periodic flooding of a river), or a process (e.g., nitrification of water) that threatens individuals and society in terms of production (making a living) or reproduction (being alive). Hazards are real, essentially inevitable, and intricately linked with the boons and benefits that people and societies accrue from nature and technology. The South Asian monsoon supports the entire summer food crop of the region but its periodic failure leads to catastrophic drought while heavy rainfall leads to flooding and loss of life. Nuclear power supplies roughly 20 percent of electrical power in the United States but its production facilities are vulnerable to dangerous failures and its waste byproducts are hazardous for thousands of years. Thus, hazards are merely the negative potentials of all things; they can be described, quantified, and discovered over time.

> **Hazard**  An object, condition, or process that threatens individuals and society in terms of production or reproduction
>
> **Risk**  The known (or estimated) probability that a hazard-related decision will have a negative consequence

Of course, not all hazards are "natural," insofar as many hazardous things are entirely anthropogenic (human caused), including chemicals like DDT or processes like nuclear power. Most hazards fall in between natural and anthropogenic, since the influence of human practices on environmental systems and the impact that these systems have on human behaviors often make such distinctions irrelevant. In either case, an understanding of the character of the environment as a hazard allows people, governments, and firms to make informed decisions about how to deal with these socio-natures.

The fact that many of these problems are intermittent or not entirely predictable makes coping with them more difficult, however. Droughts (periods with sub-average rainfall) can pose a risk for crops, for example, but do not occur every year. Do we plan pessimistically, assume a state of perpetual drought, and dampen our overall production? Do we ignore the hazard altogether and proceed ahead with our fingers crossed? Or can we develop some middle path, where we hope for the best, but create contingency plans for the worst?

## Decisions as risk

Deciding between such alternatives is to make decisions based on **risk**, which is understood as the known (or estimated) probability that a hazard-related decision will have a negative consequence. All decisions have some risk, but some risks are miniscule and others enormous. The chance of rolling snake-eyes (a pair of ones) on two dice is 1 in 36, meaning that on average, you will fail to roll it 35 out of 36 times you throw the dice. This is not a problem if you have no stake in the outcome. Betting your life savings on such a roll is a considerable risk, however. Wagering your savings on the roll of a seven, which occurs 1 in 6 times, is a better bet (though the authors recommend against it). Dealing with the environment in this light, decisions can be calculated in terms of the probability of success or failure and of good or bad outcomes.

There is always risk, for example, in planting a rain-fed crop, because droughts are always possible. In some areas, however, droughts are so infrequent as to present a minimal risk. In other areas, they are common enough for farmers to make creative adaptations to lower their exposure to the risk of crop failure by planting drought-resistant varieties mixed in with their regular crops. These adaptations may be more expensive than business-as-usual and so yield less profit from year to year, but they minimize the probability of total disaster. Such ingenious risk-averse farming adaptations to drought are common in most arid and semi-arid parts of the world.

The geographer Gilbert F. White pioneered the effort to manage risk in floodplain management. White was a public servant in a number of Depression-era Federal agencies asked by Franklin Roosevelt to assess the management of the Mississippi, still reeling from the events of 1927. Studying the problem, White concluded that it was the single-minded focus on engineered solutions (levees and river constraints) that was at the root of the problem, and that "multiple adjustments" to flood problems needed to be devised for local conditions. These multiple adjustments would take into account the specific risks posed by different parts of the problem. This included social aspects, including insurance subsidies for people living in floodplains, poor information on periodicity and location of flood hazards, bad evacuation planning, sloppy zoning, and badly managed relief systems. The river, he essentially concluded, is prone to flooding. So the question becomes: How can people working and living in and around this river accept this fact more rationally? These conclusions were discussed in his later dissertation, *Human adjustment to floods: a geographical approach to the flood problem in the United States* (White 1945), which is now a classic work of environment–society research.

Given the more recent flood disasters in places where resources are abundantly available, including Hurricane Katrina in the United States in 2005 and the disastrous flooding in the United Kingdom in winter 2003, for example, and leaving aside the horrific flooding catastrophes in the developing world in recent years, it is not clear that White's recommendations have ever been heeded fully. What sort of risk assessment was performed for determining federal and state investments in levees in New Orleans? Were the levels of commitment concomitant with the levels of risk? How might "multiple adjustments" yet be brought to bear for decision-making on the Mississippi?

By understanding environment as hazard and decision as risk, creative and improved decisions might yet be made, assuming you can calculate and consider exposure to risk by knowing with some certainty the probability, frequency and/or severity of hazards.

Environmental conditions as uncertain

Risk must be distinguished from **uncertainty**, however, which is a condition describing the degree to which the outcomes of a decision or situation are *unknown*. Uncertainty plagues most risk calculations and its sources are numerous.

**Uncertainty** The degree to which the outcomes of a decision or situation are unknown

**Risk Perception** A phenomenon, and related field of study, describing the tendency of people to evaluate the hazardousness of a situation or decision in not-always-rational terms, depending on individual biases, culture, or human tendencies

The first source of uncertainty comes from the highly uneven or unstable behavior of many environmental systems over time. People may base their decisions about water supplies, for example, on the fact that a serious drought has occurred once every 20 years in the past century. But this time frame may not be typical. What if droughts are far more common (perhaps twice a decade) over the past thousand years, and the period of the past century, upon which people are making decisions, is anomalous? Such was the case with the Colorado River Compact, the formal legal charter by which states in the western United States divided up their shares of Colorado River water in 1922. That document was written and signed during a period of unusually high rainfall and high water flow, and so it underestimated the risk of "severe drought," a condition now known to be more common to the region than ever assumed. As a result, the river is constantly *over-allocated*: the states together have legal rights to more water than physically exists in most years!

Uncertainty can also come from encounters with new hazards, for which no meaningful experience exists to help assess risks. This is especially true of technological hazards. The dawn of petrochemical pesticides in the 1930s, for example, led to decisions like the widespread use of DDT, a chemical whose accumulation in the tissues of wildlife led to the decline of many species, especially of birds. Bans and controls were imposed far later, long after their effects were felt. Thus, uncertainty affects risk calculation and draws a veil over human knowledge of complex natural systems. New information can improve risk decisions by increasing what people know, but some degree of uncertainty is inevitable.

## The Problem of Risk Perception

Risk decisions are not only hindered by objective uncertainty, according to students of hazards. They are further complicated by structured biases in the way human beings subjectively tend to see and calculate risk. Examining the behavior, opinions, and psychological characteristics of real people, researchers in the field of **risk perception** have demonstrated that the real or measurable risks of some decisions are sometimes far overestimated or underestimated in daily life. This has important implications for more rational management of hazards.

While the level of physical risk to individuals, for example, is far higher for every mile traveled by car than by aircraft, people overwhelmingly fear aircraft travel more. Fears of vaccination and of fluoridation of water are commonplace, though their statistical impact on health and role in annual injury rates may be lower than that of power lawnmowers, about which people express little or no concern. *Why?*

Researchers have sought for many decades to scientifically explore these biases in order to explain why poor environmental management and human health decisions persist in the face of better information (and by implication decreasing uncertainty) (Slovic 2000). It is increasingly apparent that many risk-related biases are shared by a large number of people, which suggests deep tendencies in human risk perception.

Hazards that are seen by human beings as involuntary, uncontrollable, and having slow or delayed long-term effects are disproportionately feared by people, relative to those hazards that involve individual and voluntary choices and whose effects are immediate. The automobile, for example, is far more dangerous than the aircraft in terms of chances of accidents and injury per mile traveled, but people tend to minimize their perceived risk of driving, typically because they *feel* artificially safer knowing *they* are behind the wheel rather than someone else. Similarly, those risks viewed as catastrophic, fatal, and high impact, no matter how infrequent or rare, tend to be perceived as disproportionately risky relative to those hazards that are common and chronic. Lawnmowers may claim more victims than nuclear power, but a nuclear meltdown seizes the human imagination far more immediately than any power tool. These tendencies are summarized in Figure 6.1, which shows the results of survey research stressing the way some risks are assessed based on characteristics not linked to actual probabilities of harm or injury.

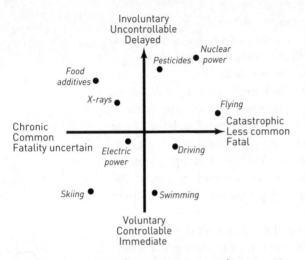

**Figure 6.1** Voluntary/Involuntary–Common/Catastrophic: A matrix for explaining what people think is risky and why. Each axis describes the characteristic of a hazard that tends to lead people, however erroneously, to assume it is or is not risky. *Source*: Adapted from Fischhoff et al. (1978).

Such biases in perception, if not offset by informational checks and balances, might lead to poor decision-making in dealing with the environment. Low-probability spectacular events may capture the attention of planners or activists for regulatory action, but the environmental and public health impacts of persistent, chronic, hazards may be equally or more widespread and damaging. Regulation of household pesticides, for example, has been a low priority for citizens and regulators both, in part because it is a "voluntary" behavior, disaggregated in small doses across vast areas. The aggregate effect of these chemicals, however, is increasingly acknowledged to be hazardous to both human health and the health of aquatic ecosystems (Robbins 2007).

Of course, many of these apparent "biases" in human perception are beneficial. How could people walk in the woods or turn on a blender if they had to stop, work through their range of options, weigh all available information, and then decide what to do? How could they even get out of bed in the morning?! "Gut checks" may not be fully rational (they are governed by emotions, what risk researchers call **affect**), but they are increas-

**Affect**  Emotions and unconscious responses to the world that influence decision-making

ingly understood as essential to making decisions, living in the world, and dealing with the environment. Risk researchers merely seek to better inform these with more complete information (Wilson and Arvai 2006).

## Making informed decisions: Risk communication

With this in mind, those who study environmental decision-making from the point of view of risk, instead of stubbornly opposing people's emotional, commonsense ways of seeing the world, have worked in recent years to find ways to take advantage of them. Working in the field of risk communication, they explore different ways of conveying information so that people can draw on their commonsense and emotional ability to decide, while decreasing their uncertainty.

This work has shown that detailed scientific data supplied in copious quantities to citizen decision-makers (lengthy technical reports, for example) does improve people's knowledge somewhat. It has also shown, however, that "values-oriented" information, conveyed in terms of local and personal priorities rather than simply scientific facts, allows people to make tradeoffs and decisions in a way that is more deliberative and that leads to outcomes people tend to be more satisfied with afterwards. Such decisions are, in a sense, more rational than those based only on cold hard facts, because people make them with a more careful view of tradeoffs and competing priorities. In other words, ironically, information that includes emotional and values-oriented components (affect) leads to less "affective" or more rationally concluded decisions. The way risk is communicated, it seems, may help counter human tendencies in the perception of environmental hazards, leading to better and more socially and environmentally sustainable decisions (Arvai 2003).

## Risk as Culture

There are significant assumptions built into this perspective, of course, worth considering before adopting a "risk perception" approach to dealing with nature. Specifically, in the same way that market approaches to the environment are rooted in the science of economics and its assumptions and methods, risk approaches to the environment are predominantly fixed in the science of psychology, and its associated presuppositions and techniques. The most important of these assumptions is that the objects of scrutiny are individual people and that the tendencies of people are universal: in human psychology, they reflect the effects of cognition and of the deep workings of the human brain.

But to some degree our sense of environmental hazards, risk, and uncertainty is learned. It comes out of school and work environments, from family, and from whether people are rural or urban dwellers, amongst a range of other influences. A century of research in disciplines like anthropology has demonstrated that many biases, perceptions, and systems of interpreting the natural world are rooted in culture, that is, the system of meanings, concepts, and behaviors that people learn from their peers and surroundings. This raises the question of whether there are universal patterns to our learned biases or whether there are distinct cultures of risk.

**Cultural Theory**  A theoretical framework associated with anthropologist Mary Douglas that stresses the way individual perceptions (of risk, for example) are reinforced by group social dynamics, leading to a few paradigmatic, typical, and discrete ways of seeing and addressing problems

In response, one of the most well-known descriptions of these cultural tendencies in risk – referred to as **cultural theory** – explains that human ways of thinking about nature and risk are neither universal (which would mean we all think the same way) nor idiosyncratic (meaning everyone thinks differently). As explained by anthropologist Mary Douglas in her book *Risk and Culture* (Douglas and Wildavsky 1983), the way people think about environmental risk is closely tied to how they think about the place of individuals within society.

Specifically, some cultures tend to stress the degree of free will and latitude of individuals in society, and the universe more generally, than others. Conversely, some cultures acknowledge or emphasize the limits and constraints that are inevitably part of living in nature and society. This range (between freedom and constraint) is referred to in cultural theory as "grid": the tendency of people in certain cultures to hold strongly that individuals tend to be constrained by social and environmental context and circumstance (for better or for worse) or to reject this and stress free agency.

Some cultures also tend to stress the desirability of solidarity and obligation between their members in order to achieve social and environmental stability, stressing the collective in the universe. Conversely, some cultures assemble the world through a notion of individuals as largely isolated and independent in a universe that is relatively robust and indifferent. This range (between prizing the collective over the individual and vice versa) is referred to in cultural theory as "group": the tendency of people in certain cultures to hold strongly that collectives or groups are important or necessary to social stability and success or to reject this and stress autonomy.

The view of risk held by each different culture, cultural theory finally suggests, is based on the convergence of specific tendencies of *both* grid and group. Significantly, neither of these cultural qualities alone describes a given culture or its risk outlook. Instead, a combination of grid and group produces a matrix of four essential outlooks (Figure 6.2). Low-grid and low-group communities or cultures are "individualists," who see people as largely unconstrained by circumstances and with little sense of group obligation or benefit. Low-grid and high-group cultures are "egalitarians" who stress voluntary associations for mutual action. High-group and high-grid cultures, whom Douglas calls "hierarchists," suggest that group solidarity and rule-constrained positions are typical or inevitable. The fourth group, "fatalists," is high grid and low group, and tend to stress constraints and context but not to see the desirability or possibility of cohesiveness and association. Notably, Douglas does not suggest that any of these groups are correct in their view of the world, only that people and cultures all operate within one of these logics, typically at the expense of other ways of thinking, and without an easy or immediate way to unlearn their ways of viewing the world.

For the purposes of understanding both risk perception and preferred risk management, Douglas argues that each of these cultures has a specific and extremely distinct way of seeing the behavior and qualities of the non-human world. For "individualists," nature is a resilient system, in which disturbances cause only temporary harm. This encourages an

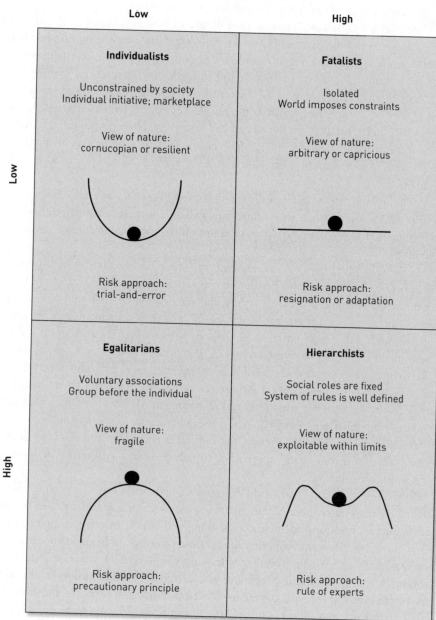

**Figure 6.2** Cultural theory: A schematic table of Mary Douglas's account of variations between cultures in terms of Group (the social solidarity between members) and Grid (the degree to which individual choices are seen as constrained). Conflicts over nature and society (for example over risks of technology) actually may reflect conflicts between cultures.

approach to risk that is somewhat open and "trial and error" oriented, with little concern for consequences that may not be far-reaching or permanent.

Conversely, "egalitarians" tend to view nature as extremely fragile and vulnerable to perturbation with little chance of recovery. Any disturbance may bring ruin to a balanced system, encouraging a risk management strategy using the "precautionary principle" – to not act without full and comprehensive information. For "hierarchists," nature is robust, but only to an extent that must be discerned through careful study and expertise. As a result, they view risk management as the purview of experts with appropriate knowledge of nature's limits. Fatalists, on the other hand, see nature as largely random, uncontrolled and uncontrollable, liable to change for reasons that may be neither manageable nor explainable. Adaptation to risks is the rule in such a culture, as is some sense of resignation.

What does this tell us about dealing with hazards? Contrary to a universal view of risk perception, there is no single structure to human bias. The same information about hazards communicated to members of different communities may be interpreted and used very differently.

Similarly, conflicts over risk management priorities, rather than being explained by a difference between those with a perception bias versus those without one, can be effectively explained by differences between representatives of different risk cultures. In cultural theory, for example, many environmental groups and their constituents are interpreted as "egalitarians," with deeply entrenched views of nature's fragility and group obligations. These are inevitably opposed to the views of "individualists" who see nature as robust and collectivity as unnecessary and undesirable. Interpretations of environmental information and decisions about uncertainty will therefore vary between groups when trying to reach any decision about proper environmental management. Debates over the introduction of genetically modified organisms (GMOs), for example, may not be amenable to more and better information, since they reflect the inevitable encounter and essential division of individualist "trial and error" approaches to nature against egalitarians holding to the "precautionary principle."

Confirming the actual real-world existence of these ideal types of groups has proven difficult. People's perceptions of particular risks vary enormously and are often tied to their socioeconomic positions or life histories, along with a host of other complex factors. At worst, therefore, this way of thinking about human culture can begin to appear a little like astrology, with each of the four types looking like a kind of sun sign (Sagittarius or Scorpio), a way of describing people's diverse preferences and behaviors in abstract clusters. Worse, even if these types proved to reflect a vast majority of people, the theory largely only *describes* risk perception but does little to explain it. Why are some groups, cultures, institutions, or countries egalitarians and others hierarchists? What makes groups change, if anything?

On the other hand, cultural theory may offer important insight to differences in the regulation of risk. This holds implications for comparing the regulatory environment in which genetically modified organisms are received, for example. GM crops, notably, are species based on existing plants that have been genetically altered to result in higher levels of productivity, resistance to certain diseases, or tolerance to cold and drought. These have not been embraced similarly throughout the world. Compare, for example, the emerging policies in the United Kingdom, where a strong regulation insists on a "precautionary principle" and where no such organisms have yet been allowed to be planted publicly or on a large scale. In the United States, on the other hand, GM crops have been treated largely like other existing varieties, with no particular increase in scrutiny towards their potential, albeit largely unknown, environmental effects. *Neither* of these might necessarily be described as irrational. Rather, these decisions could be understood as the divergence of an "egalitarian" (in the United Kingdom) versus an "individualist" (in the United States) approach to risk. Critically, these two cultures may not directly reflect those of the national populations at large, but instead only those with the power to control regulation.

At its best then, this approach provides an interesting contribution. Risk perception is not a universal brain function, but is instead located in culture, can be embedded in institutions (like environmental organizations or government agencies), and is tied to competing priorities of environmental management. Risk perception therefore is, by implication, profoundly *political*.

## Beyond Risk: The Political Economy of Hazards

Once we begin to consider the politics of hazards, however, the lens of risk, risk perception, and risk communication becomes somewhat inadequate. This is because:

1   risk is something that is sometimes imposed on others because the risk decision-maker is not the individual or group in control of decisions;
2   even where individuals make risky decisions on their own, their practical range of alternatives already has been severely diminished by political and economic context; and
3   the ability to make informed risk decisions is hampered by the political control, manipulation, and interpretation of crucial information.

Each of these facts suggests that risk analysis is a necessary but insufficient view of society–environment interactions.

### Control of decisions – the political economy of environmental justice

A key limit of risk assessment as a technique for better decision-making is that people, groups, and firms often make decisions about risk *for other people*. Consider the problem of toxic hazards and waste. More than 1,300 "superfund" sites (Federally identified cleanup locations representing hazards in violation of environmental law) are scheduled for cleanup in the United States. Roughly 11 million people (4 million of whom are children) live within a mile of these, with significant related health risks. Those living in and around hazardous sites are far more likely to come from poor and minority communities (Bullard 1990). Certainly some people "choose" to live near such sites, since they are typically lower in rent and have more easily available housing. Nevertheless, most of these sites were created by former businesses, military installations, or municipal services that are long gone, whose owners have benefited from their actions, and who never consulted local communities or people concerning their activities. This represents a situation of environmental injustice, therefore; hazards are distributed unequally over space and throughout society, with a structured bias towards vulnerable populations (see Chapter 7).

> **Externality**  The spillover of a cost or benefit, as where industrial activity at a plant leads to pollution off-site that must be paid for by someone else

We can certainly improve situations like these by providing more and better risk information to people living in and around such sites. But the problems run far deeper. To the degree that the costs of risky decisions are borne by others (and so become **externalities**; see Chapters 3 and 4), there is mismatch between the site of decision and the location of the hazard. This mismatch is deeply ingrained in the political and economic structures of communities, cities, and regions. Reconnecting the risk decision-maker to the hazard, therefore, requires more than careful or rational analysis of the hazard itself; it demands scrutiny and restructuring of the system of power relations surrounding decision-making.

## Constraints on decisions – political economy of the range of choice

Even where people find themselves in control of their risk decision-making, they are by no means in control of the social and ecological constraints that affect their decisions. Consider the case of farmers in drought-stricken West African environments. Observed in a state of crisis in the 1970s and 1980s, when many starved and out-migrated from the region, it might be possible for a naïve observer to argue that the specific weather conditions of the period were entirely unprecedented, that the society was poorly adapted to drought, that its constituent members made poor risk assessments, or – when seeing them pray for rain – that they came from a culture group of Douglas's "fatalists."

Careful examination of the longer history of their situation, however, as geographer Michael Watts (1983) asserted in his essay "On the Poverty of Theory," shows that the rules governing the economy and society of people in the region had been so fully transformed in recent decades as to make "adjustment" to the drought of the period essentially impossible. Specifically, he demonstrates that coping with severe water shortages is an historical trademark of the people of the region, and that only with the onset of colonial rule, and later regional and international commodity trading, did taxation and complex demands for cash in the economy make it hard for farmers to maintain and exchange resources in the face of crisis. In other words, changes in social and economic structures meant that the choices and assets available to farmers were severely reduced in a period of climatic stress. The result was widespread starvation in a region with hundreds of years of successful adaptation to drought.

Can a risk approach improve our understanding of such situations? Certainly, but it must be first acknowledged that the suite of resources available to people and societies facing hazards changes through political and economic processes. Addressing risks means understanding not only the characteristics of the environmental hazard in question, but also the way households, communities, and states are constrained by larger systems. These systems that govern exchange (e.g., commodity prices), access to resources (e.g., credit or social services), obligations (e.g., debt), and power (e.g., autonomy of decision-makers) are themselves part of the socio-environmental risk equation and a simple optimization or calculation of probability is inadequate to address them together (Chapter 7).

Such transformations are also by no means unique to traditional African societies. It could be argued that modern urban culture in places like the United States and the United Kingdom has itself undergone a radical reshaping of its risk landscape. Within the past few decades the traditional economic and political tools available to address risk (like state social services, for example) have decreased, exactly at a time when new risks and hazards (e.g., SARS) stemming from new technologies and interactions are on the horizon. Whole global systems of environmental risk have emerged that are closely linked to financial markets and global social systems. A typhoon in China presents a complex set of problems for financial markets in the United Kingdom, for example, which impinge on available capital for home buyers in the United States, with implications for state resources that set the conditions for levee construction in New Orleans, and so on. Living within such conditions means moving past various simple risk calculations for individual hazards and rethinking the management and distribution of hazards in a political and economic global "risk society" (Beck 1999).

## Control of information – the political economy of information

Finally, risk assessment and communication approaches to society–environment interactions have demonstrated the inevitability of uncertainty but cannot resolve how much certainty is required to act. Any decision is inevitably social and political and divergent interests are unlikely to agree. But where elite institutions (large firms, well-organized social classes, governments, or bureaucratic bodies) control the process of deriving and communicating information about risk, there is an inherent tendency to overstate the certainty of current knowledge or to understate how much information is necessary to make an informed decision. Risk decisions are often made, therefore, amidst competing accounts (or constructions – see Chapter 8) of the hazards related to the decision.

Such situations are by no means unusual! They include decision-making around many technological hazards, including nuclear power and nuclear waste. For such cases, uncertainty is inevitable and some kind of risk assessments must be made: to produce nuclear energy, for example, or not; or to bury the waste either in deep permanent storage facilities or more easily retrievable sites. But many things control and inhibit the interpretation of information in such cases. Commonly, risks are simply miscommunicated or explained in partial terms. Perhaps the most dramatic case was that of risks associated with cigarette smoking, which were long known and proven within the industry, but withheld by tobacco companies only until whistle-blowing former employees risked life and reputation to challenge them, only then leading to the release of the facts (a dramatic conflict made famous by the film *The Insider*).

In a more directly environmental case, tetra-ethyl lead in gasoline was also long known by the gasoline industry in the United States to be terrifically toxic and widespread through fumes. The delay in communicating these risks to the public slowed regulatory response, and only in 1986, after decades of exposure, were alternative additives used to replace the toxin, leading to a 75 percent reduction of lead in the blood levels of people in the United States (Kitman 2000). History has shown that the manufacturers and producers of products that pose these kinds of risks are unlikely to provide full information about these without some consumer or government coercion; after all, it is simply against their commercial interest to do so.

Beyond this, even when a full set of facts are communicated to citizens and consumers, exactly analogous to the information that government experts or private firms possess,

## Box 6.1   The Bhopal Gas Disaster

In Bhopal, India on December 3, 1984, a cloud of toxic methyl isocyanate (MIC) gas was discharged from a pesticide plant owned by the US-based Union Carbide company. The cloud quickly settled over adjacent neighborhoods, killing thousands of people immediately. By 1991, the official figures counted 3,800 dead. An event of this magnitude is extremely difficult to imagine. Consider that these figures far exceed those of the World Trade Center terrorist attack of 2001. Beyond these casualties, more than 11,000 people were permanently disabled by the release, since MIC exposure causes respiratory failure, impaired vision, and a host of crippling long-term conditions. An estimated 15,000 to 20,000 resulting premature deaths have occurred in the period since.

While a disgruntled worker was initially blamed by Union Carbide for the accident, it has since become apparent that an enormous range of dangerous errors and shortcuts were made by Union Carbide leading up to the disaster and that the event was no freak, but a very nearly predictable result of excruciatingly poor management. At the time of the accident, Union Carbide was one of the largest and most successful corporations in the world, logging $9.5 billion of sales in that year alone. How could a resource-rich modern corporation allow an accident of this scale?

In this case, Union Carbide was driven to shift from simply combining raw materials to the more hazardous process of actually manufacturing them for pesticide production (called "backward integration") at sites like Bhopal by the onset of global competition. Because of poor sales, moreover, the plant itself had been marked for dismantlement, but was kept in operation with far reduced safety procedures. An unchecked faulty valve had allowed water to flow into tanks that began the reaction. A safety refrigeration unit had been drained for use elsewhere and a gas flare safety system had been offline for months. From a risk analysis point of view, it would seem that Union Carbide did *everything* wrong.

This raises some serious questions about how risk is actually managed and distributed in a global economy. Certainly poor technical and safety procedures are the proximate causes, but the forces of global competition certainly contributed to short-cutting. So too, the globalized business model that allows the offshoring of the most dangerous parts of the production process is a significant factor in increasing risk. From the Indian side of things, it is also certainly true that the lax oversight of the plant by government officials contributed in no small way. In the rush for investment, the state looked the other way, forcing us to consider the political economy of risk as it is circulated in global capitalism. Most tellingly, the casualties of the event were mostly from the poorest strata of society, living in a slum adjacent to the plant, raising the further specter of environmental injustice, where marginal communities bear the brunt of hazards while profits are accumulated at distant sites.

### Reference

Broughton, E. (2005). "The Bhopal disaster and its aftermath: A review." *Environmental Health: A Global Access Science Source*. Retrieved September 17, 2009, from www.ehjournal.net/content/4/1/6.

there may be disagreements over the degree to which the current level of certainty is sufficient to act (Shrader-Frechette 1993). The incentives and benefits for decision-makers and firms to resolve debate quickly and to act are high, and therefore they have an interest in diminishing (or underplaying) uncertainty. The difference between state regulators/ power companies who may seek to develop nuclear waste and related waste storage, for example, and average citizens who may invoke precaution, may not be simply "cultural" in the sense Douglas suggests, therefore. It may instead reflect fundamental differences in where each group sits in a complex political economy. Both can make rational risk decisions and still not agree. Risk assessment can no more resolve such differences than it can change the nature of bureaucratic government or the capitalist economy. Though a powerful tool, the limits of a risk perspective are essential to acknowledge.

## Thinking with Hazards and Risk

In this chapter we learned that:

- By thinking about environmental problems as hazards, it is possible to more rationally consider and weigh risks.
- Risk is understood as the known (or estimated) probability that a decision will have a negative consequence.
- This is different than uncertainty, which surrounds all of the possible unknown characteristics of a hazard.
- One of the problems with assessing risk is that people's perception and estimation of risk is not fully rational and is influenced by emotion or affect.
- Indeed, differing communities or groups of people may have cultural predispositions for thinking about risks in different ways, and assigning responsibility for risk to different parties (like individuals or the government).
- Because individuals and groups are differentially exposed to risk owing to conditions beyond their control (like poverty and low political or social status), risk and hazard approaches to the environment are best considered within political and economic frameworks.

## Questions for Review

1. Using the example of hurricanes, support the statement (from page 81) that "most hazards fall in between natural and anthropogenic."
2. Do you belong to a more individualist, egalitarian, hierarchist, or fatalist culture? Support your answer with some specific examples.
3. Residents and governments of Europe have been much more critical of genetically modified crops and foods than have residents and governments of the United States. Explain this using cultural theory. (Bonus: Supplement your answer using political economy.)

4. Why do toxic waste dumps usually represent examples of environmental injustice? (Think: Who chooses to live near a toxic waste dump?)
5. Consider the long lists of precautions and risks on the label of an over-the-counter insecticide or herbicide (check one out at your local home and garden shop if you're not familiar with this). How would this list differ in the absence of governmental regulations? Why?

## Suggested Reading

Beck, U. (1992). *Risk Society: Towards a New Modernity*. London: Sage.

Bostrom, A. (2003). "Future risk communication." *Futures* 35: 553–73.

Douglas, M., and A. Wildavsky (1983). *Risk and Culture: An Essay on the Selection of Technological and Environmental Dangers*. Berkeley, CA: University of California Press.

Fischhoff, B., P. Slovic, et al. (1978). "How safe is safe enough? A psychometric study of attitudes towards technological risks and benefits." *Policy Sciences* 9(2): 127–52.

Jasanoff, S. (1999). "The songlines of risk." *Environmental Values* 8(2): 135–52.

Johnson, B. B., and C. Chess (2006). "From the inside out: Environmental agency views about communications with the public." *Risk Analysis* 26(5): 1395–407.

Kasperson, R. E., D. Golding, et al. (1992). "Social distrust as a factor in siting hazardous facilities and communicating risks." *Journal of Social Issues* 48(4): 161–87.

Kitman, J. L. (2000). "The secret history of lead." *The Nation*, March 20: 11–30.

Rissler, J., and M. Mellon (1996). *The Ecological Risks of Engineered Crops*. Cambridge, MA: The MIT Press.

Slovic, P. (2000). *The Perception of Risk*. London: Earthscan.

## Exercise: Mapping Risk

You have a choice of three different possible residences in the same town. The characteristics of the three possibilities are as follows:

|  | Choice 1 | Choice 2 | Choice 3 |
| --- | --- | --- | --- |
| Risk | No flood | 250-year floodplain | 50-year floodplain |
| Benefits | No river view/access | River view | River view and access |
| Time to pay off | 8 years | 15 years | 30 years |

Property #3 is subject to a devastating flood every 50 years, an event that can destroy your house and all of your property. Property #2 can be subject to similar flooding, but only once every 250 years. On the other hand, there is no risk of flooding to property #1.

Property #3 has a beautiful view of the river and private access to swimming and boating. Property #2 has a good river view, but no direct access to the water. Property #1 has neither a river view nor access to the river.

Property #3 will take you 30 years to pay off, while property #2 will take you 15 years. Property #1 can be paid off in eight years.

Which property would you choose and why? Compare your decision with your classmates' choices. Are they the same or different? Do you think that the differences in decisions are based on your personal backgrounds, cultures, or other factors? How would you mitigate some of the negative possibilities of your decision (either lack of amenities or high risk)?

If you chose $x$, what might make you choose $y$ instead?

# 7

# Political Economy

*Credit*: R. Gino Santa Maria/Shutterstock.

## Chapter Menu

## Keywords

- Anthropocentrism
- Commodification
- Commodity
- Conditions of production
- Eco-feminism
- Environmental justice
- Exchange value
- First contradiction of capitalism
- Globalization
- Means of production
- Overaccumulation
- Primitive accumulation
- Production of nature
- Relations of production
- Second contradiction of capitalism
- Social reproduction
- Spatial fix
- Superfund
- Surplus value

# The Strange Logic of "Under-pollution"

*"'Dirty' Industries: Just between you and me, shouldn't the World Bank be encouraging MORE migration of the dirty industries to the LDC [Lesser Developed Countries]?"*

In December 1991 World Bank Chief Economist Lawrence Summers (now economic advisor to US President Barack Obama) sent a memo with the quote above as the subject line. In the body of the memo, he laid out his argument for why it would be more efficient and cost-effective to export polluting industries to poorer countries. First, because wage structures are lowest in less developed countries, pollution would mean fewer losses in earnings due to morbidity and mortality. That is to say, because workers in poor countries are generally paid less, if they became ill or died because of pollution, the economy would suffer fewer losses to potential earnings than if the equivalent happened in more developed areas. Second, he argued that to his mind, "under-populated countries in Africa are vastly UNDER-polluted." In other words, some countries were not absorbing their fair share of global pollution. The real problem here was that, because the effects of "dirty industries" occur in specific places, there was an over-concentration of pollution in more developed countries. This fact, combined with the high cost of moving solid waste from one place to another, Summers argued, "prevented world welfare enhancing trade in air pollution and waste." Finally, he argued that the level of demand for a clean environment for either aesthetic or public health reasons depended on income. Poor people would rather have economic development than a clean environment.

The memo was leaked to a number of environmental groups and outrage ensued. Mr Summers explained that the memo was a sarcastic response to the proposal of a colleague at the Bank. Whether he was being sincere or sarcastic, though, the market logic Summers used is manifest in many of today's green development policies. Take, for example, carbon trading. This form of regulation allows for the production of waste in one place to be absorbed in another using a global exchange of pollution (see Chapters 3 and 9). In such a scheme, industries in the United States, for example, can purchase carbon credits from other firms that are not using their full allotment, so that the pollution created by one company can be "offset" by another. Effectively, we are creating a global trade in air pollution, something that Summers argued for.

Consider, also, the case of international trade in solid waste. While Summers argued that the high cost of solid waste transport would prevent it from occurring on a wide scale, he implies that this would be a more efficient way of managing pollution. Today, we have largely overcome the barriers to a worldwide market in importing and exporting waste. The international waste trade is a multi-billion-dollar business. Millions of tons of hazardous waste are transported across international boundaries every year. Many of these flows are from more developed countries to less developed countries where disposal costs are lower and environmental regulations are more lax or are more likely to go unenforced. In many ways, even if Summers' proposals were not serious, they have come to pass.

In addition to its eerie prescience, the memo raises basic questions about society's relationship to nature. How is it possible to imagine that part of the world is under-polluted? What kind of economic logic does this require and what drives such reasoning and action?

This chapter presents a perspective on environment and society that helps us to understand how the logic of the memo is an artifact of the modern capitalist world in which we live, with disturbing implications. A political economy approach to understanding environment and society argues that, at base, it is the structure of the economy and the set of power-laden relationships (worker–owner, industrialist–politician) that produce both the environment in which we live and our perceptions of it.

This chapter, therefore, seeks to introduce four basic sets of ideas. First, it aims to summarize the insights of Karl Marx for ecological problems, pointing to the way the capitalist economy is prone to *crisis*, especially ecological crisis. Second, it works to show how the natural world around us might be viewed as *produced*, explaining the strange ways we have come to live in, and consume, nature. Third, it explains how the economy attempts to resolve the problems and crises it has created by increasingly globalizing and exporting both production and waste, showing how *uneven development* and unevenly distributed environmental problems are symptoms of the contemporary economy. Finally, it shows how movements to protect the environment, as in **environmental justice** or antitoxins activism, are often rooted in issues surrounding *social reproduction*, with implications for gender and environmental action.

> **Environmental Justice**   A principle, as well as a body of thought and research, stressing the need for equitable distribution of environmental goods (parks, clean air, healthful working conditions) and environmental bads (pollution, hazards, waste) between people, no matter their race, ethnicity, or gender. Conversely, environmental injustice describes a condition where unhealthful or dangerous conditions are disproportionately proximate to minority communities

In order to understand this political economy perspective, therefore, we need to first discuss some ideas of Karl Marx. His analysis of the capitalist economy underlies much of this thought, and his work has been extended by many scholars interested in the relationship between the environment and society.

## Labor, Accumulation, and Crisis

Karl Marx, an economic philosopher of the nineteenth century, was one of the most influential scholars to study the relationship between political, economic, and social systems. While he said relatively little about nature directly, many have extended his analysis to explore the role that natural systems play in society. This section will describe some of the most salient tenets of this analysis and then discuss how these can be extended to understand contemporary relationships between environment and society. While much of the terminology of this theory may seem specialized and abstract, the basic concepts are universal and familiar. The most relevant concepts that we can draw from Marx for understanding nature–society relations are labor, accumulation, contradiction, and crisis. These elements, Marx explains, are all part of different historical economies (medieval feudalism, for example, or modern capitalism born during the industrial revolution), but they behave differently and relate to one another differently in differing eras. In today's "modern capitalism" these elements are basically arranged so that:

- people's *labor* is sold on a market;
- which allows for the *accumulation* of capital by a small number of individuals;
- this creates *contradictions* since capital becomes over-concentrated;
- which leads to disruptive financial and ecological *crises*.

In other words, in political economy, environmental problems are already built into the economy that causes them.

## Labor

The key component of Marx's thought is the role of labor in integrating nature and society. For Marx, when people mix their labor with the resources of the natural world to make a living, nature and society become bound together inextricably. In everything we use and consume, therefore, in every **commodity** (running shoes, laptop computers, golf lessons, pizzas, carpets, handguns) are both the human labor that made the object (or that made the machine that made the object) and the diverse elements of the natural world of which it is comprised. The table you are sitting at while reading this book is obviously made of wood (from trees) or plastic (from petroleum), screws (from steel and therefore iron), and shellac (derived from insect resin). But it also contains all of the human labor necessary to bring those elements together and combine them into a single object. Anyone who has ever tried to make their own furniture recognizes that there is both lumber and sweat in every such object. Labor is the act of altering nature and bringing it into the process of making the human world.

In a modern economy very few people own their own business and equipment, and very few have access to raw materials like oil or iron. Most people work for someone else, someone who owns the machinery, computers, equipment or factories (called the **means of production**) and who secures the natural raw materials (called the **conditions of production**). Put another way, in the capitalist system of today, some people (workers) *sell their "labor power"* to other people (capitalists) who put it to use however they please. They then combine the labor of these workers with raw materials and machinery to produce commodities. These objects are then sold to consumers. Usually, these consumers are also workers themselves, which means they are actually buying back the products of their own hard work. These commodities are sold on a market.

The capitalist takes a share of this sale, the **surplus value**, and keeps it for themselves (otherwise what's the point of owning a factory?) and pays the worker what is left over, perhaps setting aside some of the surplus to maintain or invest in the conditions of production: planting new trees to replace those cut down, for example. Crucially, if workers are paid the full value of their labor or natural systems are reinvigorated at the same rate they are drawn down, there is little or nothing left for the capitalist. Capitalists have to "underpay" labor

**Commodity**  An object of economic value that is valued generically, rather than as a specific object (example: pork is a commodity, rather than a particular pig). In political economy (and Marxist) thought, an object made for exchange

**Means of Production**  In political economic (and Marxist) thought, the infrastructure, equipment, machinery, etc. required to make things, goods, and commodities

**Conditions of Production**  In political economic (and Marxist) thought, the material or environmental conditions required for a specific economy to function, which may include things as varied as water for use in an industrial process to the health of workers to do the labor

**Surplus Value**  In political economic (and Marxist) thought, the value produced by underpaying labor or over-extracting from the environment, which is accumulated by owners and investors

**Figure 7.1** The secret of surplus value, in a nutshell. By the same principle, the environment must be worked harder and underinvested in order to sustain surpluses. The artist, Fred Wright (1907–84), made panel-cartoons and animated shorts for the United Electrical Workers Union throughout his career. *Credit*: Fred Wright Papers, 1953–1986, UE 13, Archives Service Center, University of Pittsburgh.

and/or nature or there cannot be capitalism. This presents a fundamental challenge to the social or ecological sustainability of capitalism (Figure 7.1).

## Accumulation

Marx viewed capitalism as a relatively new kind of economy, a recent product of the eighteenth and nineteenth centuries. Previously, people combined their own labor with the natural materials of the world around them and either made all of the goods they needed themselves, or traded with their neighbors to meet their needs. In other words, goods were **exchanged for their value** as useful things, measured as the value of these objects to the person making and trading the products of their own work. This was possible, at least in part, because people controlled their own means of production (a cobbler owned the basic tools required to make shoes, for example), and the raw materials were largely available in the world around them.

In such a view, there is nothing inevitable about organizing an economy in a way where workers and their products are separated from each other by a market and a bunch of capitalists. It has been organized in other ways in the past and could be organized differently in the future. How, then, did the economy get this way? Why are people forced to sell their labor instead of employing it themselves? How did capitalists come to control the means and the conditions of production?

Looking back at economic history, there are clear periods where it was made more difficult for people to utilize their own labor and to gain access to natural resources. Marx specifically refers to the example of the Enclosure Laws in England. These laws, mostly enacted between 1780 and the 1820s, privatized communal areas and forests and therefore pushed many independent smallholders off of the land. This was an essential building block of modern capitalism, because it not only allowed for a select group of people to own and control the means and conditions of production (in this case, land), but it also took away the ability for people to feed themselves independently. With no land on which to grow food or gather materials to produce goods for trade, many people were left with nothing except their own labor power. They had to sell this labor power in order to survive and were thus forced into wage work. This process of enclosure and appropriating previously communal or free land is sometimes referred to as **primitive accumulation**.

From this historical point forward, accumulation would be the central driving imperative of the economy, with serious implications for the environment. Once a set of social relationships like these (described as the **relations of production**) is established between a few people in the world who are capitalists and many people who are wage-earning workers (whether they cut meat, write code for computers, wash cars, sell household products, or perform financial services), a set of imperatives is introduced into the economy. To produce a surplus (and so make a living without applying their own labor to the manufacture of goods), capitalists must keep capital

**Exchange Value**   In political economy (and Marxist) thought, the quality of a commodity that determines the quantity of other goods for which the commodity might be traded at a given moment. Compare to use value

**Primitive Accumulation**   In Marxist thought, the direct appropriation by capitalists of natural resources or goods from communities that historically tend to hold them collectively, as, for example, where the common lands of Britain were enclosed by wealthy elites and the state in the 1700s

**Relations of Production**   In political economic (and Marxist) thought, the social relationships associated with, and necessary for, a specific economy, as serfs/knights are to feudalism and workers/owners are to modern capitalism

(money that is invested in the system to make more money) in circulation, constantly investing in labor power, in the means of production, and in the conditions of production. Also, since competition between capitalists continues to drive down surpluses over time, capitalists must frantically 1) keep innovating production techniques to squeeze more surplus out of the same amount of labor, 2) keep accelerating the process of buying and selling to maximize the number of profitable transactions, and 3) keep cutting some of the value put back into workers and the environment. These ongoing pressures demand the constant circulation of capital and the continuing expansion of the system, which are both critical for the survival of capitalism and capitalists. This account of the economy as an accelerating treadmill (Foster 2005), constantly racing to keep itself in place and devouring labor and materials to survive, raises further questions about the sustainability of capitalism.

## Contradiction and crisis

One can well imagine that such a system might crash if it was faced with a slowdown in buying and selling of goods or a scarcity of resources, or some combination of these. Consider the socioeconomic events of 2008 in the United States and the United Kingdom, a time when resource scarcity (fuel and housing) squeezed the capacity of workers to buy goods (houses, cars, and even food), leading to downturns in the rate of turnover and surplus, leading to further pressure on wages (and massive layoffs) and on nature (aggressive exploration for energy resources in environmentally sensitive areas), resulting in … further crisis.

In contrast to a free market theory of the economy (Chapter 3), a political economic approach posits that such crises happen *all the time*, that they are endemic to the modern economy, that they define the history of social and political adjustments of the past two centuries, and that they inevitably lead to perverse and undesirable social and environmental conditions. Though we like to think of such situations as aberrations, as cases where the system has failed to work, for Marx, such crises are simply an expression of the internal contradictions of the capitalist system itself. Crisis – with layoffs, homelessness, environmental destruction, and firms scrambling to stay in business – is actually the system working.

Marxists have argued that a contradiction lies at the heart of these crises: **overaccumulation**, the over-concentration of wealth in the hands of a few capitalists. On the one hand, accumulation is necessary to the continued expansion of the capitalist system. The more owners earn, the more they can put back into production and the more surplus value they can appropriate. This tends to lead to a concentration of wealth in fewer and fewer hands. On the other hand, accumulation is bad for the system as a whole because it limits the capital available to circulate, especially when a few wealthy people and large firms control most of the capital. Such a condition is certainly evident in the global economy. Exxon Mobil Corporation's 2005 profits were $36.13 billion (the highest ever for a US company), making it bigger than 125 of the 184 economies measured by the World Bank (Blum 2005).

**Overaccumulation**   In political economy (and Marxism), a condition in the economy where capital becomes concentrated in very few hands (e.g., wealthy individuals) or firms (e.g., banks), causing economic slowdown and potential socioeconomic crisis

This situation of overaccumulation leads to strange symptoms in the economy, crises caused by either overproduction or underconsumption. In the first case, there is simply too much stuff being made to be bought in a reasonable amount of time. This can be due, for example, to improvements in technology that allow for extremely high levels of production, resulting in gluts. Underconsumption, on the other hand, means that consumers are not buying products at a sufficient rate to clear the market. This is generally the result of workers earning insufficient wages to buy all of the goods available to them. This is also unsurprising, especially considering the simultaneous effort on the part of capitalists to keep workers' pay as low as possible when workers are the main body of consumers.

Of course, overproduction and underconsumption are two sides of the same coin. The problem is that each product that is not sold or bought represents a value that is no longer circulating through the system. Because the circulation of value is so critical, something has to be adjusted to ensure that the system as a whole does not collapse. During the Great Depression, for example, overall global consumption declined and the system ground to a halt, with catastrophic results for producers and workers around the world. Unemployment was at high levels, due to drops in production. In order to spur investment in the United States, New Deal policies enacted by the Roosevelt administration put people to work on public projects (bridges, roads, buildings) so that they could earn wages. This, in turn, would help to stimulate production. A similar logic also produced the 2008 tax rebates as part of an economic stimulus package aimed at getting people to spend more.

As crises accelerate, the need to increase production while cutting costs leads to "heroic" efforts on the part of capitalists, companies, and boards of directors, all of which lead down a path of diminishing returns. Massive wage cuts can be introduced or companies can move production and jobs to places where lower wage rates and benefits are normal. But, if you put all of your potential buyers out of work, or pay them lower wages, who will buy your goods? And, if you exploit workers to the point that they are not as productive as they could be, they might have more sick days or work more slowly. Or, radically, workers might get fed up with this situation and walk out. If this happens on a significant scale, the system is in danger of collapsing, bringing about a transition to some other form of economy, with different social relationships. This drive to accumulate and its accompanying transformative results are sometimes referred to as the **first contradiction of capitalism**.

**First Contradiction of Capitalism**   In Marxist thought, this describes the tendency for capitalism to eventually undermine the economic conditions for its own perpetuation, through overproduction of commodities, reduction of wages for would-be consumers, etc., predicted to eventually lead to responses by workers to resist capitalism leading to a new form of economy. Compare to the second contradiction of capitalism

Consider, though, that another, related, way to maintain accumulation, aside from placing further stress on workers, is to draw heavily on natural resources, which are often free "common" goods (Chapter 4). What are the implications of trying to forestall a social crisis by igniting an environmental one?

## The second contradiction

According to the theorist James O'Connor, Marx said only a few things about nature directly, but they have significant implications for environmental sustainability. First, Marx argued that periodic bad environmental conditions can result in an economic crisis. This means that scarcity of raw materials due to drought, hurricane, or other natural disasters might cause a crisis in the production system. Similarly, Marx acknowledged that there may be some absolute limits to production imposed by nature. So, even though capitalists are continually looking for ways to produce more stuff, they might be somewhat limited by the availability of raw materials, for example. Either way, if raw materials are unavailable, or simply too expensive to mine or harvest (see Box 3.1 on Peak Oil), this means that the costs of producing goods will go up. In turn, production of certain commodities could be reduced, leading to slowdowns in circulation as described above.

More disturbingly, Marx argued that capitalist forms of production/extraction, especially in the areas of agriculture and forestry, were bad for nature in a way that parallels the exploitation of human labor power. To maintain surplus, value must be overdrawn from either labor or nature:

> the increase in the productivity and the mobility of labour is purchased at the cost of laying waste and debilitating labour itself. Moreover, all progress in capitalist agriculture is a progress in the art, not only of robbing the worker, but of robbing the soil. (Marx 1990: 638)

The more severely you exploit the land (or workforce), the less able the land (or workforce) will be to sustain production over the long run, again leading to crisis. To make his argument, O'Connor draws on the work of Karl Polanyi. Rather than thinking of scarcity as a problem caused by rising demand (Malthusian overpopulation – Chapter 2), or as a technical issue to be managed within the system (risk analysis – Chapter 6), or as an inevitable result of the separation between humans and nature (deep ecology – Chapter 5), Polanyi argues that scarcity is produced by capitalism, which extracts resources necessary to make commodities. Scarcity, then, is not a natural condition, but one that is created by marketing of nature (as we will discuss in the case of bottled water, Chapter 13). So, capitalism creates its own limits, not only because it changes humans into cogs in a machine, but because it also extends into, and uses up, nature.

If this is true, the environment may present a limit to capitalist growth. As capitalism relies on the exploitation of nature as a resource, the propensity to destroy nature could be another path toward the disintegration of the economic system and the rise of alternatives. This ecological theory of crisis and social transformation is often referred to as the **second contradiction of capitalism**.

Summarizing this argument, O'Connor explains that environmental conditions are not just a limit to economic production, but that the pollution of nature by capitalist enterprises also threatens the health and wellbeing of the earth and the workers who live on it. This leads to a struggle over **social reproduction** (see also below), understood as the process of procuring the basic needs required to keep people healthy and productive. In this way of thinking, the labor power required to keep the economy going is undermined by capitalist production. If laborers cannot reproduce themselves effectively, labor power is decreased. Since capitalism is constantly under-

**Second Contradiction of Capitalism** In Marxist thought, this describes the tendency for capitalism to eventually undermine the environmental conditions for its own perpetuation, through degradation of natural resources or damage to the health of workers, etc., predicted to eventually lead to environmentalist and workers' movements to resist capitalism, leading to a new form of economy. Compare to the first contradiction of capitalism

**Social Reproduction** That part of the economy, especially including household work, that depends on unremunerated labor, but without which the more formal cash economy would suffer and collapse

mining its own conditions to save costs – producing waste, unclean air, bad water, and other conditions that can threaten workers – this is a predictable outcome (recall the incredible 75 percent reduction in blood lead levels after imposing limits on the gas industry explained in Chapter 6!).

Paychecks, then, are not the only reason for groups to organize against capitalism. Some of the most important contemporary agents of social change are not labor unions, after all, but rather "new social movements." These movements focus on workplace safety, toxic waste disposal and production, etc. (see the section on environmental justice below). Here, instead of seeking better wages or work conditions, people assert the basic right to a clean environment. In other words, rather than focusing energy and attention on taking over factories and workplaces, new social movements are aimed at ameliorating the effects of the economy on public health and the environment. These new social movements also differ from more traditional worker movements in that they are often wary of technology and traditional notions of development. While Marx thought that technology would eventually free labor from the factory by replacing labor power with machines, a "green" political economist of nature would argue that many of these technologies endanger the environment and human health. Rather than look to increased economic and technological development as the answer, then, some new social movements view development as part of the problem (Figure 7.2).

Part of what these new social movements demand, therefore, is a more environmentally friendly way of producing necessary commodities. As new demands are made on the capitalist system to be less polluting, capitalists are no longer allowed to treat nature however they want, and are subject to scrutiny on the part of the public and regulatory bodies. In theory, this new level of public accountability and scrutiny would set us on the path to socialism, or some other, more sustainable and transparent society and economy. O'Connor's conclusion is that continuing exploitation of nature and people will eventually

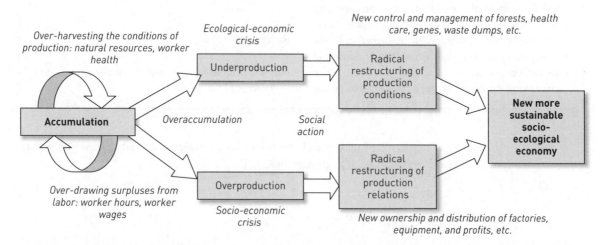

**Figure 7.2** Schematic representation of the possible contradictions that capitalism produces and the social and environmental responses they engender, possibly leading the way to a more sustainable and transparent society. *Source*: Adapted from O'Connor (1988).

## Box 7.1 Love Canal

Between 1942 and 1952, the Hooker Chemical Company of New York dumped more than 20,000 tons of hazardous materials, including highly toxic chlorinated hydrocarbons, in a clay-lined canal in a Niagara Falls neighborhood. When Hooker was done, they covered the site with dirt. Years later, the local school board obtained the property and built on the site.

By the late 1970s, health problems were rife throughout the community, including epilepsy, asthma, and urinary tract infections, as well as miscarriages and a range of birth defects. Seeking a thorough investigation of this crisis, Lois Gibbs, a local woman with no previous political experience, brought national attention to the problem through direct activism, protest, and wrangling with state authorities. Her efforts resulted in a full disclosure of problems at the site, an evacuation of residents, and a federal cleanup.

In the years after, defenders of Hooker Chemical have insisted that they gave warning concerning the presence of the chemicals to the local school board and so the company should not be held liable. Nonetheless, it is clear that school boards are not well versed in issues surrounding hazardous waste and that even if they were complicit in this disaster, the chemical company might well still be accountable for producing the hazard in the first place. A central problem therefore, beyond the horrible exposure of toxic chemicals to unsuspecting residents, lay in determining *who was legally liable* for damages and human health problems and *who was responsible for cleaning* the mess? Specifically, the question was whether or not liability ceases when a contaminated property changes hands, whether or not

there has been disclosure of risks. At the time of the case, these matters were not clear in US law. This matter was settled, to a great degree, by the landmark passage in 1980 of the Comprehensive Environmental Response, Compensation, and Liability Act (CERCLA – sometimes known as Superfund). That law insists on "joint, strict, and several" liability, which means all of the parties involved are responsible, creating a serious disincentive for any such dumping in the future, since once a mess is in the ground, its creator is in part responsible for it, no matter who buys the land, under whatever circumstances.

As a result, Love Canal established several critical precedents and made a symbolic international breakthrough as a result: 1) average middle-class citizens might be exposed to extremely hazardous materials in and around their homes, 2) companies free to operate outside of serious scrutiny have the capacity to put enormous numbers of people at risk with little legal recourse or transparency, 3) citizen activism can galvanize national attention and support on environmental problems through direct action, and 4) women often represent the central leadership on issues of environmental justice. More radically, the case paved the way for a comprehensive revision of how one might think about risk and responsibility in environmental law

### Reference

Blum, E. D. (2008). *Love Canal Revisited*. Lawrence, KS: University Press of Kansas.

lead to a "rebellion" of nature and the rise of social movements that demand an end to "ecological exploitation" (O'Connor 1988: 32).

## Production of Nature

Given the state of environmental and working conditions, from a political economy perspective it is reasonable to ask why such a rebellion and reformation of the economy has not occurred. While O'Connor sees a high potential for nature to set limits to capitalist production and to incite social action and undermine the system itself, there are other, more pessimistic interpretations of the relationship between the environment and society. Most prominently, the **production of nature** thesis suggests that *nature itself* is often made and remade through economic processes and that people have come to consume it as a commodity.

Geographer Neil Smith (1996) points out a paramount example of this problem. In the 1990s, The Nature Company (most recently acquired and renamed by the Discovery Channel) made significant profits selling petrified wood, butterfly nets, and books on penguins to consumers hungry for green products. Simultaneously, however, the store's parent company was cited as a party responsible for pollution at a **Superfund** site and further violations related to hazardous waste treatment and storage. This appears at first to be simply an

> **Production of Nature** In political economic thought, the idea that the environment, if it ever did exist separate from people, is now a product of human industry or activity
>
> **Superfund** The environmental program established to address abandoned hazardous waste sites in the United States
>
> **Commodification** The transformation of an object or resource from something valued in and for itself, to something valued generically for exchange. In Marxist thought, the rise of the exchange value of a thing, over its use value

unpleasant irony. Smith suggests, however, that it is exactly because nature can be treated as an object of *consumption* – as a commodity – by consumers that social and ecological issues in *production* are disguised, hidden, or forgotten.

Specifically, scholars concerned with the production of nature explore our ideologies of nature. They argue that modern worker/consumers hold two contradictory notions of nature at the same time, which together serve continued exploitation. The first of these ideas is that nature is something *external* to ourselves and society. By seeing human society as separate from nature, we have allowed humanity the apparent right to use it in whatever manner we see fit. In the case of capitalist society, this generally means the exploitation of nature as raw materials for commodity production, or the **commodification** of nature itself. If we want to buy or sell something on the market, after all, we must be able to see it as an object, external to us.

At the same time, however, we hold an *internal* conception of nature that suggests that all humans and non-humans are subject to natural processes. Here, we are internal to nature and therefore at the mercy of its laws. Because of this notion of nature, we view human society and the capitalist system as natural entities guided by the rules of nature. This means that the current state of society is simply an outcome of natural and inevitable rules. By universalizing the economy and society, economic competition and other concepts that are quite specific and recent, seem to be timeless and inevitable. This latter view has made it difficult to question the current socio-ecological system or imagine alternatives (see Chapter 8).

These contradictory viewpoints (nature is separate from people/people are part of nature) are the key to understanding capitalist ecology. On one level, as Smith argues, this is realized in the way that "green" companies sell people products aimed at making us feel closer to nature. Nature has been made an object (a bird feeder, a microscope, a remote controlled shark …) in order to market it for consumption. At the same time, though, we view this consumption as a way of immersing ourselves in, or taking ourselves back to, nature.

The process of commodification, however, takes other forms. Firms and investors are always trying to commodify nature, in order to maintain accumulation as described before.

If natural goods remain free to everyone, there is no potential for profit. But, if someone can claim ownership of a resource, then he or she can charge others for the use of it. The move towards privatization of natural resources like water is a prime example. In Bolivia over the past decade, attempts were made by a multinational corporation, Bechtel, to commodify water delivery. The state would sell the rights to the water supply of the city of Cochabamba to the company, who would, in turn, sell the water to residents through unit pricing. Here, a previously free good, water, having been commodified, becomes a source of accumulation for Bechtel. A political economy approach stresses the degree to which many forms of nature we all share and experience have increasingly come to be commodities we exchange.

Taking the production of nature seriously leads to two conclusions about contemporary environmentalism. First, we should be suspicious of any attempt to use market principles to solve environmental problems, especially those that seek to turn natural materials into tradable goods. Second, we must be critical of any environmental activity or activism that treats nature as external to politics and the economy. Though such activities may be able to save a few key areas of wildlife habitat or impose sanctions on individual companies for polluting, they cannot address head-on the contradictions at the core of the environmental crises that created habitat loss and pollution in the first place: the perverse relationship between nature and society which is inherent to contemporary capitalism.

Instead, according to the production of nature argument, a viable approach to the environment must follow a very different route. First, it must assume the inevitability and creativity of the social relationship with nature. This means that we should surrender to the fact that the nature that we know has been altered through its inevitable entanglement with human labor, and vice versa. We must acknowledge the implausibility of an autonomous nature, something that we might find in an essential and pristine condition, including romantic ideas about maintaining or going back to "wilderness" (see Chapter 8). Only by eschewing this nostalgia is it possible to have meaningful debates over what kinds of produced natures are most desirable.

Second, a production of nature approach means that environmentalism must keep in mind the global system of domination inevitably embodied in the economy. While capital can never fully control nature or people, it will always try. Given the accumulation of power and wealth inherent in capitalism, we must pay attention to the ways human relationships with nature vary among people differently situated in the economy, whether they work in a sweatshop in Indonesia, a rural ranch in Montana, or a grocery store in London. These people differ in their knowledge and experience of nature precisely because they work in different parts of the economy, in different relationships to capital. To create a meaningful environmental movement, these many economic and ecological positions will have to be united, bringing together people from very different places and backgrounds. No small task!

## Global Capitalism and the Ecology of Uneven Development

This is made all the more difficult by the spatial strategies employed in capitalist production. Specifically, as Summers' memo suggested, we now have the ability to export hazardous and solid waste internationally, meaning that many of the byproducts of the high-consumption lifestyle of the United States are absorbed by poor people half the world away. From the point of view of capitalism, Africa *is* under-polluted, but perhaps, not for long.

To this point, we have largely discussed the growth of capitalism and the contradictions that emerge from it in the developed world. Political economy approaches to the environment–society relationship, though, also help to explain the increasingly global reach of capitalism and its worldwide impacts on people and nature.

The capitalist economic system, as discussed above, requires sustained accumulation of capital and its continued circulation through the system. The ability to create economic growth can be endangered by contradictions inherent in the system itself which lead to periodic crises. Crises usually stem from overproduction and/or underconsumption, as we saw earlier. In either case, goods are not moving quickly enough through the system to sustain accumulation and produce profits. One way to resolve these periodic crises is to use the increasingly global reach of trade to subdivide the world of production and consumption. If workers in your own country are too expensive because of regulation, or if environmental regulations prevent you from getting cheap raw materials, why not move production somewhere else? Alternately, if your workers have already bought all of the goods that they can, why not find new markets abroad?

Both of these are examples of what geographer David Harvey has called the **spatial fix** (Harvey 1999). The crises caused by the treadmill of accumulation can be avoided, in theory, as long as capitalist production and consumption can be extended to new places around the world. This is the logic that leads to offshoring (the practice of transnational companies (TNCs) who move production facilities abroad) in an attempt to lower production costs and to create new consumers. Often production is moved to countries with fewer labor and environmental regulations. Since the 1970s many North American and European companies have moved production to Latin America and Asia. This has had profound impacts on environmental quality in these regions.

**Spatial Fix**   The tendency of capitalism to temporarily solve its inevitable periodic crises by establishing new markets, new resources, and new sites of production in other places

Some have argued, though, that such processes might be good for the host countries. Production means jobs. Jobs mean money. Money equals economic growth. Economic growth means development. Development is good for everyone. This resembles some of the logic behind free market accounts of the relationship between environment and society (Chapter 3). A political economy perspective, however, would raise questions. In addition to questions about the specific quality of life for workers in these new factories and the important environmental issues that arise, critics might take issue with the characterization of development as always and everywhere being a good thing. Development, particularly when measured in terms of economic variables like gross national product (GNP), often means further exploitation of workers and nature. Further, in a capitalist system, development is always uneven. While workers in the system might earn money, it is very little in comparison to the surplus that is flowing back to factory owners. While capitalists from Europe and North America are earning money, hand over fist, the environments of Latin America, Asia, and Africa are being destroyed, privatized, and commodified to further increase profits. Moreover, because international companies are removed from these effects by distance, they have little incentive to protect resources or even to provide for the continued health of their workers. They can always move. Workers and nature become disposable.

This tendency for profits to flow and be collected in some parts of the world while other parts are exploited is not new. Uneven development (Smith 1984) has long characterized the capitalist system. Colonialism, practiced in Latin America, Asia, and Africa by European countries (which some would argue was itself a spatial fix for dwindling domestic markets), provided the Global North with practically free resources for industrialization. Moreover, these political and economic relationships have had lasting consequences that must be considered when thinking about contemporary development practices.

> **Globalization**   An ongoing process by which regional economies, societies, and cultures have become integrated through a globe-spanning network of exchange

They might also be viewed as the foundations of **globalization**, the tendency for capitalist production and consumption practices to spread to the global scale. This process is a logical extension and result of the spatial fix since transnational companies often have lower production costs and can drive out more local smaller producers (see Chapter 12). Converging global consumption patterns also follow from this geographic strategy, as for example in the worldwide spread of fast food (see Chapter 14), which has meant increasing demands for industrial-scale food production, usually in the form of single crops that require intensive inputs of pesticides, fertilizers, water, and energy to grow and process, and which leave behind clear-cut forests (Chapter 10) and heavy loads of carbon in the atmosphere (Chapter 9).

## Social Reproduction and Nature

All of these externalized environmental damages, whether in the United Kingdom or Mexico, the United States or India, have real impacts on people's daily lives. In this regard, political economy approaches, even as they often focus on the problem of production (how, where, and under what conditions stuff is made), also have an interest in how people live, and are sustained (what is sometimes called social reproduction). To maintain an economy, any economy, its lifeblood must be preserved, after all. This includes natural resources, like air, water, and raw materials, but also, as we saw above, people themselves. The daily lives of people and their families, what Cindi Katz calls the "fleshy, messy, and indeterminate stuff of everyday life" (Katz 2001: 711), are thus of great importance to the political economy of nature. Given that many of society's interactions with the environment are precisely in this sphere (consider: you encounter more wild species in the forms of microbes and bacteria in kitchens and bathrooms than anywhere else in your daily life), the notion of social reproduction is clearly important for a political economy of nature.

For Katz, we can only understand society's relationship to the environment if we come to terms with how people live, how they interact to survive, and how they might work to improve or change the conditions of their lives. The environmental aspects of this sphere of the economy are the conditions in which people live, which can indeed be degraded in pursuit of surpluses. If a company, for example, dumps its toxic wastes in the ground (as at Love Canal, explained above), they have not only harmed the environment, but they have also endangered their workers' abilities to do their jobs (sick workers are less productive). Alternatively, as in the Bolivian case, if water is privatized and commodified, this also limits workers' access to the essential resource. In either case, there are possibilities for environmental change, if people demand that the company clean up pollution or if citizens organize to stop the privatization of water.

## Environmental justice

That this kind of collective action does not happen more often, as suggested by the production of nature thesis, may be because people have come to think of the environment as a commodity that can be bought and sold, rather than an intrinsic part of people's daily lives. It may also be a question of the differential spatial impacts of pollution. That is, some people may enjoy the consumption of a commodity produced far away, where the nasty effects of its production are more directly felt.

Environmental justice activists point directly at this problem, and highlight the persistent tendency for environmental hazards, waste, and noxious facilities to be sited in and around minority and low-income communities. The observation that poor, black, and immigrant communities are disproportionately put at risk from exposure to things like lead, silica dust, and pesticides is in fact an old one, with activists in the early twentieth century working in vulnerable communities amongst people who were most socially and politically disempowered (Gottlieb 1995). Some of the earliest environmental activists in the United States were urban social workers, primarily women, agitating for regulation of dumping and sewage and pressing for reform of public health and safety.

In the past hundred years, evidence has mounted of a simple, reliable, and direct correlation between proximity to dangerous things, like lead smelters and toxic waste dumps, and race and class. Low-income, black communities are consistently more likely to be close to hazards than whiter, wealthier ones. This spatial correlation between toxic facilities and minority populations was brought to broader public attention in the 1980s by Robert Bullard, who studied the distribution of waste facilities in the American South. In his book *Dumping in Dixie: Race, Class, and Environmental Quality* (1990), he famously concluded that

> Growing empirical evidence shows that toxic-waste dumps, municipal landfills, garbage incinerators, and similar noxious facilities are not randomly scattered across the American landscape. The siting process has resulted in minority neighborhoods (regardless of class) carrying a greater burden of localized costs than either affluent or poor white neighborhoods. Differential access to power and decision making found among black and white communities also institutionalizes siting disparities.

Though many activist groups and scholars agreed with Bullard's assessment, a debate emerged about whether the phenomenon was based on race or class. That is, are certain people living with toxic waste and other hazards because they are minorities or because they are poor? There is still disagreement on this point today, although it is clear that both are factors. Another debate in the field hinged on the question of whether vulnerable populations lived in certain areas where hazardous facilities were then sited, or if they moved into these areas after the establishment of such facilities, owing to lower housing costs. The answer to that question is, of course, both.

Both of these debates are arguably "red herrings." Rather than arguing about the timing of waste disposal and residential growth, or whether discrimination was based on race or class, it would be more useful to think about how these are related to larger processes of economic growth and disenfranchisement. Whether or not people lived in an area before dumping of hazardous materials occurred, the current structure of land development and ownership means that poor people and minorities were bound to be located in the same, marginal areas. This is partly due to land rents and partly due to the relative power of white wealthy neighborhoods to resist any proposed disposal sites or other toxic facilities near them. This differential ability of some people to resist having hazards dumped in their neighborhoods is sometimes referred to as NIMBYism (Not In My Back Yard-ism). A political economic approach to environmental justice goes beyond this parochial thinking to suggest how differential exposure to hazards is inextricably linked to the economy and the power structures within it.

### Gender and the political economy of environmental activism

**Eco-feminism** Any of a number of theories critical of the role of patriarchal society for degrading both the natural environment and the social condition of women

In addition to questions about class and race, many people interested in environmental justice study the role of gender in environmental politics, reform, and crisis, an approach often referred to as **eco-feminist**. Consider that, in a startling number of cases, protests against hazardous materials have been spearheaded by women. Women make up 60–80 percent of the membership of mainstream environmental organizations, but they figure even more prominently in grassroots organizations addressing environmental health and related issues (Seager 1996). The antitoxins movements of the past half century, by and large, have been dominated by women, including high-profile scientists like Rachel Carson as well as citizen activists like Lois Gibbs and Erin Brockovich.

Why would this be? Many argue that since responsibility for the health and maintenance of the household traditionally falls to women in many cultures, they are the first to notice, and act to oppose, environmental dangers that are products of capitalist production (water, air, and soil pollution that affect human health, for example). This position in the household is not, however, based on any inherent characteristics of women. Rather, it is a social circumstance with roots in the modern economic system. Historically, capitalism exploited people's labor in the workplace but also their labor at home. Someone had to work to produce the workers, after all. While historically men were taken out of the domestic sphere to perform wage work, women were left in the home and given the responsibility to provide food, clothing, and other basic (reproductive) necessities for the entire household.

At the same time, however, and as a painful irony, the opinions, observations, and concerns of women in this capacity have been at times woefully overlooked by experts.

Women often launch initial complaints about what are only later acknowledged to be major environmental crises. In the case of Love Canal (see Box 7.1), a notorious New York housing development fraught with asthma, epilepsy, and urinary tract infections, which was later disclosed to be sited atop chemical waste, women activists were dismissed as "hysterical housewives." This characterization, by the media and by the companies involved in dumping, was used to argue that the women activists lacked sufficient knowledge, expertise, or rational judgment (Seager 1996).

As a result, these become struggles not just about pollution, but also about expert knowledge. Traditional scientific understandings of nature pride themselves on objectivity and an external relationship to nature. This scientism is challenged by activists, often women, who hold more experiential views of what types of environmental problems society faces. This kind of relationship to the environment means that women often notice subtle changes in the environment (often ones that affect the health of loved ones and community members) that may be missed by big environmental groups more concerned with issues like wilderness or biodiversity preservation. Environmentalism, in this sense, needs to be a political economic movement.

## Environments and Economism

Such an analysis of environmentalism is compelling, because it causes us to think rather differently about environment and society. By exploring the relationship between nature and capitalist production, such approaches open up new areas for regulation, possibilities for environmental organization, and grounds upon which to be skeptical of the claims of neo-Malthusians (Chapter 2), free market evangelists (Chapter 3), and institutionalists (Chapter 4). While this certainly opens new paths to understanding environment and society, there are potential pitfalls.

Specifically, political economy approaches risk the extremely **anthropocentric** view of nature solely as a resource for production and social reproduction (with its attendant limits, as per Chapter 5). By tying together the society and environment through the analysis of economic processes, this approach makes other ways of thinking about non-humans (in terms of animal rights, for example) difficult.

> **Anthropocentrism**  An ethical standpoint that views humans as the central factor in considerations of right and wrong action in and toward nature (compare to ecocentrism)

By declaring the environmental movement ineffectively "bourgeois" in its unwillingness to confront the economy, for example, this perspective abandons some avenues of politically organizing around the interest of non-humans. This closes down an enormous number of potential allies and regulatory approaches for managing nature.

Additionally, there is potential for these approaches to reify the economy, meaning that this way of thinking often treats "capitalism," which is merely an abstract conceptualization of complex relationships, as if it were a solid thing with a concrete existence. This provides less room for seeing or imagining other existing productive social and ecological relationships in the world around us. By pinning all explanations for environmental problems and issues on capitalism, moreover, a range of other causes, relationships, and issues might be left off the table in an effort to solve problems.

In many of its iterations, the political economy approach also defers progress on environmental issues to a point *after* economic ones are solved. We may not be able to afford such a luxury in a time when society faces an enormous range of ecological problems and capitalism shows no immediate sign of vanishing.

## Thinking with Political Economy

In this chapter we have learned that:

- An influential school of thought stresses the way politics and economics are merged throughout history and the way human interactions with nature are mostly mediated through work, labor, and the economy.
- By observing the history of economies since the industrial era, the political economy approach concludes that the global economy tends towards accumulation of capital in very few hands, but that it is also highly prone to periodic and spectacular crises.
- The implications of this for the natural environment include the contradictory tendency for the economy to overexploit natural resources, leading to further economic problems but also to the rise of environmental movements.
- A political economy approach also focuses on the way nature becomes commodified and the way it is unevenly developed and degraded.
- It further points towards the way that many of these environmental and economic problems are worked out and experienced in everyday life (or social reproduction), leading to environmental justice movements, often led by women.
- While they are extremely powerful tools for understanding the relationship between nature and society, critics have suggested that these approaches tend to be anthropocentric and economistic.

## Questions for Review

1. Using political economy, describe a contemporary common private sector job (say, a nurse at a private hospital or a retail clerk). Include the terms "exchange value" and "surplus value" in your description.
2. What is primitive accumulation? Why was/is it necessary for the development of capitalism?
3. Why (according to a political economy perspective) do capitalist economies inevitably tend toward crises caused by overaccumulation?
4. Discuss the first and second contradictions of capitalism. Be explicit as to why (again, according to political economy) these are unavoidable contradictions inherent to capitalism.

5. Women's environmental experience and knowledge is sometimes different from that of men, owing to the political economic structures. Describe why this might be so and provide an example.

## Suggested Reading

Athanasiou, T. (1996). *Divided Planet: The Ecology of the Rich and Poor*. Boston, MA: Little, Brown, and Co.

Marx, K., and F. Engels (ed. H. L. Parsons) (1994). "Marx and Engels on Ecology." In C. Merchant (ed.) *Ecology: Key Concepts in Critical Theory*. Atlantic Highlands, NJ: Humanities Press.

Perrault, T. (2008). "Popular protest and unpopular policies: State restructuring, resource conflict and social justice in Bolivia." In D. Carruthers (ed.) *Environmental Justice in Latin America*. Cambridge, MA: MIT Press, pp. 239–62.

Singer, P. (2001). *Marx: A Very Short Introduction*. New York: Oxford University Press.

Wheen, F. (2007). *Marx's Das Kapital: A Biography*. Boston, MA: The Atlantic Monthly Press.

## Exercise: Commodity Analysis

The contemporary economy produces increasing masses of "e-waste" – computer monitors, processors, cell phones, televisions, and other goods that, while still functional, are thrown away or occasionally recycled. These typically become hazards, as they contain mercury and other heavy metals that enter waterways and the ecosystem. A political economy approach to this issue would insist that such planned obsolescence of "e-waste" is not an incidental or accidental product of the system, but instead an essential one. Explain why this might be so. What are the tendencies of capital that make such waste necessary? How is surplus produced so that waste is inevitable? What environmental justice issues might arise in the disposal and recycling of these goods?

# 8

# Social Construction of Nature

*Credit*: Lane V. Erickson/Shutterstock.

## Keywords

- Concept
- Constructivist
- Co-production
- Discourse
- Ideologies
- Narrative
- "Nature"
- Power/knowledge
- Race
- Relativism
- Signifying practices
- Social construction
- Social context
- Wilderness

## Chapter Menu

## Welcome to the Jungle

Imagine the following scenario: You have embarked on the journey of a lifetime!

Perhaps you have a great interest in wild places. Maybe you have canoed the Boundary Waters in Canada, or hiked Scafell Pike in the Lake District of the United Kingdom, or walked part of the Appalachian Trail, Yellowstone, or the Grand Canyon in the United States. Even summers and winters at home may have natural phenomena of interest to you: ice forming on lakes, migrating birds, blooming trees. You may be an avid viewer of nature films, like *March of the Penguins*, or television programs on *Animal Planet* or maybe you read magazines like *National Geographic*.

For you and people who share an interest in nature, few places captivate the imagination as much as the jungles of Borneo. Borneo is an enormous island in the South Pacific, split between the nations of Malaysia, Indonesia, and tiny Brunei. The *National Geographic* essay featured Borneo's justifiably famous wildlife – Sumatran rhinoceroses, clouded leopards, pygmy elephants, gaudy birds like hornbills and barbets, and of course the critically endangered orangutan.

And now imagine you get to experience this nature first hand. Arriving in Borneo you fly into Kota Kinabalu, Malaysia, a bustling city of some 700,000 people. From there you take a one and a half hour bus ride, followed by a jeep taxi to your first destination, a remote and rustic inn. You appreciate the authentic feel of the inn. Not only can you sleep under mosquito netting in an open-walled room, but the thatch-roofed inn seems to melt into the surrounding forest. You may also be happy that the people working the inn (cooking, cleaning rooms, serving tea) are all native Borneans – at least, they appear to be.

On your first full day on Borneo, you hire a local guide for a hike deep into the jungle of a nearby national park, hoping to glimpse some of the island's famous wildlife. After a disappointingly long line at the entrance gate to the national park (so many tourists!), your guide zips past the larger bus groups in his jeep and heads all the way up the road into the cloud forest, stopping at a small, nearly unmarked turnout. "This is where we will first hike. You have your binoculars, right?" he reminds you. So you head out onto the trail. After about 15 minutes the sounds of birds, insects, and running water have overtaken the noise of buses on the main park road. People have disappeared from view. At the same time, the trail is surprisingly well maintained, and between your guide, your guidebook, and the interpretive signs (in English, no less!) all along the trail, you feel safe, well informed, and *finally*, your dream come true, deep in the wild jungle of Borneo.

You might recognize that this story relates to some of the themes already presented in this book. There is, for example, a political economy at work here. The bustling eco-tourism gives a select few locals an opportunity to make a decent wage working, but at the same time, as more land is set aside in nature reserves, constraints are put on where indigenous people can hunt in the forest. Your choice of vacation destinations – a rustic inn and a national park, rather than a beach resort or cruise ship – is influenced by your personal environmental ethics and belief in the "sustainability" of low-impact eco-tourism.

But something more may be at work here too. Many parts of this forest have been inhabited and used by people for many thousands of years. Where did they go? Your experience benefits from the carefully maintained paths and the signs that point out important species or views. Who put them there? While the place you are observing is of course profoundly *natural* – it is extremely remote, has insects, wild animals, and jungle plants – it is also undeniably *social*. Your social priorities determined where you would go and what you might expect. A conscious decision has been made to remove people from the landscape to preserve the jungle's condition for your experience. The helpful signs are scientifically accurate but they also inevitably reflect human and social choices about what (of the literally billions of things you might observe in the forest) is most important.

In this sense, what you see in the forest and what you *think you see* are both influenced by other people. Your ideas about Borneo come from other sources (movies, television, family) just as the Borneo presented to you on your trip combines indigenous and scientific representations of the forest. These ideas, images, and assumptions are all **socially constructed** – made up piece by piece – elsewhere and by other people.

What would it mean, then, to say that the nature that you have long imagined and that you are now experiencing is socially constructed? At first glance, this claim might seem patently absurd. Isn't this as *authentic* as nature gets? A tropical rain forest, a *protected* treasure trove of biodiversity: This is **wilderness**, right? And, as wilderness, must it not be fully separate from its opposite – society?

While the answers to these questions seem self-evident, a **constructivist** perspective starts by looking behind these answers, asking, for example: What unspoken assumptions are buried within the ideas of authenticity, protection, and wilderness? Once we begin to scrutinize our experience of the environment in this way, we begin to see how your nature, and everyone else's, is a product of social processes, beliefs, ideologies, and history.

**Social Construction**   Any category, condition, or thing that exists or is understood to have certain characteristics because people socially agree that it does

**Wilderness**   A parcel of land, more or less unaffected by human forces; increasingly, wilderness is viewed as a social construction

**Constructivist**   Emphasizing the significance of concepts, ideologies, and social practices to our understanding and making of (literally, *constructing*) the world

**Nature**   The natural world, everything that exists that is not a product of human activity; often put in quotes to designate that it is difficult if not impossible to divvy up the entire world into discrete natural and human components

The "**nature**" in this story – a tropical cloud forest – is socially constructed in at least two ways. First, our *ideas* of "pristine" and "authentic" tropical nature are not simply raw mental images that come into our minds from the air. Rather, our ideas of the tropics as Edenic paradise, our personal valuation of biodiversity, even the idea of wilderness itself: these are all the products of culture, media, education, and the like. But it is more than just our ideas about nature that are social constructs (again: *products of social processes*). The national park itself is a construct, rather than a fragment of raw, asocial nature, captured and protected, frozen in time. The entrance fees, maintained roads, hiking trails, limits on indigenous uses, interpretive exhibits, and forest management practices – all of these together *construct* the very nature that we expect to see when we visit such parks. Even before the park itself was established, the indigenous residents actively managed this area for millennia, selecting plant and animal species, planting and setting fires. Like the pine forests of colonial New England and the rainforests of the Amazon, the cloud forests of Borneo are not simply the end result of the physical processes of brute nature at work.

They are, rather, the products of biophysical processes, a specific physical environment, *and* a long history of human occupation and management.

In this chapter, we will examine what it means to scrutinize our ideas about nature and environmental knowledge, as products of history and social processes, as *social constructs*. A logical place to begin this overview is with an examination of the word "nature" itself.

## So You Say It's "Natural"?

As Raymond Williams famously wrote nearly 40 years ago, "nature is perhaps the most complex word in the English language" (Williams 1976: 184). As Williams points out, the word nature has at least three common usages:

(i)    the essential quality and character of something;
(ii)   the inherent force which directs either the world or human beings or both;
(iii)  the material world itself, taken as including or not including human beings.

Since these three definitions are closely related, and tend to overlap in our thinking, discussions of "nature" do not always adequately specify the way in which the term is being used. If we stated, for example, that peasant societies live "closer to nature" than urban ones, are we saying that they are more in touch with some "true human nature" (perhaps that urban dwellers have become removed from)? Or are we saying that they are more moved by natural laws? Or perhaps they have a better sense of these natural laws? Or are they simply closer to material nature itself: soil, rain, and wind? The lack of precision typical in the use of the word "nature" can lead to confusion and misunderstanding.

Beyond this, nature presents a further problem. In all of the definitions presented above, and in most common speech and thinking, nature is understood as that state, condition, or quality that is before, separate from, or outside of society, human history, and volition. For example, the first definition provided above describes the assumed timeless, universal, and basic quality of something. Bears eat berries, for example, it's in their nature. Nothing people might say or do is likely to change this underlying and immutable characteristic.

Or so it would seem, anyway. Difficulties ensue when we begin to examine how we arrived at our knowledge of these properties. If we examine our knowledge of these essential characteristics, hints of society and human history frequently intrude. Even when natural properties of things are derived from authoritative methods (such as the scientific method) or sources (such as religious texts), for example, these methods and sources themselves rely upon social concepts, constructs, and a **social context**. For scientific facts to exist, for example, scientists must *agree* on the characteristics of the natural world. These characteristics must be described in socially developed languages and words. And these concepts all have histories, moments of invention and political contexts that make them seem normal and taken for granted, at least for a time. They also have places and periods when they become less stable and convincing.

**Social Context**  The ensemble of social relations in a particular place at a particular time; includes belief systems, economic relations of production, and institutions of governance

**Race**  A set of imaginary categories distinguishing types of people, typically based on skin color or body morphology, which varies significantly between cultures, locations, and periods of history

Take the troubled concept of **race**, for example. For well longer than two centuries, it was believed by many educated Westerners to be scientific fact that there were different biological races of humans. By the turn of the twentieth century, many people believed, moreover, that the different races represented different levels of biological evolution. Many people (scientists, clergy, and lay people alike) placed some groups (such as Africans and Australian aborigines) on a "lower level" of evolutionary hierarchy and believed them to be more closely related to apes, that is, *less human*, than others. Such an understanding justified a huge range of ugly historical practices (the slave trade among them) and arbitrary forms of discrimination and oppression. This concept, however appalling to us, was quite convenient for many people for a long time.

Needless to say (or so one would hope), the notion of different biological races of humans has been comprehensively debunked in the past century. Actual evidence of essential racial differences is nonexistent. Genetic analysis has advanced in recent decades, moreover, to demonstrate that similarities between all people, despite superficial outward appearances, are overwhelming. In genetic and physiological terms, there simply are no essential or definitive races of people.

Yet racial categories indeed remain very "real" for many people and places. The racial categories we know, even while they have no natural physical basis, are rooted in a power-laden historical social consensus. Stated another way, supposedly factual information (about the "*nature* of humanity") has been shown to be a thoroughly social construct, one supported by non-innocent concepts and practices, and rooted in a very particular historical context (Western European imperial expansion). More than this, the power of racism as a way to order society has been, at least in part, bolstered by the apparent "naturalness" of racial differences. People did not *appear* to have invented these racial categories (though it turns out they had!), instead the categories seemed to preexist any social decision. The apparent "naturalness" of such social constructions makes the pernicious categories on which they are based harder to challenge.

This raises some important and disturbing questions about what we know to be true about the world and how we come to know it. More than this, it suggests that understanding something as natural, pre-given, and non-social may be problematic in itself and indeed undesirable.

To say that some claim or concept that we think of as "natural" actually might be "socially constructed," therefore, is to ask at least one of the following questions:

1   Is this claim or concept natural, inevitable, timeless, and universal?
2   If not, at what point was it invented? Under what conditions?
3   What are the social, political, or environmental effects of believing that this claim or concept is true, natural, or inevitable?
4   Would we be better off doing away with the concept altogether, or rethinking it in a fundamental way?

Such a set of questions can be applied to a great many things (Hacking 1999), with implications for how we think about problems of the relationships of nature and society.

## The social construction of New World natures

Consider the case of "pristine" pre-Columbian America. For hundreds of years, explorers, writers, teachers, politicians, and average people have operated under the assumption that the landscapes of North and South America in the time preceding 1492 were largely unaffected by human activity. They were, in a word: pristine. It was assumed that, while native people certainly used the resources of the land, their relatively small population size and crude tools and technology meant that they had little impact on their environment. Some forms of the argument further suggest that native peoples maintained a land ethic that made their ecological footprint especially light; being closer to nature and more in touch with environmental systems and flows, the impact of the many peoples of the Amazon, the Mississippi, the California coast, and the Yucatan was assumed to be relatively minimal and benign. When Europeans arrived, and from that time forward, they believed themselves to be visitors in, or tamers or invaders of, a wild natural landscape.

Even at the time of initial contact between Europeans and native people, however, there was plenty of evidence to the contrary, including complex land use patterns, large cleared regions, and effective forest thinning through use of fire. Centuries of historical research and archaeology, moreover, have consistently demonstrated that native impacts were quite heavy in many places as species were aggressively selected, forests were felled, and large areas were planted and burned. In Chapter 14, for example, we see that the Incas had long since domesticated potatoes by the time Columbus set foot on the shores of the Americas, and they had fully transformed the mountainous terrain of the Andes in the process, to grow and harvest this food.

If this is true, what made it so hard for Europeans to see, and eventually acknowledge, the facts in front of them? In part, European perceptions were deeply influenced by their biblical notions of the earth. They placed their Eden, as a real geographic place, on the far side of the Atlantic. The forests, rivers, and plains of the New World appeared to them as unchanged since the time of creation. The facts that many of the forests were carefully cleared by natives for hunting, that the plains were heavily worked cornfields, and that the rivers were rich with silt from the construction of native cities, markets, and monuments could not penetrate the preexisting social construction of nature that settlers carried with them in their Bibles.

As much or more than religion, however, political imagination played a crucial role. European belief in a non-transformed New World nature was perfectly congruent with territorial ambitions. For colonial Spanish, French, and English settlers, a central guiding assumption about the relationship between labor and natural rights was that land and resources are possessed only insofar as they are put to use (see discussion of John Locke in Chapter 6). Unused land was, for Europeans, unowned land, available for improvement under colonial rule. Productive land was expected to look a certain way, like typical farms or managed forests in the Old World. Because of these preconceived notions, lands that had been systematically used and improved by native peoples, before the conquistadores arrived in Vera Cruz, Mexico, came to be defined and designated as "baldios" or "yermas": wastelands. Wasted land was available for improvement, indeed by implication it *demanded* improvement (Sluyter 1999) and therefore appropriation by the colonizers.

This construction of a pristine nature in the Americas made not only systematic land uses, but also complex and productive New World cultures, completely invisible. This was a belief that matched European assumptions about the backwardness of native peoples (and indeed the characteristics of their race, as per above). Contemporary stories that insist Mayan temples or Nazca desert lines had to have been made by space aliens rather than sophisticated native architects do much the same. Either way, the natural conditions of the New World as they were understood by later generations of people were prefigured by a social construction of native landscapes. This social construction helped to establish and justify European rule and rendered the native peoples of the Americas invisible and mute.

The point of examining the social construction of nature in this case is not simply to deny the truth of an incorrect claim (that people had not impacted the land prior to Europeans), although this is, of course, important. It is to ask a further set of questions: If this claim about the Americas is not true, what were the effects of assuming it to be true? How did this claim remain "true" and widely believed even in the face of evidence to the contrary? Whose interests are served by its remaining "true"? Who becomes more powerful and who less powerful in the establishment of a certain set of environmental facts as the "true" ones?

The answers to these questions shed light not only on the history of the Americas but also on how we think about environment–society relations more generally. They suggest that there is a relationship between environmental *knowledge* – which claims and ideas about the state of the environment are known to be true – and environmental *power* – what groups and interests control the environment and its resources. To approach environmental issues from the point of view of the social construction of nature, then, is to question what we know about nature and how we came to know it.

## Environmental Discourse

So far we have shown how both a) nature itself (the material world we live in, study, and explain) and b) our knowledge of nature and the world can be understood as socially constructed. Employing the notion of "**discourse**," we can take this understanding further to explain how and why specific ideas of nature come to be normal, taken for granted, or inevitable, when they are not.

Geographers Barnes and Duncan (1992: 8) define discourses as "frameworks that embrace particular combinations of narratives, concepts, ideologies, and signifying practices." This definition is a little dense, but its pieces can be made plain. A **narrative** is a story. Examples of environmental narratives include "the tragedy of the commons" (Chapter 4), for example, though there are countless others. In a narrative, there is a story line with a beginning and end. A **concept** is a single idea, usually captured in a word or a phrase. "Carrying capacity" (Chapter 2) is a concept, as is "market" (Chapter

**Discourse**   At root, written and spoken communication; thicker deployments of the term acknowledge that statements and texts are not mere representations of a material world, but rather power-embedded constructions that (partially) make the world we live in

**Narrative**   A story with a beginning and end; environmental narratives such as "biological evolution" and "the tragedy of the commons" aid our comprehension and construction of the world

**Concept**   A single idea, usually captured in a word or a phrase

3), or even apparently natural things, like "risk" (Chapter 6) or "forest" (Chapter 10). **Ideologies** are normative, value-laden, world views that spell out how the world is and how it ought to be. That all living things have natural rights (Chapter 5) is ideological, as is the notion that people are naturally free and that the role of government is to realize and enable that freedom. **Signifying practices** are modes and methods of representation, the actual techniques used to tell the stories, introduce and define the concepts, and communicate the ideologies: geographic information systems, newspapers, television advertising, scientific papers, are all signifying practices, but there are many more.

> **Ideologies**  Normative, value-laden, world views that spell out how the world is and how it ought to be
>
> **Signifying Practices**  Modes and methods of representation; the techniques used to tell stories, introduce and define concepts, and communicate ideologies

Discourses put these elements together into powerful, coherent, mutually supporting frameworks, which are persuasive and tend to stand the test of time. Consider the "tragedy of the commons" from Chapter 4. This discourse has a powerful and ironic story at its core: two herders compete for a common pasture and destroy it in the process. It is bolstered by a set of interlocking concepts, like the "Prisoner's Dilemma" and "free-riding." Ideologically, it is powerfully connected to the notion that people make free and rational choices but that they may not always act in their own enlightened self-interest. The story is conveyed in an enormous range of media, from scientific papers on the topic to environmental magazine articles to textbooks (like this one!). Many of these elements are borrowed from elsewhere and cobbled together from earlier persuasive concepts and systems of ideas. Once these pieces are in place, they set the terms of discussion, suggest imagery and outcomes, and direct our ways of thinking. In the process, the assumptions on which the ideas are based, the places from which the concepts are borrowed, and the history of the ideas are often lost or obscured. This transition makes it more and more difficult to remember that discourses are stories; the social context of their invention disappears, and they become true.

Once such a transition has occurred, it becomes difficult to think outside the internal logics of a discourse. If the set of concepts at your disposal for thinking about the world includes "carrying capacity" (Chapter 2) but not "induced intensification" (Chapter 3), it becomes hard to think about population as anything except a drain on a finite natural world. As such, discourses set the hidden rules that guide what can and cannot be said (and done) in particular places, times, and contexts.

Environmental discourse analysis is a method for examining these constructions of nature that attempts to reverse this trend towards the stability of discourses by placing them under scrutiny, unveiling their histories and assumptions, and attempting to foreground the ways in which *social context* influences their production and maintenance. This represents a critical departure from traditional approaches because it often focuses on the concepts (e.g., health, sustainability, hunger) and sources of knowledge (e.g., government experts, scientific laboratories) that other accounts take for granted and assume to be independent and authoritative. A social constructionist approach to the environment is not simply interested in discourses only, therefore, but the social institutions to which they are tied.

Historically, discourses, including discourses about nature, have been propagated and reinforced by important social institutions, like community patriarchs, the church, or the monarchy. In modern society, environmental discourses are increasingly tied to institutions such as governments, schools, hospitals, and laboratories. These institutions are sites of power that are in the business of managing the relationship between environment and society, not so much by telling people what can and cannot be done, but instead by cementing specific understandings of what is and is not true.

### Power/Knowledge

**Power/Knowledge**   A theoretical formulation associated with the philosopher Michel Foucault, which holds that what is known and held as true in a society is never separate from power, such that knowledge reinforces relationships of power but also that systems of power are associated with their own specific regimes of knowledge

This notion that discourses are always embedded in relations of power – that discourses should never be presumed innocent – is an insight indebted to the influential French theorist Michel Foucault. Foucault argues that there can be no knowledge outside of power. Therefore, he upends the singular (and seemingly simple) concept of "knowledge," replacing it with "**power/knowledge**" (Foucault 1980). "Power" here indicates the thing lost or displaced in dominant, uncritical usages of the term "knowledge." Stated another way, there is no knowledge innocent of (without) power: all knowledge is power/knowledge.

For Foucault, that "power" is not something that only governments, corporations, or certain social classes possess; it is not something simply used by the powerful against the powerless. Rather, power is diffuse and omnipresent (think of food coloring dropped into a glass of water): people (all people) are always and inevitably operating within a field of power. This is not to imply that all people everywhere are equally empowered. Do not, in other words, confuse Foucault's "field of power" with a mythical "level playing field." What is important to remember here is that power is the result of relationships between people, rather than something possessed by individuals who can wield it over others. Think, for example, about your last visit to the dentist. Did you pay extra attention to your teeth before you went? Maybe brushing more? Flossing twice a day? Did you do this because you thought the dentist would punish you by having you arrested by the fluoride police – or banished to gum disease hell? More likely, you were afraid that, because of his or her authority over you, in the area of dental health, your dentist might gently scold you or simply remind you that you should take better care of your teeth. This kind of subtle admonishment has power over you, not because of who your dentist is as a person, but rather because he or she, by going to dental school (a social practice), has been invested with authority.

This is not to say that such power is absolute. Because it relies on social relationships, it can always be challenged. Foucault's theorization of power, therefore, serves an analytical and political purpose. By showing that power pervades all levels of society, and produces knowledge as discourse, marginal individuals and groups might regain a political presence within (rather than against) power by destabilizing authoritative discourses.

In a social constructivist approach to the environment, therefore, the object is not simply to valorize or debunk a specific environmental story, but also to understand how the process of truth-making (or the "truth effect" in Foucault's words) is linked to processes that establish social power. By carefully, painstakingly, and empirically demonstrating the history of specific ideas, their relationship to institutions, and their social and environmental effects, social construction seeks to make problematic discourses less stable, reveal the interests they serve, and so make space for other ways of thinking about nature. According to its adherents, by showing that things we take to be true, immutable, and timeless have histories and social contexts, it allows us to imagine alternatives and possibilities.

**Figure 8.1** Power to the people: An environmental protester holds a huge cardboard cutout of Australian Federal Climate Change Minister Penny Wong in front of a planned "regeneration fire" in Tasmania. This is an act of empowerment: By tagging the national government's top climate official a "climate criminal," these protesters are destabilizing the dominant governmental discourse of climate stewardship. *Credit*: Peter van der Pasch.

In environmental politics, for example, when an activist protesting the over-cutting of national forests dresses up as Smokey Bear (the symbol of the United States Forest Service), but wearing a flannel shirt, suspenders, and wielding a chain saw, they are destabilizing the government's favored image as stewards of public land and blurring what is supposed to be a strict distinction between federal managers and corporate loggers. They are reading and rewriting discourses of forest management (see also Figure 8.1). Beyond activist efforts in the manipulation and inversion of imagery, discourse analysis represents a demanding area of environmental research in itself.

## The discourse of North African desertification

Consider the case of desertification in North Africa. As recently as 1997, a UN report concluded that the Sahara Desert is on the advance northwards and that the region, once the Roman Empire's "breadbasket," had become a desert. Radical measures would be required, by implication, on the part of governments, the international community, and non-governmental organizations, to restore the region's lost potential. It is increasingly clear, however, that this conclusion is one based on old ideas, texts, and claims originally made by colonial officers. As geographer Diana Davis demonstrates in an exhaustive and detailed historical analysis of the "desertification discourse," there has been a long-standing and well-established assumption that the region was once heavily forested and that its degradation came from overgrazing, especially in the period following what was called the "Arab Invasion" in the past thousand years. As part of this assertion, maps of the region were drawn up over the eighteenth and nineteenth centuries laying out what kind of land cover ought to occur in the region, based on assumptions about what the "climax" communities should be: how the landscape should look, if properly managed. In this case,

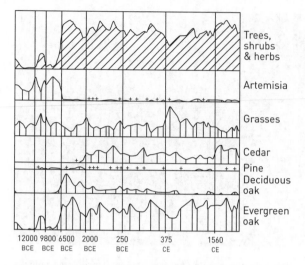

Trees, shrubs & herbs

Artemisia

Grasses

Cedar

Pine
Deciduous oak

Evergreen oak

12000 9800 6500  2000      250      375     1560
BCE  BCE  BCE   BCE       BCE      CE      CE

**Figure 8.2** Pollen evidence from Morocco over 14,000 years. Despite the ongoing dominant discourse that post-Roman North Africa experienced "desertification" and loss of tree cover following "Arab Invasions," pollen data shows no significant decrease in tree pollen in the past thousand years. *Source*: Davis (2007), p. 11.

climax communities were assumed to look like the forests of Europe. In contrast then, photos of the open desert were used to capture a sense of forlorn emptiness. These kinds of claims and documents can be dated back to the earliest surveys of French colonizers in Morocco, forming the basis of an important environmental story (Davis 2007).

Pollen core data from the region positively refutes any such story, however. These data, which show the relative abundance and diversity of flowering plant species across the area for thousands of years (including trees, grasses, herbs, and shrubs), suggest a significant increase in tree cover around 8,000 years ago, and then a long period of shifts, up and down, but without any evidence of loss of trees, tree species, or vegetation more generally. Put simply, accounts of post-Roman desertification and an expansion of the Sahara following an "Arab Invasion" have no basis in anything like contemporary scientific knowledge (Figure 8.2).

There are a few ways to think about this apparent misunderstanding. One can certainly argue that pollen data and other paleoecological information simply were not available to French colonizers. Their misunderstanding was based on scattered information and ancient texts with little scientific validity. Basically, we might conclude that this was an "honest mistake."

A constructivist research effort goes further, however. It asks who benefited from a discourse of desertification? How does such a discourse take form and gain momentum? How is it tied to institutional, political, and even ethnic power relationships? As a discourse, the structure of these elements can be viewed as internally powerful and coherent, accounting for some of its longevity. It contains a compelling narrative ("paradise lost"), key concepts ("climax vegetation"), an ideology of progress and restoration, and signifying practices in the form of government documents, landscape photography, and scientific papers. As far as power goes, the discourse of desertification clearly suited the needs of colonial officers. It has long been a crucial justification for land use controls and investment in land improvement mechanisms, especially the settlement of nomads and other native peoples. Assuming that a vast forest had been destroyed by irrational indigenes provides a powerful justification for land control and improvement.

None of this is to say that land degradation has not occurred in places in North Africa, or that massive overgrazing might not lead to a depletion of grass cover, or that people do not impact the environment in arid regions. All of these things may be true and yet the evidence for a massive anthropogenic creation of the desert in this region is minimal. What exists in its place is a strong, coherent, and compelling discourse of desertification, linked to colonial and state authorities. Nor is any of this unique to North Africa.

## Wilderness: A troublesome discourse

In his landmark essay "The Trouble with Wilderness," environmental historian William Cronon argues that "the time has come to rethink wilderness" (Cronon 1995: 69). The dominant view of wilderness, he suggests, as a land untouched by humans, a pre-historical and pure nature, is troublesome for a number of reasons. First, the feelings and experiences we have about, within, and around wilderness areas are learned, inherited, and at bottom, "quite profoundly a human invention" (p. 69), or what we have been calling in this chapter a social construction. So what does it mean, exactly, to say that wilderness is a social construct? For one, it forces us to recognize that the word itself has a culturally and historically specific meaning. Wilderness – as pure nature, landscapes untouched by humans – is a construct most specific to Western European-based, modern cultures. Indeed, most indigenous and non-Western cultures do not even have a word for lands untouched by humans and do not distinguish between "wild" and "non-wild" landscapes. Wilderness is a *culturally specific* construct. But even within this culture of wilderness, the meaning of the term itself has changed enormously within its brief history, even within the past 250 years.

Wilderness today is prized. It is mapped, sought out, hiked through, and protected in reserves. But you do not have to go back far in history to find quite different meanings:

> As late as the eighteenth century … to be in a wilderness was to be 'deserted,' 'savage,' 'desolate,' 'barren' – in short, a 'waste,' the word's nearest synonym. Its connotations were anything but positive, and the emotion one was most likely to feel in its presence was 'bewilderment' – or terror. (Cronon 1995: 70)

Much of this comes from the Bible itself. Adam and Eve, after all, were banished from a garden (a cultural landscape, paradise) to the wilderness. It was nothing less than the God-given project of human history to reclaim the wilderness for humanity. Even into the nineteenth century, the project of American Western expansion was often viewed (and explicitly portrayed) as bringing civilization to the wilderness (Figure 8.3).

By the turn of the twentieth century, however, this meaning had been turned on its head. Wilderness had become scarce and with expansion now coast to coast, Americans sensed that the vast untrammeled (and supposedly unpeopled) landscapes of the "American frontier" were gone, forever. The debate over

**Figure 8.3**  John Gast, "American Progress," 1872. Lady Liberty brings light (and farmers, plowed fields, telegraph lines, trains, etc.) as she drives the darkness (and the "Indians" and wild animals) out of the *wilderness. Credit*: Library of Congress.

the damming of Hetch Hetchy Valley in Yosemite National Park (as outlined in Chapter 5) serves as the classic example of this new, highly valued conception of wilderness. As Cronon puts it: "For [John] Muir and the growing number of Americans who shared his views, Satan's home had become God's own temple" (p. 72). In this way, wilderness as a concept is troubling because it disguises human priorities and values as "natural" conditions.

More problematically, as noted previously in this chapter, the very lands of the New World that were perceived to be wilderness by arriving European settlers were not empty of people in any way. Some 4–12 million indigenous Americans at the time of Columbus's arrival had radically modified the land for cultivation, hunting, and other uses. They are lands with a history. Even Yosemite Valley and the High Sierra mountains – paragons of wilderness and wildness – were inhabited, managed landscapes long before the arrival of European Americans in the early nineteenth century. Most of the very "wildest" landscapes that conservationists have fought to protect for the past 150 years are very much the *products of human labor*.

To make them into wilderness, therefore, requires a form of violence. The very natives that managed the vegetation of Yosemite Valley with controlled, seasonal fires had to be forcefully removed before a wilderness could be created. This is not, mind you, just a matter of history. To this day, members of the Blackfeet Nation in Montana are cited for "poaching" in Glacier National Park, on lands to which they were guaranteed hunting and fishing rights *in perpetuity* through their treaty with the US government. In this way, wilderness is troubling because it has enabled and even necessitated violent social and political relationships.

Additionally, Cronon argues that the idea of wilderness has a more subtle yet equally problematic effect on the American environmental movement as a whole. The US environmental movement has long relied on the idea of wilderness as one of its foundational constructs. A century ago, when protection of large tracts of (only apparently) untouched land was the central priority, the emotional value of wilderness was undoubtedly useful for the creation of national parks and other reserves.

But what does it mean to valorize wilderness today in the same way? The implication of wilderness environmentalism, according to Cronon, is that the natures most deserving of preservation "must also be pristine – remote from humanity and untouched by our common past" (p. 83). Yet environmental history teaches us that few of those places exist now, if they ever did. Placing wilderness at the center of environmentalism serves to disguise and minimize the range of important natures all around us, in cities, vacant lots, and old farm fields. An environmental movement aimed only at preserving wilderness, therefore, will largely ignore places we live, work, and play (see "environmental justice" in Chapter 7).

Like all other discourses, wilderness merges stories, concepts, and ideologies that make it possible to see and do certain things, while at the same time disguising and hiding other things. To argue that wilderness and nature more generally are social constructions is not to argue against preserving this or that tract of land. It is, rather, to demand a more honest and reflective examination of our beliefs and actions, and the role of social power in creating the nature we think we know.

## Box 8.1    Wild Horses

What's in the power of a name? Consider this: Which of the following would you consider worthy of conserving, a population of wild horses or one of feral horses? On one hand, the question presents a false dilemma, because in several places across North America (as well as in many others worldwide), both names refer to the same animal, *Equus ferus caballus*, the domestic horse. On the other hand, the fates of local populations of these animals often rest in which construction, wild or feral, wins out in the end. To better flesh out the (literal) power of contested social constructions of nature, we will briefly examine a case study of one controversial band of wild/feral horses.

The Ozark National Scenic Riverways (ONSR), a unit of the US National Park Service located in the southern Missouri Ozarks region, contains a small population (usually between 20 and 35 individuals) of free-roaming horses. The presence of the "wild" horses, which are believed to be primarily descendants of Depression-era draft horses abandoned by poor farmers, predates the park, which was established in 1964. Beginning in 1970, federal legislation mandated that national parks manage designated "natural areas" primarily for ecological integrity over cultural or recreational considerations. In the official park management plans, ONSR managers began referring to the horses as "feral horses," alien denizens of an otherwise "natural" ecosystem. Park documents made no reference to the cultural significance of the horses. Rather, they repeatedly cited the dangers that "alien" species pose to "natural" ecological processes. Indeed, by the 1990s managers were considering removing the herd under the guise of ecosystem management.

The proposal to remove the horses from the park appalled many locals. In 1992, the Missouri Wild Horse League formed to fight for permanent protection of the band of horses. For many area residents, the horses were a pleasing and noteworthy part of the landscape (they remain, after all, the only free-roaming horses in the US Midwest). But they were also much more than this. The horses represented a symbolic and material link to their communal, agrarian history. Indeed, Ozark local history (as it is in many places) is infused with narratives of loss of control to outside forces: loss of farms to the Depression, loss of forests to timber barons, even loss of common lands to the strictures of a federally managed park. The horses were survivors, of all this. They *belonged* there.

So which side was "right"? Well, both, really. From the Park Service perspective, *yes*, the horses do alter the local ecology. Even acknowledging that equids are native to North America, *these* equids are not. Moreover, "native" North American equids lived in a very different ecosystem, which included (among other differences) equid predators like cougars and wolves. On the other hand, the wild horse advocates are "right," too. The horses are a beloved feature of the local culture and landscape. And they *are* "wild," right? They are a self-sustaining population of animals, after all. In the end, the wild horse advocates won out, courtesy of a piece of federal legislation granting permanent protection to the horses.

So who's to say, anyway, what is "natural" and what isn't, what belongs and what doesn't? Please note: this is not a rhetorical question. Someone *will* define these boundaries. And others *will* disagree. What (or *who*) is allowed to stick around is often decided by clashes of contested social constructions.

### Reference

Rikoon, S., and R. Albee (1998). "'Wild-and-free,-leave-'em-be': Wild horses and the struggle over nature in the Missouri Ozarks." *Journal of Folklore Research* 35(3): 203–22.

## The Limits of Constructivism: Science, Relativism, and the Very Material World

That said, not everyone has embraced the turn toward constructivist accounts. Among the problems identified by critics is the challenge of science, the threat of relativism, and the need to take seriously the more-than-constructed nature of the world.

### What about science?

A fundamental part of our understanding of environmental problems comes from scientific investigation. Ever since the Enlightenment, it was assumed (at least by scientists) that scientific knowledge offered a transparent perspective on nature, "a uniquely privileged vehicle to true knowledge about the world" (Demeritt 2001: 26). Science, as such, is a method for finding truths already "out there," "natural laws" or the "natural properties of things."

A constructivist perspective turns this fundamental assumption of science on its head, arguing that "scientific knowledges are made in historically specific, socially situated practices, rather than 'found'" (Castree and Braun 1998: 27). Moreover, many constructivists argue that the very "things" we take to be the natural objects of scientific inquiry (e.g., quarks, DNA, soil samples) are not found (or discovered) either, but rather socially constructed through the process and practices of science, reflecting more about the social world of science than the natural world. Researcher Donna Haraway has demonstrated, for example, that successive ideas and paradigms in primate research (on topics like parenting behavior, sex, resource use, and warfare) have tended to reflect the historical social concerns and anxieties of the period of their investigation, and were often imbued with sexist, racist, and colonial discourses (Haraway 1989). Even and especially scientific knowledge, in this way of thinking, is social and political.

And yet the practice of science is largely congruent with many of the environmental goals and values that constructivists espouse, including scientific investigation of the hazards of modern industry and its demonstration of the threat of global warming, among countless examples. Indeed, constructivist research, with its amassing of evidence and logical forms of argument, is often itself a form of scientific inquiry. How can treating science as discourse square with seeing science as a unique way of getting at important truths?

### The threat of relativism

For some, by retreating from the universal power of science to adjudicate debates about nature, social construction leads to the problem of **relativism**. In theory, if knowledge of the "natural" world is rooted in "social" constructs, stories, and ideologies, it becomes difficult or impossible to establish reliable information and knowledge upon which to act. When disputes occur, what and who are we to believe? On what basis can we argue for or against various human behaviors: dumping in rivers, using

**Relativism** Questioning the veracity of universal truth statements, relativism holds that all beliefs, truths, and facts are at root products of the particular set of social relations from which they arise

nuclear power, cutting trees? In its stronger forms, relativism argues that all statements and knowledge are *only* relative to their social context. This strong form of relativism can, in theory at least, devolve into *nihilism*, the "view that nothing is knowable" (Proctor 1998: 359).

In response to this, constructivists argue that not all forms of relativism are as debilitating as others. In its weaker forms, relativism simply reminds us that all statements about reality are conditioned and recognized (written and read; said and heard) in a social and political context. No statement can present a "mirror of nature" (Rorty 1979). Even so, most constructivists take a path of moderation, adhering to the idea that while environmental science is never free of social and contextual influences, these alone do not determine scientific findings in any simple way. Certainly, acknowledging the validity of constructivism makes it more challenging to make bold statements about reality and about truth. But it does not hold that one is a relativist simply because one admits to the complexities of the relationships between environmental knowledge and environmental power.

## Constructivism in a material world

This still leaves a difficult point. There is plenty of reason to believe that the environment and the objects of the world around us influence our social and political context as much as the social and political context influences our knowledge of these objects. Global warming is a scientific construct *and* a globally observed temperature phenomenon (Chapter 9). Wolves are a politically constructed site for ideological rancor *and* a species of wild animals that operates with its own rules (Chapter 11). At what point do the material qualities of nature become important to explaining social and environmental outcomes? Can we admit to a constructed nature and still take seriously the sticks and stones of a complex material world? Many critics insist not.

We would suggest, however, that the concept of **co-production** offers a route out of the perceived constructivist impasse. By co-production we mean the inevitable and ongoing process whereby humans and non-humans produce and change one another through their interaction and interrelation. Humans and non-humans are

> **Co-production** The inevitable and ongoing process whereby humans and non-humans produce and change one another through their interaction and interrelation

discrete, but always entangled. As Donna Haraway observes: "beings do not pre-exist their relatings" (Haraway 2003: 6). Humans constantly remake the world but are remade in the process. Social discourses and narratives are fundamentally "material" since they are crucial in directing human understandings, impacts, and behaviors, which in turn act to change landscapes, exterminate or generate new species, emit gases, and change the flow of water on the surface of the earth. At the same time, however, these landscapes, species, and gases are "social" in that they act on and confront the taken-for-granted in dramatic ways, reworking social systems, discourses, and ways of understanding the world.

Consider the case of Hurricane Katrina, which made landfall in 2005 in the largely unprepared city of New Orleans, with an enormous death toll, especially amongst poor and minority populations. The disaster is widely recognized to be the result of engineering failures in regional infrastructure of dams and levees, and it brought to light deeply socialized differences in resources and opportunities amongst the city's black and white residents, which had been heavily erased by representations of the city in the past. News sources during the event cast the storm in a range of perverse lights (consider online images of black people with salvaged goods described as "looters" and white people described as "finders" – see http://politicalhumor.about.com/library/images/blkatrinalooting.htm).

Here was a disaster exacerbated by risk decisions, filtered through racially tainted coverage, interpreted through complex discursive lenses and perhaps even increased in severity by anthropogenic global warming. In all these ways there are social and constructed elements to the horrible events of that storm. But it would be analytically useless and definitely bizarre to claim that the cyclonic phenomenon itself, its high winds, and its deluge of water were a discourse. Rather we might observe that politics and wind, language and water, ideology and levees, discourses and storm surges together produced an outcome that remade the world.

In every interaction, with every object on earth, such outcomes are produced on a tiny scale every day. The lesson of constructivism is to take seriously the complexity of the conditions under which we understand those interactions and make sense of the world around us.

## Thinking with Construction

In this chapter, we have learned that:

- Many things or conditions we have thought of as normal or "natural" have later been shown to be social inventions or merely ideas (for example: race).
- "Social constructions" typically take the form of taken-for-granted concepts or ideas that direct our thinking or action, often unbeknownst to us without critical examination.
- This holds implications for many key concepts we use to think about society and environment, including ideas like "wilderness" or "desertification."
- Environmental stories or narratives formed from constructions of this kind can hold enormous political and social implications by placing blame, directing policy, or developing solutions that may be inappropriate, undemocratic, or environmentally unsustainable.
- By analyzing environmental discourses, stories, and narratives, we can learn where our taken-for-granted ideas come from and how they might be resisted or changed.
- Critics suggest that taking this approach too far may lead to relativism or dismissal of science.
- Reconciling the material reality of the environment with the powerful social constructions that influence our thinking is therefore a major challenge.

## Questions for Review

1. Is a national park a more "natural" place than a region dominated by farms? Try to answer this question using a constructivist perspective (in which the correct answer is not a simple "yes").

2. Describe how history and science expose the concept of "race" as a social construction.
3. When Europeans explored the New World, how did they perceive the "nature" they came into contact with? How did these perceptions legitimate their taking of the land?
4. Foucault posited that "knowledge" cannot be understood without taking "power" into consideration. Explain.
5. Discuss the profound difference between viewing North African desertification as an objective phenomenon and as a *discourse*. (Bonus or alternate: Do the same thing for "wilderness.")

## Suggested Reading

Braun, B. (2002). *The Intemperate Rainforest: Nature, Culture, and Power on Canada's West Coast*. Minneapolis, MN: University of Minnesota Press.

Castree, N. (2005). *Nature*. New York: Routledge.

Cronon, W. (1995). "The trouble with wilderness or, getting back to the wrong nature." In *Uncommon Ground: Rethinking the Human Place in Nature*. New York: W. W. Norton and Co., pp. 69–90.

Davis, D. K. (2007). *Resurrecting the Granary of Rome: Environmental History and French Colonial Expansion in North Africa*. Athens, OH: Ohio University Press.

Demeritt, D. (1998). "Science, social constructivism and nature." In B. Braun and N. Castree (eds.) *Remaking Reality: Nature at the Millennium*. New York: Routledge, pp. 173–93.

Foucault, M. (1980). "Truth and power." In *Power/Knowledge: Selected Interviews and Other Writings 1972–1977* (ed. C. Gordon). New York: Pantheon, pp. 109–33.

Haraway, D. (1989). *Primate Visions: Gender, Race, and Nature in the World of Modern Science*. New York: Routledge.

Robbins, P. (1998). "Paper forests: Imagining and deploying exogenous ecologies in arid India." *Geoforum* 29(1): 69–86.

Sluyter, A. (1999). "The making of the myth in postcolonial development: Material-conceptual landscape transformation in sixteenth century Veracruz." *Annals of the Association of American Geographers* 89(3): 377–401.

Wainwright, J. (2008). *Decolonizing Development*. New York: Blackwell.

## Exercise: Analysis of Energy Discourses

Consulting magazines, newspapers, television and the internet, identify several advertisements from energy-producing companies (e.g., Exxon, BP, etc.) and several publicity ads for major environmental groups (e.g., Sierra Club, Wilderness Society, etc.). What discourses (narratives, concepts, and ideologies) are embodied in the pictures and words in these advertisements? What stories do they tell or imply about nature and society? How are the two sets of discourses similar? What work do they do for each of the interest groups involved?

# Part II

# Objects of Concern

In the first part of this book, we reviewed diverse ways of thinking about people and the environment, in the abstract. Our goal was to show that starting from different assumptions can lead to startlingly different conclusions, and that while many competing ideas exist, not all of them are universally persuasive. There may be a marketplace of many ideas, in other words, but some ideas are more valuable than others. We also showed how thinking about the world in some terms makes it harder to think about it in other ways. Market-based thinking, for example, is somewhat incompatible with viewing the environment in political-economic terms.

In the second portion of the book, we provide some grounded examples of the way such ideas are animated by considering some real-world objects. In this section, we want to stress that each object presents a peculiar puzzle for people, owing to its own characteristics. The way carbon dioxide circulates globally in the atmosphere makes it different than many other pollutants, for example, as does its specific historic relationship to humanity through the process of combustion. Certain species of tuna, similarly, tend to school with dolphins, presenting a different kind of puzzle for society, at least as long as people are interested in eating tuna and sparing dolphins. Water in a bottle differs drastically from water from the tap or from a river, though *not* in its health characteristics or contaminants as we shall see, but instead by how it circulates, relies on energy inputs, and connects to the economy. We hope to show in this handful of examples how these basic agreed-upon puzzles (everyone agrees tuna school with dolphins) can lead to highly divergent ways of thinking about things as environmental "problems" (is tuna production an ethical problem or rather one of market failure?).

Our selection of objects is deliberately eclectic and wide-ranging, but they are by no means random. What could carbon dioxide have in common with French fries? Both represent things that are produced and created through human activity but also things that – once sent into the world – act back upon us in complex ways, forcing us to rethink our relationship to the broader environment. Moreover, by selecting these diverse things, we want to encourage you to *avoid* thinking about human–environment relationships as a laundry list of already-known "problems" (like global warming, desertification, or food scarcity) and instead consider how human social, political, and economic life mixes with the many animals, plants, chemicals, and minerals of the world around us to create outcomes and situations, many of which – but by no means all of which – are profoundly undesirable (species loss, hunger, forest fires, or disease, for example).

These become problems of very different kinds, however, when we apply differing ways of thinking. Solutions to the problems posed by society–environment relationships require us to take seriously the specific nature of *things* in the world around us, therefore, as well as *ideas* that both open and limit our thoughts and actions.

# 9

# Carbon Dioxide

*Credit*: Alexander Ruesche/EPA/Corbis.

## Keywords

- Cap and trade
- Capital accumulation
- Carbon cycle
- Carbon sequestration
- Coase theorem
- Collective action
- Command-and-control
- Emissions trading
- Greenhouse effect
- Greenwashing
- Photosynthesis
- Surplus value
- Uneven development

## Chapter Menu

### Stuck in Pittsburgh Traffic

At 4:45 on a hot summer afternoon, commuter traffic approaching the Liberty Tunnels outside of Pittsburgh, Pennsylvania comes to a halt, as it does in cities all over the world during rush hour. Perhaps 100,000 cars will pass through these tunnels in a single day, a reliable daily reflection of the commuter culture of cities. The average American worker commutes approximately 45 minutes every day, with many commuters traveling twice as long or more. Bumper to bumper for 20 minutes, this traffic jam is an expected part of the workday for everyone involved. It is a normal part of life and the economy, if an unpleasant one.

As cars crawl through the "tubes" (as they're known locally), with air conditioning making the temperatures bearable for drivers, a complex set of exchanges is occurring. Fluids and gases are flowing through the vehicles, maintaining the pace of traffic, however slow. These cars are propelled by a process of combustion, which combines oxygen from the air with gasoline in the tank to make tiny explosions that break apart the bonds of the fuel (which is basically a tangle of carbon and hydrogen atoms) and release energy that makes it possible to drive. What's left over after the reaction? Some $H_2O$ (water) and a bunch of $CO_2$ (carbon dioxide). Raw fuel and oxygen go in one end and water and carbon dioxide come out the other. Notably, for every two molecules of fuel that go into the process, at least 16 molecules of waste $CO_2$ come out, as well as a range of dirty particulates, and a number of other pollutants and gases.

The $CO_2$ that escapes from these tailpipes is entirely invisible. It rises into the atmosphere, mixes with other gases and apparently disappears. Regrettably, what is out of sight and out of mind is by no means out of action. The carbon dioxide emitted by these cars today may reside in the atmosphere for a century or far longer, being only slowly reabsorbed into the world's oceans and vegetation over long time periods. In the meantime, those molecules will float through the air, trapping heat energy coming from the earth and raising the overall temperature of the planet, with potentially disastrous consequences.

Surely though, what harm could come from 16 tiny molecules of $CO_2$? A typical car, regrettably, produces considerably more than that. The average vehicle emits ~150 grams of carbon dioxide per kilometer traveled. Since the average American driver covers ~19,000 km per year, the average car is actually producing around 2,800 kilograms of $CO_2$ every year, significantly *more than its own weight*.

This calculation, moreover, excludes the carbon required to produce the energy to make the car in the first place as well as the energy to make and maintain the roads and to extract the fuel and raw materials that go into the car's tank, tires, and body. This is also beyond the carbon required to transport and power the rest of a commuter's life, from the food on their breakfast table to the heating and cooling of their home and work, indeed even the energy that built their house and workplace in the first place. All of these things required the emission of carbon dioxide (along with a great many other gases and waste products, of course). How did a tasteless, odorless, invisible gas become so fundamentally tied to our lives, and to the foundations of our society? And at what cost to the planet and the economy?

# A Short History of $CO_2$

The element carbon actually makes up only a tiny fraction of the earth. Most of that (perhaps 99 percent) is buried in the crust, relatively near the earth's surface (the earth's crust is only 60 km or so thick), but locked up in sedimentary rocks. The rest of the world's carbon is in a state of continuous motion. Moving between the oceans (where most of it resides), the atmosphere, plants, and soils, carbon circulates, occurring most notably as solids (graphite and diamond) and gases (carbon is in methane and carbon dioxide). It moves beyond these states into a huge range of organic molecules, however, including fats and sugars. In this way, it moves in and out of plants and animals and forms the fundamental building block of life on earth.

$CO_2$ – the most common gaseous form of the element – is constantly circulated into life forms and also recaptured in oceans and the atmosphere. The process of **photosynthesis**, which is the basis of all life on earth, essentially combines the sun's energy with $CO_2$ and water to make sugars that form the tissue of plants, with oxygen left over. In this way plants draw carbon from the soil and atmosphere as they grow but emit carbon back into the air and put carbon back into the soil when they decay. The same is true of animals (including you and me), which further participate in the **carbon cycle** (Figure 9.1).

A small proportion of carbon escapes this cycle, as it drops from the ocean floor and from deep soils back into the earth's crust. Sedimentary rocks, which contain lots of carbon, are formed from this slow deposition of layers of silt and organic material on the ocean floor, hardening over millions of years to become part of the earth's crust. On the other hand, by removing sediments from the earth, including coal (carbonized wood) and petroleum (a transformed fossil of marine life), and setting them on fire, people release large quantities of carbon into the atmosphere very quickly, and (quite significantly) *at a rate far faster than they are naturally returned to geological storage.*

**Photosynthesis**   The process through which plants use the sun's energy to convert carbon dioxide into organic compounds, especially sugars that are used to build tissues

**Carbon Cycle**   The system through which carbon circulates through the earth's geosphere, atmosphere, and biosphere, specifically including exchanges between carbon in the earth (e.g., as petroleum) and the atmosphere (as $CO_2$) through combustion and back again through sequestration

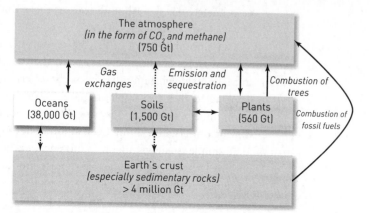

**Figure 9.1**   Carbon on earth. A Gt is a gigaton, or a billion tons. Ninety-nine percent of carbon is contained in the earth's crust. Most of the flow of carbon is between the oceans, the atmosphere, and soils and plants. Modern industry has removed large quantities of carbon from the earth's crust and, through combustion, has released them into the atmosphere as $CO_2$.

## The changing $CO_2$ content of the atmosphere

The amount of $CO_2$ in the atmosphere is by no means fixed. Indeed, it has changed dramatically over the past several hundred million years. At the beginning of life on earth, some two billion years ago, microbiota teemed across the planet. Tiny prokaryotes thrived thanks to an atmosphere consisting of very little oxygen, one that instead contained a great deal of carbon dioxide. These tiny, simple life forms took many shapes but all lived by breaking down the inorganic compounds around them to build their bodies. They were especially good at using abundant water, breaking it apart to use hydrogen atoms, and turning sunlight and $CO_2$ (both of which were plentiful) into energy. The incidental byproduct – the *waste* – of all these processes was oxygen, released into the atmosphere as $O_2$. As these "cyanobacteria" continued to thrive, the atmosphere slowly changed during the Precambrian period, from one with high proportions of $CO_2$ to one with a more significant proportion of oxygen. Today the earth's atmosphere (though still dominated by nitrogen) now contains around 20 percent oxygen and less than a tenth of a percent $CO_2$ (about 381 parts per million (ppm) of $CO_2$, to be more exact). New life forms came to take advantage of the now available oxygen (as we ourselves do) as these older ones began to vanish. For these older bacteria, an atmospheric cataclysm occurred: these older life forms were *poisoned by the very atmosphere they had produced*. Anaerobic bacteria, the descendants of earth's first life, still persist but they live in different environments and smaller areas than they used to, including mud flats and your own intestines (Margulis and Dolan 2002).

There is a lesson here. Living things have the ability to influence the biochemical characteristics of the earth, often in ways that change the conditions to which they must adapt in order to survive.

Certainly, what is true of ancient prokaryotes is still true for human societies. We are as intimately tied to carbon dioxide as those tiny bacteria were. Indeed, about 18 percent of your body, by mass, is carbon.

Ever since the advent of fire, however, our relationship to carbon has intensified profoundly. *Homo erectus* likely mastered the use of fire more than 400,000 years ago. Ever since, human civilization has been characterized by ever-increasing demands for energy. Cooking food, for example, requires a considerable amount of energy, as does staying warm under cold conditions (or cold under hot ones!), but so does building structures or moving objects, everything from automobiles to water.

In simpler societies, much of this energy is provided by manual labor. As societies have become larger, more complex, and more productive, however, they have come to depend on supplemental energy, which in the modern world comes primarily from burning fossil fuels: coal, oil, and natural gas most notably. The process of burning these liquids and rocks results in the release of enormous amounts of energy, but also in the emission of carbon, which flows into the atmosphere in the form of $CO_2$. As contemporary economies have learned to build larger buildings and cities, move freight around the globe, and grow more food by inputting fertilizers and pesticides (themselves made from petroleum), more energy is required, and so the combustion of fossil fuels increases apace. Modern civilization, put simply, is carbon civilization.

The results of this growing carbon-based economic activity are now quite apparent from measures of the atmosphere over time. Since the 1950s, scientist Charles Keeling and his colleagues have been measuring concentrations of carbon dioxide in the atmosphere from an observatory atop Mauna Loa in Hawaii. Graphed over time, the data show $CO_2$ (in ppm) content as a steep curve, steadily increasing (Figure 9.2).

This clear recent trend is only the tail end, moreover, of an earlier change towards a carbon-centered civilization. That major shift was one from human and animal labor power to machine power, with a massive concomitant rise in supplemental energy demands to drive machines. Steam engines that depended initially on wood and later upon coal came to dominate manufacturing and transport after 1800, before which objects (from shoes to shovels) were made largely by hand and vehicles (from rickshaws to military vessels) were powered by people, animals, or the wind. With the addition of each new engine in the 1800s and 1900s came a simultaneous act of combustion, whereby trees, lumps of coal, or barrels of oil were set ablaze to make heat and steam, to spin turbines for electricity, and to drive the motion of equipment and vehicles. Our car stuck in Pittsburgh traffic moves forward only by dint of burning hydrocarbons, pushing metal pistons in the engine, spinning the axle and rolling the tires so we can crawl along the freeway, bumper to bumper. Each act of combustion and each tiny explosion releases $CO_2$.

Other modern economic activities alter the carbon content of the atmosphere as well. By clearing large areas of forest for development or agriculture, for example, people perform a double impact. Destroying forests whose photosynthesis captures (or **sequesters**) carbon reduces the power of the trees to capture that carbon. Burning them then releases the carbon stored in each tree back into the atmosphere. It also leaves fewer plants on earth to recapture that carbon.

Although our direct measurement of $CO_2$ in the atmosphere begins only in the 1950s, we have records at our disposal from far earlier that show the unmistakable fingerprint of this machine age transition. Tree rings and sediment can be used to reconstruct climate conditions in the past thousand years or more, and the gases locked up in deep and ancient ice cores from the Antarctic and Greenland can be analyzed to tell us about the atmosphere when the ice was first laid down (Figure 9.3).

Evidence from the past millennium is compelling. There is certainly some slight variability from decade to decade over the period spanning the Middle Ages, a time of feudalism, peasant production, and religious hierarchy, to the 1700s, a period of global contact, scientific expansion, and colonialism. But these minimal variations are dwarfed by the massive influx of carbon dioxide into the atmosphere accompanying the industrial revolution after 1800 (Intergovernmental Panel on Climate Change 2007).

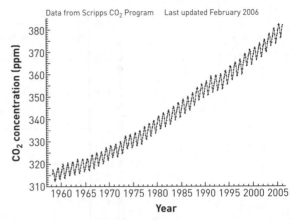

Data from Scripps $CO_2$ Program    Last updated February 2006

**Figure 9.2**   The Keeling curve: Atmospheric concentrations of $CO_2$ since 1958. The overall trend is a steady upwards curve. Annual variability (smaller up and down oscillations) is a result of the seasons in the northern hemisphere; plants green up in the spring, taking $CO_2$ out of the atmosphere, and die-off in the Fall, re-releasing that carbon. *Source*: http://www.aip.org/history/climate/xMaunaLoa.htm.

**Carbon Sequestration**   The capture and storage of carbon from the atmosphere into the biosphere or the geosphere through either biological means, as in plant photosynthesis, or engineered means

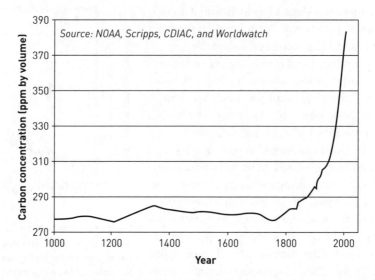

**Figure 9.3** Atmospheric concentration of carbon. Over the past thousand years carbon in the atmosphere has hovered at around 280 parts per million. With the explosion of industrial activity starting in 1800, more parts of human activity depend on combustion of fossil fuels and atmospheric carbon has increased exponentially as a result. *Source*: http://www.earthpolicy.org/Indicators/CO2/2008_data.htm.

## From carbon loading to climate change

This ongoing transformation of the atmosphere would receive little comment or attention if it did not have profound implications for the environmental and social systems of our planet. The central problem is that carbon dioxide, along with a number of other gases that are part of the atmosphere, plays a major role in regulating the temperature of the earth.

This is because much of the incoming short-wave radiation of the sun that strikes the earth is absorbed into earth surfaces and is transformed into long-wave radiation, which we experience as heat, and which is re-radiated back into space. Some portion of that re-radiated energy does not escape the earth, however, and instead is absorbed in the atmosphere that cloaks the planet. This atmospheric energy is trapped by a set of specific gases, including carbon dioxide. And make no mistake, this is good news! The result of this naturally trapped heat is a temperature on earth that can support life; without this effect – the **greenhouse effect** – the planet would be very cold indeed. But if a buildup of gases increases the level of heat-trapping over time, it can be reasonably predicted that global temperatures will rise.

**Greenhouse Effect**   The characteristic of the earth's atmosphere, based on the presence of important gases including water vapor and carbon dioxide, to trap and retain heat, leading to temperatures that can sustain life

This process, linking gases in the atmosphere with global temperatures, has occurred since long before human beings arrived on the scene. It has also proven extremely sensitive to changes in the gas content of the atmosphere. We know, for example, that the decline of carbon dioxide in the atmosphere during the ancient Precambrian period (discussed above) had the effect of cooling global temperatures dramatically. So the buildup of these gases in the atmosphere over such a short period of recent history raises the prospect of serious environmental change. The short-term global climate record reflects an overall pattern of warming. This warming has led to indirect effects as well, including the rise of sea levels resulting from the melting of ice, as well as overall decline in global snow cover (Figure 9.4). On average, the world is indeed getting warmer.

Carbon dioxide is only one of a number of greenhouse gases that act in this way. The others include naturally occurring gases like water vapor, methane, and nitrous oxide, along with entirely artificial industrial compounds, including chlorofluorocarbons (CFCs). Despite the long list, $CO_2$ is problematic not only because its concentration is high and rising quickly but also because it has high "radiative forcing," which means that per unit, it has a high effect on trapping heat. All of these gases are currently building up in the atmosphere and temperatures are indeed rising.

In this sense, "global warming" is an accurate description, but it is an incomplete one. In terms of *warming*, we do know that, because these gases reside for long periods in the atmosphere, there will be ongoing warming and that the

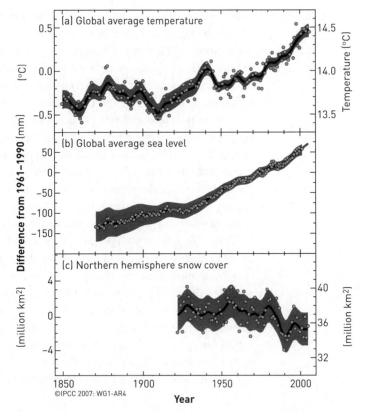

**Figure 9.4**  Global average temperatures, sea level, and snow cover. Over precisely the period when there has been a radical increase in the emission by people of gases that are understood to trap heat in the atmosphere, global temperatures have risen steadily, sea levels have increased, and snow cover has declined. *Source: Climate Change 2007: The Physical Science Basis*. Working Group I Contribution to the Fourth Assessment Report of the Intergovernmental Panel on Climate Change. Figure SPM.3. Cambridge: Cambridge University Press.

average global temperature will likely rise between 1 and 4°C in the next century. But this also sets into motion a large range of possible regional changes in climate and weather patterns. The myriad effects of this change in the global energy balance may include drying of some areas and increased rainfall in others, increased drought in some places and increased floods in others, decreasing sea ice, rising sea levels, and even the collapse of circulating currents that regulate the warm and moist conditions of some areas, possibly leading to isolated regional cooling.

The effects of this change will be *global* in the sense that everyone in the world will be affected. But here too the variability of impact will be enormous. Given a rise in sea level of a meter or more, countries in the island Pacific (actually home to 14 million people) are likely to disappear altogether. Drought in sub-Saharan Africa, where a large proportion of the population survives on rain-fed agriculture, means massive starvation and migration. Wealthy people all over the world may be able to direct a larger proportion of their budgets to air conditioning and food provisioning but the poor will not. Without some change in the trend of atmospheric concentration of $CO_2$ and related gases, therefore, the ironic effect of massive carbon-fed economic development during the past century may be an environmental situation that challenges and dismantles the very possibility of further economic growth and that punishes those who have contributed least to the problem.

Beyond people, the effects of such an environmental shift are just as profound. With rapidly changing temperature and precipitation, an enormous number of plant and animal species are likely to become suddenly endangered; many will be quickly lost. While high-profile cases like the polar bear are already receiving attention, the problems are much more serious if species more critical to the global food and energy web are eliminated. Such species (especially insects, plants, and marine plankton) are crucial to the survival of a range of others, meaning the potential collapse of whole ecosystems.

These changes in the global system could include "positive feedbacks" that might further enhance warming (melting of ice sheets reduces the reflectance of the sun's radiation back into space, for example) and some "negative feedbacks" that might decrease warming (increasing cloud cover that might conversely increase that reflectance) (Pearce 2007). Closer to the question of $CO_2$, however, an increase in warming will likely result in human adaptations that demand more energy (air conditioning, for example, or chemical inputs for agriculture), and which may increase the demand for more fossil fuels, and so increase emission of $CO_2$. Despite the complexity of the system, then, the high and increasing rate of carbon dioxide emissions *in particular* has become a matter of global concern.

## The puzzle of carbon dioxide

This brief review of the natural history of carbon dioxide reveals many things about the relationship of environment to society. First, it suggests that the history of civilization is necessarily also a history of gases, that molecules ($CO_2$) and elements (carbon) are as much a part of society as people, and that humans and non-humans influence each other in complex ways. It further suggests that the state of the environment is given to dramatic changes over the long term and that there is not necessarily a balanced, eternal, or inevitable state of nature. Greenhouse gas content in the atmosphere has changed over the eons, as have global temperatures. On the other hand, it also points to the profound influence and difficult-to-reverse effects that a specific form of society (industrial society) can have on complex earth systems, with implications not only for sustaining that form of society, but also for sustaining the current diversity of life on the planet.

More specifically, this review stresses two things about carbon dioxide that make it a specific kind of object:

- First, it is ubiquitous; carbon is everywhere, in all of us, and in a state of constant flow, making it hard to capture, pin down, and isolate and making its impacts often distant in time and place from its sources.
- Second, it is extremely sensitive to economic activity: the history of modern economic growth is completely intertwined with carbon dioxide, making a simple delinking of

human social formations and the content of the atmosphere difficult, like an operation on conjoined twins rather than a mere amputation.

Carbon dioxide therefore provides lessons for understanding where humans are positioned in the world, but it also presents a *problem*: given its ubiquity and economic sensitivity, how might atmospheric carbon loading be stemmed or reversed? Differing perspectives provide differing insights into this question, with regrettably somewhat contradictory answers.

## Institutions: Climate Free-Riders and Carbon Cooperation

In Chapter 4 we established that problems associated with the environment were commonly problems of **collective action**, where there is a necessity for joint action by many people but where there is an incentive for each of them to not cooperate. Controlling and diminishing the flow of carbon dioxide into the atmosphere, as mentioned previously, is exactly this sort of problem. Using an institutional perspective that stresses these features of socio-environmental problems provides some insight into the question, therefore.

> **Collective Action**  Cooperation and coordination between individuals to achieve common goals and outcomes

### The carbon prisoner's dilemma

Consider, in a simplified case, that two countries emit carbon. Both would benefit from halting global warming by "biting the bullet" and bearing the costs of reducing their own emissions. There are myriad reasons, however, why each would not act.

First, of course, is the inevitable disadvantage each country would suffer if they were to act while the other country did not. While both countries would benefit from a lower carbon load in the atmosphere, only one country would have suffered the sacrifice of making it so. To complicate matters, there are inevitable uncertainties surrounding global climate change and its possible effects. Without knowing the precise costs of inaction, it is difficult for either country to gauge their own commitment to potentially expensive action, especially if the effects will be uneven, with some countries bearing greater costs of inaction than others. Similarly, it is difficult to monitor whether another country is fully meeting any commitments to decreasing greenhouse gas emissions. While burning forests might be seen from space, the $CO_2$ coming out of every power plant in a country is difficult to measure. Each of these is a classic element of the *Prisoner's Dilemma*, that metaphoric explanation that stresses why cooperation and collective action so often fail. But there is even more than this!

Carbon's common role in a fully globalized economy also makes it a difficult problem to solve. One country might impose limits on carbon, forcing entrepreneurs to develop ingenious but expensive methods to make products that emit less carbon. But these products compete on a global market where the firms of the inactive nation market their cheaper carbon-dependent products for far less. One country acts aggressively, while the other "free-rides," enjoying both the benefits of the first country's reduced emissions and the better market position from having not imposed carbon limits.

At the same time, not all countries start their bargaining from the same point. The industrial revolution that occurred in Europe and the United States turned these regions into major carbon emitters for hundreds of years. Over that time, Europeans and Americans enjoyed greatly increased standards of living but at tremendous costs in terms of atmospheric change. Poorer countries may have large and/or more rapidly growing populations but their per capita carbon emissions are a tiny fraction of those of their counterparts in wealthy nations (Chapter 2). Why should a poorer nation, with its low per capita carbon emissions, agree to stringent controls when they have not yet enjoyed the fruits of a carbon economy? Why should a richer nation sacrifice its economy to reduce emissions while poorer nations, with their larger populations, stand to become the largest emitters over the medium or long term? Carbon's mobility exacerbates this conflict since it disconnects its places of emission from its locations of impacts: one person in the United States enjoys the benefits of emitting carbon through daily car commuting, while the costs are borne in a failed crop harvest amongst farmers and countries in West Africa, a half-world away. One country has an incentive to act, the other does not.

Finally, add to this the fact that there is no effective higher authority than the individual nations themselves to enforce any agreement: a de facto "state of anarchy" (Thompson 2006). While the United Nations is a powerful deliberative body, it does not represent a central power that can easily force international compliance with a regulatory statute. Instead, the model of international law must be followed, and the negotiators themselves must enforce the rules. Taking all of this into account, it would be hard to believe that international agreement of any kind might be reached. And yet one has been reached.

## Overcoming barriers through flexibility: The Kyoto Protocol

When the international community sat down in the late 1980s to work out a strategy to address climate change, the United Nations Framework Convention on Climate Change, they were faced with all of the hypothetical barriers to cooperation described above and several more. Eventually they established an international agreement, the Kyoto Protocol, a binding document of international law, which addresses and seeks to overcome these barriers and imposes emission reductions (of all greenhouse gases including carbon) on signatory countries. The solution to each and every one of these problems was to include elements of flexibility in the treaty that made it easier for signing countries to agree.

The details of the agreement are extensive, however, and a closer look at them reveals the complexity of overcoming the barriers to cooperation inherent in the climate problem. First among these barriers to cooperation were problems of uncertainty. Not knowing the potential costs of *either* implementing climate change controls through a treaty *or* those of not doing so, the international community was forced to negotiate, not really knowing what the future would bring. As a result, Kyoto contains specific provisions for alteration, update, and expansion over time. Article 9 specifically calls for ongoing revision and review of the protocol "in the light of the best available scientific information and assessments on climate change and its impacts, as well as relevant technical, social, and economic information."

Kyoto also establishes graduated goals, giving countries time to develop technologies and enforcement strategies to deal with emissions. The protocol imposes national limits on carbon emissions and increasingly stringent reductions demanded for signatories over time, with the intermediate goal of forcing countries to eventually reach emission levels equal to 5.2 percent below their 1990 levels. By developing a treaty that could change over time, the international community made it easier for hesitant and uncertain nations to sign on, knowing that the details are not entirely fixed. For nations that favored more aggressive action (those of the European Union, for example), it was possible to sign on to an unsatisfactorily weak treaty, knowing it could expand and become stronger in the future.

The second central barrier was the concern over economic costs and competition. For wealthier countries and those nations worried about the way compliance with emission reductions might make them economically uncompetitive, the protocol contains a wealth of provisions to lessen the impact and increase national flexibility. Most obviously Kyoto does not tell countries *how* they must reduce emissions. Countries can impose draconian restrictions, flexible carbon taxes, cap and trade systems (see below), or any other policy instrument they choose to meet their goals. This flexibility enhances national "buy-in" by allowing each country to tackle the problem in its own way. This approach was further enhanced by an explicit **emissions trading** provision between signatory countries.

> **Emissions Trading**  A system for exchanging the right to emit/pollute limited amounts of determinant materials (like greenhouse gases). These rights or credits are exchangeable between emitters, but subject to a total regulatory limit

The final barrier to cooperation was that of "distributive conflicts" of the kind outlined previously, where wealthy, low population, high per capita emitting nations (e.g., European countries) debated with poorer, higher population, low per capita emission nations (e.g., India) over the share of responsibility and obligation to act. Kyoto's framework deferred this conflict by establishing that countries have "common but differentiated responsibilities."

This means an agreement that wealthy countries are the largest current emitters, that their development has been predicated on carbon emissions over long periods, and that they have the obligation to act first on emission reductions. In terms of the protocol, this essentially means that most of the provisions of the treaty apply only to "Annex I" nations, wealthier industrial and post-industrial countries like France and Russia. "Annex II" countries, like China and India, are signatories to the protocol, but do not face the same obligations. Rather, they can be beneficiaries of technology transfer through the "Clean Development Mechanism," a mechanism where Annex I countries can meet their reduction goals in part by developing carbon emission reduction schemes and technologies in Annex II countries. These projects can include activities like tree planting or construction of new lower-emission power plants, or other interventions that help Annex II nations "leap frog" carbon industrialism and develop their economies without a period of high emissions, as occurred in the wealthier industrial world previously.

Annex II countries can also develop their own, independent emissions targets and goals (as China has in recent years), reconciling their need for economic growth with the eventual need to reduce and control emissions. From an institutional point of view, this again represents a key element of compromise that made it possible for the treaty to be signed and eventually come into force. The principle of common but differentiated

**Table 9.1**  Why is Kyoto a "shallow" treaty? There were a number of obstacles to cooperation around the Kyoto Protocol. The result of these conflicts and complexities is a document that is necessarily flexible and therefore somewhat "shallow," but one that sets the groundwork for more aggressive cooperation and action in the future

| Obstacles to cooperation | Flexible components |
| --- | --- |
| Uncertainty and differential commitments | Treaty rule alterations over time and graduated goals |
| Competitiveness concerns | Policy flexibility and emissions trading provisions |
| Distributive conflicts | Common but differentiated responsibilities |

*Source*: Adapted from Thompson (2006)

responsibilities, however, came to be seen by the United States as the main barrier to joining the protocol, and it is still not a signatory to Kyoto.

A range of other ad hoc concessions also made the document more flexible. The treaty could only come into force if 55 percent of all global emissions came under its mandate. Thus, the sum of all emissions of all participating countries had to add up to a majority of the earth's carbon emissions. Once the United States withdrew from the treaty, with its enormous share of carbon emissions, Russia became the necessary signatory to propel the protocol into force. That country, with its massive standing forests, insisted that existing forest stock be counted for some form of carbon credit. This concession, and others like it, became a part of the protocol.

In sum, the Kyoto Protocol represents an institutional innovation that attempts to overcome enormous barriers to cooperation through flexibility. These elements are summarized in Table 9.1. Such flexibility, of course, also means that the agreement is not as strong as many would prefer. As political scientist Alexander Thompson has pointed out, these elements of flexibility, which together make the Kyoto Protocol extremely "shallow", are the hallmarks of many international treaties at their nascent stages, most of which later became far stronger agreements (The Law of the Sea, for example). Because the Kyoto Protocol is designed to evolve over time, moreover, the possibility that it can grow sharper "teeth" definitely exists. Even if the protocol is eventually scrapped, its elements and format serve as templates for climate governance and the trust the treaty built, along with many institutions (like carbon markets – see below), form the core of future elements of cooperation.

An institutional perspective (as summarized in Chapter 4) reinforces this view. Specifically, institutional theory stresses that if collective decision-making is encouraged and rules can be established, however tentatively, agreements are reached that build trust. With trust and experience between parties (what institutional theorists call social capital), it becomes possible to make more aggressive, more far-reaching sets of rules over time, to continue to tackle commons dilemmas. Perhaps the best treaty is one that starts out life as "shallow," a highly flexible agreement that works to get as many signatory nations as possible, even if it makes less comprehensive reduction demands.

## Beyond Kyoto: Towards less shallow institutions or something else entirely?

There is some evidence that broad shallow treaties, having achieved consensus and leadership, are shifting to deeper forms. Notably, successive periodic meetings (or "convening of the parties" – CoPs) from the United Nations Framework Convention have continued to bring more stringent and comprehensive controls to the discussion table. A pre-meeting for the 2009 CoP in Copenhagen, held in March of 2009, suggested more comprehensive initiatives and targets, including a larger market-oriented component (see discussion of REDD below). By bringing a larger number of signatories to the table in successive rounds, and slowly ramping up stipulated goals, Kyoto mimics forms predicted by institutional theory.

Signatory countries, especially in Europe where leadership on climate change has been strong, have also been comparatively highly successful in limiting their greenhouse gas emissions, at least compared to non-signatories. According to the Global Carbon Project (a scientific forum that tracks international emissions, among other things – www.globalcarbonproject.org), European Union signatories decreased their emissions by 2.6 percent between 1990 and 2004, while emissions in the United States increased by roughly 20 percent. These reductions, it is important to note, might have occurred without the Kyoto Protocol, assuming there was an inevitable ongoing trend in European Union technology and regulation. Nevertheless, Kyoto shows the possibilities of a shallow treaty taken seriously over time.

Having said this, a great deal of activity on climate change appears to be occurring that poorly resembles the kinds of things predicted and advocated in institutional thinking. In the United States, for example, the "Mayor's Climate Protection Agreement" (Box 9.1) has initiated activities across the country, specifically without many of the necessary elements outlined by institutional theory, including any form of enforcement. Here, highly local and somewhat uncoordinated activities appear to have led to real reductions in emissions in cities, each with its own potential climate problems and strategies. This suggests that while a "post-Kyoto" period in managing $CO_2$ and other greenhouse gases may contain strong, negotiated, international efforts at dealing with the atmosphere as "common property," as predicted by institutional theory, other kinds of more ad hoc, local, and regional efforts, based on entirely different logics, may be more important than anyone has previously predicted.

## Markets: Trading More Gases, Buying Less Carbon

Many critics of the Kyoto approach to managing emissions suggest that the treaty is inefficient and relies too heavily on state mandates and too little on trade mechanisms to determine priorities and techniques for carbon reduction. One way of thinking about this problem that takes these concerns seriously, market-based thinking, invites us to consider the carbon flowing into the atmosphere, along with its potential negative effects, as externalities of market transactions and economic activity (Chapter 3). Viewed this way, we

## Box 9.1   Mayor's Climate Protection Agreement

Climate change is almost universally assumed to be a "global" problem. This is because carbon doesn't stay put and circulates through the atmosphere, having effects (e.g., global warming) far from its source. This view also leads to an understanding of the problem as largely one of international treaty-making to solve an international common property problem.

It might at first seem unclear, therefore, why the smaller political jurisdictions in the world – cities – would be leading actors in initiating a coordinated response to climate change. And yet, by May of 2007, over 500 US mayors had signed an agreement to meet or exceed Kyoto Protocol targets within their jurisdictions (a 7 percent reduction from 1990 levels by 2012). The Mayor's Climate Protection Agreement stresses that municipalities use the tools at their disposal, including planning, transportation, forest plantation, and public education, to decrease greenhouse gas outputs, especially carbon. Originally launched by Seattle Mayor Greg Nickels, the agreement now spans the country, and includes cities in every state. While for some municipalities the agreement may be symbolic, for most it has led to radical action and real reductions. The city government of Seattle, for example, has cut its own greenhouse gas output by more than 60 percent of 1990 levels and is moving quickly to bring the city as a whole into compliance. Taking advantage of obvious capacities in urban governance, like new density zoning and the provision of public transportation, cities are far more effective than individual people in changing regional climate footprints (Mayor's Climate Protection Center 2009).

But why did cities act? In some cases, it is clear that the potential effects of global change on specific munic-

ipal resources (water, for example) prodded cities into action. Some coastal cities face sea-level rise, others depend on ski tourism that depends on snow pack. There is some instrumental logic, therefore, for city action to precede that of US Federal government, which, at the time of writing, had not yet signed and ratified the Kyoto Protocol. So too, many of the efforts of cities to meet these targets – reducing the energy demands of municipal buildings for example – represent win–win cases, where the city benefits in cost savings by meeting emission goals.

This case also suggests some of the limits of strictly instrumental logic for explaining environmental governance behaviors, however. These cities all have clear disincentives to participate. Transition to a climate-friendly city can be expensive. Institutional common property theory does not necessarily predict rapid cooperative action from disconnected parties like cities, especially when they are under no coercive obligation to act. And yet they did act. Cities are not the only such actors, moreover. US states have set themselves binding targets, as have interstate compacts. Clearly the political will to address climate change is broader and more universally distributed than might have been guessed. The Mayor's Climate Protection Agreement represents a very real and also happily surprising effort to stop the runaway train of climate change.

### Reference

Mayor's Climate Protection Center (2009). Retrieved March 19, 2009, from www.usmayors.org/climateprotection/.

understand the potential effects of warming merely as the hidden costs of production that have not been appropriately borne and paid for by emitters of greenhouse gases reaping the benefits. After all, the potential costs of global warming are predictably economic as well as social and ecological. New costs of cooling homes can be predicted from warming, for example, as can failed crop harvests resulting from undesirable and disastrous weather events. Even a massive loss of biodiversity resulting from global climate change represents a kind of hidden cost, insofar as new and unknown species may be economically valuable now or in the future.

From a market-centered way of thinking, the problem here is that the economy is working inefficiently if those emitting the carbon (including commuting drivers, factory-owners, or forest-clearing loggers) are not paying the "true" costs of their activities, which are instead visited upon other people or society as a whole. This hidden subsidy (of sorts) needs to be remedied one way or another.

Regulating emission reductions, as in Kyoto, is one way of managing the problem. It is possible to argue, however, that this approach may be terribly inefficient and therefore may lead to unnecessary sacrifices because the benefits of reduction are far lower than the costs. How do we know when the benefits from carbon reduction are worth the costs required to achieve them? How do we know what techniques are most effective in managing the problem? Who should pay for managing the problem and how should the responsibility be shared? Kyoto answers these questions, as we have seen above, by negotiating a set of rules between diverse parties and imposing the resulting framework on the negotiators. It is democratic, in that sense, but is it efficient?

In contrast, the way of thinking summarized in Chapter 3 suggests that the most efficient way of dealing with such a problem is to discover the true costs and benefits and determine the best actors to shoulder them in a market. Using a market approach might 1) let those experiencing costs of climate change *contract* with those causing the nuisance and pay them to cease, 2) insist that those who demand carbon reductions alter their own *consumption* behaviors to create new supplies of carbon-neutral goods and services, and/or 3) cause emitters to *internalize* these externalities so that the costs are borne by the producers, providing them with an incentive to lessen or halt the impacts. The first approaches are largely consumer-driven, while the latter is producer-driven. All should achieve the same result, an efficient diminution of undesirable environmental harms, but only insofar as people are willing to pay for that outcome.

## Consumer choice: Green carbon consumption

If people are worried about temperatures rising, they can pay the owner of a factory some fee in exchange for the reduction of their emissions. This contractual approach to the problem follows from the logic of the **Coase theorem**: no matter who the polluters or the aggrieved parties are, the most efficient solutions come from market transactions. If maintaining the climate is truly important and failing to do so visits serious costs on some people, then people should be willing to pay to forgo negative outcomes.

> **Coase Theorem** A thesis based in neoclassical economics, holding that externalities (e.g., pollution) can be most efficiently controlled through contracts and bargaining between parties, assuming the transaction costs of reaching a bargain are not excessive

As Coase also pointed out, however, the efficiency of a contractually negotiated outcome requires minimal transaction costs, those expenses associated with actually negotiating the deal. On that front, turning this idea into a reality faces serious obstacles. If someone in Guam is threatened by sea-level rise owing to my driving a half-ton truck in California, how can we come to reach a negotiated settlement? How can a contract be applied that requires five billion people to negotiate with hundreds of thousands of factories and hundreds of millions of automobile drivers?

Barring this, how can the real cost of carbon be determined? Approached through the market, a more attractive solution emerges through consumer choice. If people do not want carbon to be emitted in industrial production, they are free to vote with their wallet, and buy only those products that are "climate-safe." Products might be labeled with per-unit carbon emissions to allow consumers to comparison-shop for the most climate-friendly products. Those products that do not provide such options would lose revenue and be forced out of the market, or their manufacturers compelled to invest in technology that makes them competitive in a new consumer culture. Green labeling on an enormous range of products already follows this logic.

A second solution is more direct. Consumers have the option of directly purchasing back their carbon emissions through offsetting. Carbon offsetting typically involves a consumer sending money to support some form of activity that "retires" carbon from the carbon cycle, by sequestering it in growing trees or otherwise facilitating an activity that reduces carbon (replacing a coal power plant with a solar powered one that produces an equivalent amount of energy using a lower amount of carbon).

According to the organization "Carbonfund," for example (www.carbonfund.org/), an individual seat on a 7,792-mile flight from Chicago, Illinois to London, England contributes 3.82 tons of carbon to the atmosphere. Carbonfund will accept a check for $38.53, which they will transfer into projects including some combination of renewable energy ("supports clean energy development, such as wind, solar, and biomass"), energy efficiency ("reduces existing energy use, much of which comes from coal, oil, and natural gas"), or reforestation ("absorbs existing $CO_2$ emissions, which helps to reduce the excess greenhouse gases that humans have added to the atmosphere"). The results of these activities, they assure, will sequester almost four tons of carbon from atmospheric circulation. And for your trouble, you also receive a certificate, a bumper sticker, a window decal, and a pen.

In theory as well as practice therefore, people's willingness to pay has been harnessed into new markets in this way, and directed towards effectively reducing individual carbon footprints, consuming our way to a greener world.

## Producer-driven climate control: Carbon markets and cap and trade

At the other end of the commodity chain, companies controlling the production of goods and services (from airlines to pulp and paper mills) can be considered equally culpable in carbon emissions as consumers. As pointed out in Chapter 3, one approach to carbon emissions is to impose rules on companies, telling them how much carbon they can emit and even the technologies they need to use to lower their rate of emissions.

A market approach eschews such **command-and-control** thinking under the assumption that firms will innovate solutions better on their own, albeit only if some financial incentive is attached to doing so. Here the solution lies in creating an incentive for the emitters of carbon dioxide to invest in new technology on their own, and find the least costly way of doing so. This requires an imposed collective limit of emissions on a group of participant members. Such a limit may be designed to decrease over years or decades, moreover, leading to reduced emissions over time. Each individual firm within the market owns a specific number of

**Command-and-Control** Forms of regulation that depend on government laws and agencies to enforce rules, including such things as regulated limits on pollution or fuel efficiency standards; contrasts with market-based or incentive-based approaches

**Figure 9.5** The strange logic of carbon offsets. *Credit*: *Joy of Tech* comic courtesy of geekculture. com.

units of emissions, which represent the amount of pollution they can produce. Those who release fewer emissions than their allotted units may sell their surpluses to firms who cannot meet their limits (Figure 9.5).

Such a carbon market is by no means hypothetical. The European Climate Exchange (or ECX) opened for business in 2005 and, in the United States, the Chicago Climate Exchange (CCX) launched in 2003. In both markets, participants face annual greenhouse gas emissions targets. Firms able to reduce their emissions below these thresholds through innovation or change in production levels can bank their "surpluses" or sell them to other firms.

The two markets differ from one another in a significant way, however. ECX operates in a political context in which internationally negotiated limits on national carbon emissions have already been imposed through the Kyoto Protocol. CCX, on the other hand, operates within a truer free market context in the United States, a country that is not a signatory to Kyoto. Nevertheless, CCX lists hundreds of members of the market, including major chemical companies, food processors, and power generators, as well as state, county, and municipal governments. Each contracted member must achieve a not-insignificant 6 percent reduction from a 2003 baseline by the year 2010. Those above this limit can use banked credit from previous years or bid for available carbon, which was trading at more than $100 per 100 metric tons of $CO_2$ equivalent at the time of writing (go to www.chicagoclimatex.com/).

The problems with such an approach are many, of course (see below). Primarily, voluntary markets, to date, have enrolled only a tiny portion of the total number of emitters and emissions in the United States. The United States produces roughly six billion tons of carbon per year and CCX trades perhaps 1 or 2 percent of that total. Most of these participants have joined for symbolic reasons, to demonstrate a "green" awareness and commitment, or otherwise reap benefits through public relations. Certainly most current corporate participants in these markets are doing so because stricter regulation is likely to come in the future. Participating in CCX allows them to get "ahead of the curve." Either way, most emissions continue to escape voluntary controls.

**Cap and trade** has also become a potentially more important part of whatever international regime follows Kyoto (as described above). A program called "Reducing Emissions from Deforestation and Forest Degradation" (or REDD) is now proposed that will stress purchasable offsets of forest, targeted especially at marginal communities, indigenous groups, and other, typically poorer communities, in forested and deforested areas. The mechanisms of this instrument are still under debate and discussion, but the basic elements are increasingly clear, and would include credits purchased by emitters to maintain and enhance carbon sequestration, with capital benefits of the program ostensibly directed at communities who protect and develop forest cover. As we will see below, the history of various forms of offsetting, especially amongst marginal communities, is filled with failures. Nevertheless, the appealing logic of the approach – offsetting carbon emissions in wealthy industrial countries by paying people to protect their own environments elsewhere – matches the theoretical leanings of market proponents: discovering and paying the real price of the environment.

> **Cap and Trade**  A market-based system to manage environmental pollutants where a total limit is placed on all emissions in a jurisdiction (state, country, worldwide, etc.), and individual people or firms possess transferable shares of that total, theoretically leading to the most efficient overall system to maintain and reduce pollution levels overall

In sum, market mechanisms for rethinking and reworking human relationships to greenhouse gases are more than theoretical. Between carbon-sensitive products, carbon offset services, and carbon markets, it would seem the most active area of addressing the problem of climate change is happening through capitalist systems of exchange. But there are reasons to treat these approaches with profound skepticism. As adherents to political economic approaches to the environment ask: If capital accumulation got us into this mess, why should we have such confidence that it can get us out?

## Political Economy: Who Killed the Atmosphere?

As we explained in Chapter 7, a political economy approach to nature–society relationships stresses that the roots of social and environmental crises are in the economy. For political economists, reliance on markets, or even on rationalized management institutions, represents a distraction (or worse a deception) from the underlying drivers of the problem. For these theorists, most of the benefits of carbon emissions accrue to a tiny minority of people and corporate entities, who have little or no interest in allowing institutions to control their behaviors. Instead these interests actually seek to propel *further consumption* as a route out of the "carbon trap."

## Green consumption is still consumption

From an economic point of view, the voluntary nature of the green transaction is the advantage of market mechanisms. People will only participate in an environmental effort, after all, if they get value from their investment, even if this is merely a sense of satisfaction. In this case, aggregated consumer willingness to pay effectively reveals both social preferences and the value of carbon.

From a *political* economy point of view, however, volunteerism is the great drawback of market mechanisms as a long-term solution. As pointed out earlier, those with the greatest ability to pay for offsets are typically those *least* directly affected by the actual climate impacts of carbon dioxide emissions and vice versa. Some green consumers in France, the United Kingdom and the United States may be emotionally driven to pay for offsetting. Because of **uneven development**, however, it is the farmers in Côte d'Ivoire (whose income may be falling as a result of climate change) who are most poorly positioned to economically drive a change in the system. These poor, indebted, and nearly politically powerless farmers do not have the luxury of communicating their preference (for survival!) in a carbon market transaction. Money-valuation of a global problem in this way is both inherently undemocratic and ecologically ineffectual, since most "votes" and power rest with very few people, most of whom are distant from the impact of their own behaviors.

**Uneven Development** The geographic tendency within capitalism to produce highly disparate economic conditions (wealth/poverty) and economic activity (production/consumption) in different places

And the mismatch between sites of emissions and sites of impacts may be considerable. Small island nations throughout the South Pacific, most notably, face total inundation from sea-level rise associated with melting ice caps. Agriculturally-dependent tropical countries, including India and other South and Southeast Asian nations that depend on monsoonal rain, are likely to experience more frequent, lengthy, and potentially more severe droughts (Cruz et al. 2007). These kinds of effects will have far more severe consequences in states with little money available for relief and emergency provisioning.

In all of these examples, heavily impacted areas will likely consist of countries, populations, and economies that have contributed a small or negligible fraction of the total greenhouse gas load in the atmosphere. It is certainly true that wealthier carbon dioxide producing countries may face considerable climate-related costs and disasters. The destructive wildfires that swept across Australia in 2009 were likely conditioned by lengthy drought that is part of a climate change trend. Yet nations of the "Global South" – a term used roughly here to describe poorer, historically colonized nations throughout the world, with economies that have been underdeveloped through their relationship within global trade – have contributed little to global climate change (at least so far) but are faced with potentially devastating impacts.

**Capital Accumulation**   The tendency in capitalism for profits, capital goods, savings, and value to flow towards, pool in, and/or accrue in specific places, leading to the centralization and concentration of both money and power

**Surplus Value**   In political economic (and Marxist) thought, the value produced by underpaying labor or over-extracting from the environment, which is accumulated by owners and investors

More fundamentally, a political economy perspective on carbon dioxide stresses the fact that **capital accumulation** demands sustained consumption and the production of **surplus value** by shedding of labor and environmental costs. Recall that, for *every* seller, there is a strong and ever-present incentive to make sure that people buy, utilize, dispose of, and repurchase as much as possible, as often as possible. Green consumption is still consumption, after all, and economic growth and sustained return for investment require growth of consumption (or continued cutting of costs by externalizing more waste). So too, political economy suggests that the demands of competition make it impossible for profit-driven firms to unilaterally reduce their subsidies from the environment (for example, by increasing costs to contain emissions) and still survive in a global market. This is especially true as economic growth means competition between firms and factories all around the world, including places like China and India, where environmental regulations are historically more modest. In sum, given 1) disconnections between sources and impacts that characterize *both* the carbon cycle and uneven economic development itself, 2) the gap between the power of those affected and that of those causing the effect, and 3) the importance of sustained consumption to keep production costs low, market-based approaches based on consumption have serious limitations.

To summarize, we can isolate at least five general and persistent factors that limit the overall effective potential of consumer-based solutions to carbon emissions:

- Companies/firms *must* strive for continuously increasing consumption levels.
- Companies/firms *must* strive to keep production costs low.
- The carbon cycle is marked by extreme disconnections between sources and impacts.
- Uneven economic development (as its name suggests) is also marked by disconnections between sources and impacts.
- Those who have the most power to change the market (the affluent, generally) are far removed from those who are going to bear the brunt of the negative effects of climate change (the poor, particularly poor residents of the "Global South").

## Critique of carbon trading and other markets

Critic Larry Lohmann (2006) has extended the political economy perspective to include problems with the institutional mechanisms designed to capture and trade carbon outlined previously. Institutionalized solutions (offsetting, cap and trade, etc.), however attractive, have profound limits and pitfalls. First, there are basic problems in information, confirmation, and transparency. How do I know that this transaction *actually* results in a carbon reduction? Assuming the money goes to planting trees, nothing has actually been offset if those trees were to be planted anyway or if $15 worth of trees fails to grow in such a way as to sequester 1.5 tons of carbon. Who confirms and counts this carbon trail? What

assumptions are made in the math and how is the accounting overseen? How much carbon is required in the process of confirming the
transaction of managing and marketing the firm that oversees it?
Given the prevalence of **greenwashing** throughout all consumer

**Greenwashing**  The exaggerated or false
marketing of a product, good, or service as
environmentally friendly

goods and the blatant exaggerations, misinformation, and downright lies that typify
marketing more generally, this obscure procedure must be treated with skepticism.

Tracing dozens of failed experiments in offsetting, for example – where wealthy Northern
countries and firms pay poorer Southern countries to plant trees – Lohmann and a large
number of other on-the-ground observers show that local communities get little value
from the investment. Time and time again, the commitment of land for such projects has
resulted in loss of resources for the world's poor, and very little carbon is actually seques-
tered in these projects. Take the case of carbon forestry in Uganda, where a Norwegian
energy company (that burns coal) paid the government to plant trees, rather than reduce
their own power plant emissions. The results were typical (Bender 2006; Lohmann 2006).
The company (and its forestry partners):

- bought up a massive 20,000 hectares of land but planted merely 600 hectares in trees,
  mostly fast-growing, ecologically less desirable, and exotic species;
- provided far lower real carbon sequestration results than anticipated on paper, although
  the value of those assets, since they allow firms to not expend on industrial equipment
  for carbon reductions, is on the order of US $10 million;
- paid a fee to the government that declined rapidly as inflation increased in Uganda,
  totaling less than $110,000 for use of the land for 50 years;
- evicted hundreds of families of farmers and herders from the land and provided 43 jobs
  in the process.

The clearly colonial implications of such an outcome are evident. Costs are visited on
poorer countries, benefits accrued in wealthier ones, and limitations in real environmental improvements are hidden behind well-intentioned rhetoric. Since most of these
transactions are obscure, under-regulated, and built into already existing economic relations, a political economy approach would predict such underperformance. Political
economy highlights how, rather than being a step in the right ("green") direction, this
project was actually a very savvy *investment* for the producers, with the end-result being
a typically uneven landscape of development (in Norway) and underdevelopment (in
Uganda).

## Climate policy in political economy

Ultimately, from a political economy perspective, both market and institutional approaches
have made little headway in dealing with the carbon problem but they have, ironically,
extended the power of companies and elites to own or control more of the world, including
both land resources and the atmosphere itself. These approaches have stressed sustaining
production and consumption by merely shuffling the costs and effects of the economy
around in a sort of shell game.

As an alternative, political economy thinking tends to privilege "structural" drivers behind the emission problem, favoring massive public investment in green infrastructure (solar and wind energy production), changes in subsidy structures (favoring tax breaks for the installation of home solar panels rather than subsidies for ethanol production), strong regulatory systems (setting direct limits on industrial $CO_2$ emissions), along with green taxes and other non-market economic incentives and global agreements that favor fixed and inflexible regional limits on greenhouse gases. These approaches are the same ones, it should be pointed out, that market-oriented thinking finds to be most inefficient. From a political economy point of view, however, they direct attention back to its source: *contradictions* in the economy that favor sustained growth over sustaining the conditions that make the economy possible (the atmosphere). As the effects of $CO_2$ emissions become increasingly evident, the debate over which of these approaches can produce the most tangible and immediate results is far from academic!

## The Carbon Puzzle

In this chapter we have learned that:

- Carbon dioxide in the atmosphere has changed over millennia and that the conditions for global life can be altered by living things and people on the earth.
- The rise of industrial production meant a parallel rise in an entirely carbon-dependent economy and society.
- Carbon dioxide concentrations in the atmosphere (along with other greenhouse gases) contribute to serious but unpredictable global climate change.
- Institutional approaches to the problem stress achieving international cooperation through careful compromise and effective rule-design.
- Market approaches to the problem stress the efficiency of green consumption and transferable rights to the atmosphere.
- Political economy perspectives stress the inherent limits and undemocratic implications of economic carbon reduction solutions to carbon production problems ultimately rooted in the economy itself.

Which brings us back to the slow-moving traffic crawling through "the tubes" in Western Pennsylvania. Can international rules be designed to encourage governments to reduce the emissions of these cars in time to forestall serious problems or are global climate treaties inherently too "shallow" to propel needed local action in places like Pittsburgh? Can consumer demand for a smaller carbon footprint alone lead to replacing those 100,000 cars with ones that run on electricity and that are fueled by solar and wind power? Can wholesale public infrastructural change eliminate the Pittsburgh commute altogether or is such an approach too expensive or draconian?

The puzzle of carbon dioxide clearly demonstrates the way contemporary society is deeply intertwined with environmental systems; the global climate system can no longer be understood separately from the global economic system.

## Questions for Review

1.  How has modern society dramatically altered the carbon cycle?
2.  Without the greenhouse effect, life on earth would not exist. But too much of a good thing can spell trouble. Explain.
3.  How is the "free rider" problem so potentially vexing to the challenge of mitigating climate change?
4.  How does a cap and trade system for carbon emissions differ from the traditional regulatory approach?
5.  Describe the example given of the market-based "solution" to climate change that exacerbated global uneven development. Did this program at least achieve its stated environmental goals?

## Suggested Reading

Intergovernmental Panel on Climate Change (2007). *Climate Change 2007: The Physical Basis. Contribution of Working Group I to the Fourth Assessment Report of the Intergovernmental Panel on Climate Change*. S. Solomon, D. Qin, M. Manning et al. Cambridge: Cambridge University Press.

Lohmann, L. (2006). "Carbon trading: A critical conversation on climate change, privatization, and power." *Development Dialogue* 48.

Pearce, F. (2007). *With Speed and Violence: Why Scientists Fear Tipping Points in Climate Change*. Boston, MA: Beacon Press.

Sandor, R., M. Walsh, et al. (2002). "Greenhouse-gas-trading markets." *Philosophical Transactions of the Royal Society of London Series A – Mathematical Physical and Engineering Sciences* 360(1797): 1889–900.

Thompson, A. (2006). "Management under anarchy: The international politics of climate change." *Climatic Change* 78(1): 7–29.

## Exercise: The Ethics of $CO_2$

In this chapter, we have reviewed how the puzzle of $CO_2$ might be addressed by markets, institutions, and political economy. Explain how you might understand this problem using an ethics framework (as described in Chapter 5). How might an anthropocentric approach differ from an ecocentric one? Do polar bears have intrinsic value? How might pragmatism and utilitarianism be employed to consider options for the control of carbon? What are the limits of an ethical approach to $CO_2$?

# 10

# Trees

Credit: P. Phillips/Shutterstock.

## Chapter Menu

## Keywords

- Acid rain
- Anthropocentrism
- Biodiversity
- Climax vegetation
- Disturbance
- Ecocentrism
- Ecosystem services
- Forest transition theory
- Induced intensification
- Market response model
- Preservation
- Primitive accumulation
- Reconciliation ecology
- Second contradiction of capitalism
- Secondary succession
- Social reproduction
- Succession

## Chained to a Tree in Berkeley California

On September 8, 2008, municipal authorities, riding in the basket of an enormous cherry picker, removed four men from the upper bough of a redwood tree on the University of California campus in Berkeley, and put them under arrest. This ended a long stand-off, in which the trees of the surrounding grove had been inhabited by protesters continuously for almost two years. Zachary Running Wolf had kicked off the protest when he climbed into the boughs of a tree 21 months before, in an effort to block the university's plan to build an athletic facility on the site. An organization of protesters, local activists, and environmental groups rallied around the cause and engaged in three lawsuits against the university. Calling themselves "Save the Oaks at the Stadium," they named themselves for the 200-year-old Heritage Oak and the more than 30 mature live oaks (among many other trees) that would have to be removed to make way for construction. One of the protesters explained: "There are no other oak groves like this. Some of [the trees] have been here longer than the school has. There's no reason to cut them down when they could just as easily adjust their development plans to avoid destruction of the grove" (McKinley 2008).

The tree sitters were supported by a large group of people providing logistical support, including food and water, throughout the event. Over the months of occupation, the grove of trees was surrounded by chain link fence by the university. Civic leaders also joined the debate, weighing in both for and against the protest and the university's actions. While city law prohibits the removal or destruction of live oaks larger than six inches, the university claimed exemption as an independent authority and a state agency.

In the end, the protest was a failure and the ancient grove was removed to make way for the university's construction. In the intervening period, however, the protest was conveyed through national and international media, and gained the attention and vocal support of an enormous range of people around the country and the world. Doug Buckwald, a spokesman for Save the Oaks, suggested that the protest was one with global resonance: "When people see people standing up and taking a stand, it gives them the courage to envision possibilities in their own places. And I saw that in the oak grove" (Delcourt 2002).

Why did this protest so fully capture the public imagination and enjoin such dramatic conflict? Besides the spectacular nature of the feat – people living aloft in a redwood tree for two years – similar protests have occurred in places around the world, from Canada to India, where the fate of people and trees are entwined through economic development, political opposition, and ethical wrangling. Trees are fundamentally *symbolic* for people, after all. Consider the Christmas tree or the tree in the Garden of Eden, George Washington's cherry tree, or the story of Newton beneath the apple tree. And yet trees are also a very *material* part of human history. Trees fueled civilization for the millennia prior to petroleum and their disappearance from the earth's surface has paralleled the extension of agriculture and cities. In many ways, the tree is the core marker of the complex relationship of environment and society. As Buckwald observes above, the trees of Berkeley's oak grove do seem to stand in for something larger than themselves.

## A Short History of Trees

In this chapter, we will introduce the puzzles that trees present society, stressing that the emotional role of trees for people is as significant as civilization's deep historical need for land clearance and forest destruction. We will stress that the reproductive capacity of trees makes forests (like many ecosystems) capable of recovery and creation, even while they are liable to ongoing destruction. We will also review three radically different approaches to thinking about tree-cover decline and recovery: population/markets, political economy, and ethics.

Notably in this chapter, like the others, we attempt to stress the difference between the objects in question (in this case trees) and the systems and problems to which they are intimately tied (in this case forests and deforestation). We do this to encourage readers to think about the specific characteristics of trees, and the importance of these for considering trees for contexts and problems beyond those few we review here. For example, how do trees fit in urban environments, in arguments between people over property boundaries, or in the development of new kinds of agriculture? So while forest cover is a central concern around the world, there are more to trees than forests. In this sense, we think it useful to start from the tree itself.

We define a tree here as a perennial plant with a woody structure. This broad definition includes more than 100,000 species worldwide (though it is impossible to know for certain), encompassing a quarter of all living plant species. The number, extent, and diversity of trees across the surface of the globe between the first appearance of trees and the present day have been highly dynamic, with eras during which leafy forests likely covered much of the earth's land mass to others in which all but some conifers retreated before glaciers and other global climatic shifts. At the height of the last ice age, for example, more than 10,000 years ago, those areas of North America and Europe not covered in ice had relatively fewer trees, almost exclusively conifers. Only in the most recent warming period of the past 10,000 years have the deciduous trees we associate with temperate regions (oaks, elm, ash, etc.) come to appear and dominate (Cohen 2004).

More dramatically, there was a time on earth entirely without trees. The first trees appeared a little more than 300 million years ago (MYA), an arguably recent event on a six-billion-year-old planet. These giant ferns only later gave way to non-flowering conifers (like our contemporary pine tree), and deciduous flowering trees appeared most recently, approximately 140 MYA.

### Trees and civilization: A complex relationship

Nevertheless, for human beings, who themselves only made a recent appearance on the scene, trees and forests have typically appeared ancient and timeless, owing to their size and longevity. In a time before buildings or other large human-made monuments, some of the largest things around would have been trees. Given their enormity and their crucial role in providing food for humans, construction materials for shelter, and habitat for other species of importance to people, it is unsurprising that trees have long been a central symbol and site for worship and notions of the divine. And the history of human religious culture is filled with trees, including among countless examples the tree cult of Diana Nemorensis in Rome, the Bodhi tree beneath which Buddha found enlightenment, the Tree of Knowledge from the Bible, or Yggdrasil, the tree upon which the world sits in Norse mythology. Linked to these religious associations, trees and forests have often stood in for nature, the environment, and the non-human more generally, both as dangerous alien places, and sites for nurturing primeval truth (Figure 10.1).

Earliest uses of the word "forest" (or Old French: *forêt*), for example, specifically refer to wooded areas *outside* of the walls or fences of a park; an exterior space, associated with "wilderness" and "waste" (see Chapter 8 on the social construction of wilderness). Forest, in such an understanding, is a place away from culture. This ideology certainly made it all the easier to cut trees and clear forests for cultivation and urban expansion over the past thousand years.

When forests and trees are venerated or romanticized, on the other hand, tree-centered thinking can emerge, for better and for worse. When forest destruction was noted in Europe and America in the early 1800s, romantic poets and philosophers began to associate trees and forests with virtue and freedom separate from the vices of civilization. Such feelings were later cultivated intentionally by government and industry to offer tree-planting as a panacea for social and environmental problems. The American holiday Arbor Day, for example, was advanced by interests in the forestry industry, whose activities are in part responsible for tree-cover decline in old-growth forests (Cohen 2004).

**Figure 10.1** *Sequoia sempervirens*, the genus in the cypress family Cupressaceae. Sequoia national park. California. USA. Trees are some of the largest and longest-lived organisms around which human beings live. Their influence on human culture has been enormous. *Credit*: urosr archive/Shutterstock.

These ideas, both fearful and romantic, have heavily influenced, in often problematic ways, scientific and popular assumptions about the impact of people on trees and the historical trajectory of forest-cover change. As noted in Chapter 8, for example, theories that North Africa was heavily wooded in the recent past and that its forest-cover loss was the result of "uncivilized" indigenous behavior have proven false on both counts, raising questions about assumptions concerning the relationship between civilizations, forests, and the history of global tree cover. This reminds us that the often ideological associations between trees and people need to be kept in mind as we sort out social and ecological explanations for actual forest-cover loss and recovery.

## Climax, disturbance, and secondary succession

Chief among the tools available to ecologists in explaining and exploring tree-cover change are the concepts of **climax vegetation**,

**Climax Vegetation**  The theoretical assemblage of plants arising from succession over time, determined by climatic and soil conditions

**Disturbance**   An event or shock that disrupts an ecological system, thereafter leading either to recovery of that system (e.g., through succession) or movement of the system into a new state

**Succession**   Ecologically, the idealized tendency for disturbed forest areas to recover through stages of species invasion and growth, progressing from grassland, to shrubs, and eventually back to tree cover

**disturbance**, and **succession**. These three linked concepts, and their continued evolution in the field of forest ecology, have guided thinking about forest dynamics for a century.

The concept of climax vegetation was developed by Frederic Edward Clements, who worked at field stations around North America during his career as an ecologist in the early twentieth century. The concept suggests that broad physiographic and climatological conditions account for a "normal" or average vegetation type, "best suited" to local conditions, if things are left to themselves.

For example, throughout the arid plains of Central Asia, it is to be expected that the typical land cover will be hardy perennial grasses. On the other hand, in Eastern North America we would expect various forms of deciduous forests, including oak–pine forest for example, or beech–maple.

When a periodic event occurs in such a region, like a massive rock fall, a fire, a hurricane, or a human-induced clearance, large areas of this land cover might be eliminated or dramatically altered. Hurricanes knock down trees. Forests clear whole areas, and so on. Such an event is typically referred to as a disturbance. Disturbances are viewed as natural but also as unusual or atypical deviations that lead to only temporary changes. Over time, areas cleared of climax vegetation experience the colonization of new plants, which slowly give way to others, and eventually lead to the restoration of the climax community. After a fire, for example, grasses invade burned patches, followed by various shrubs, which are slowly shaded out by trees. This is referred to as succession.

The implications for conservation and management of trees and forests from this understanding are clear: Minimize disturbances whenever possible and allow areas to recover through stages of succession after such events by preventing human intervention or pressure.

More recent ecological science has raised questions about the simplicity of this model, especially in terms of the role of disturbance. Specifically, contemporary ecological science has demonstrated that some important species (and often typical or dominant ones) often actually *require* disturbance to reproduce and function. Many pine trees of North America, for example, have evolved with fire and so require fire events to open their cones and clear the land for their successful germination. Some perennial grasses co-evolved with animal grazing and so require some animal disturbance activity to thrive.

The implications of this for trees and forests cannot be overstated. Rather than assuming that an *absence* of disturbance is necessary for a successful or diverse ecosystem, it may sometimes be the case that keeping such events from occurring (by suppressing fire, for example) may do more harm than letting them occur. Preventing disturbance in forests may actually alter their trajectory away from the "climax" community desired by people.

A related criticism of traditional climax theory emphasizes, on the other hand, that certain scales and types of disturbances may be dramatic enough to make it impossible for ecosystems to return to their original state. Allowing some forms of disturbance, therefore, may lead to permanent changes to existing ecosystems.

These scientific debates over the role of human beings in maintaining or retarding forest cover echo the larger cultural and historical debates about trees. Are forests things apart from society that should be left alone to thrive? To what degree is human intervention necessary for creating and fostering certain kinds of tree cover? Where should we expect forests and where might we create them? By preserving or creating forests, are we enhancing or actually limiting other kinds of ecosystem development? Ecological questions about the role of humans and the environment mirror far older questions about culture and nature.

## How much forest is there now?

Of course, these debates occur in a context where total global forest cover has long been in dramatic decline. The total area of forests and woodlands around the world is estimated to have declined from 6,215 million hectares in the year 1700 to 5,053 million hectares by 1980, a loss of almost a fifth of the world's forest over a relatively short period (Grainger 2008). It is also important to keep in mind that where forests did not disappear, their composition was transformed dramatically. Where did all of these trees go?

Most available evidence points to deforestation being closely associated with the expansion of human industry, urbanization, and agricultural expansion. Forest is cleared to produce food and cash crops and to make way for urban development. Many secondary effects of human activity also hamper forests, including climate change and atmospheric pollution effects, like **acid rain**, which has damaged and destroyed forests throughout North America and Europe.

**Acid Rain** Deposition of rain or snowfall with unusually high acidity, resulting from the emission of sulfur dioxide and nitrogen oxides into the air, typically from industrial emissions. This form of precipitation is harmful for plant life and aquatic ecosystems

Yet, there is not necessarily a linear relationship between population growth and loss of tree cover. Consider, for example, that in many places where population continues to grow (for example, Europe), forests are in a state of recovery. At the same time, urbanization actually leads to an increasing density of people in smaller spatial areas, which in theory makes room for more forests. Agricultural intensification, finally (see Chapter 2), enables the use of less land for the same amount of food production. There are reasons to believe, therefore, that forests have the possibility of recovery in a world full of people.

To better determine what causes forest-cover decline and recovery, more detailed analysis is therefore essential. Obtaining reliable numbers to estimate rates of deforestation and reforestation turns out to be extremely complex and difficult, however, especially for tropical areas, where many of the world's trees remain. Sources of error are numerous, even though technologies (like satellite remote sensing) are improving all the time. Since forest cover changes quickly in many areas, with rapid cutting and rapid regrowth both occurring in different places and at varying rates, estimates of regional forest cover can vary wildly. This makes tracking and explaining the fate of trees all the more difficult (Delcourt 2002).

The United Nations Food and Agriculture Organization's 2005 Global Forest Resources Assessment, a commonly cited source, estimates that current global forest cover is 3,952 million hectares, or about a third of total land area. Table 10.1 shows the change in forest cover in the period between 1995 and 2005. Some caution should be used in examining these figures. Clustering world regions is always arbitrary and this skews statistics somewhat. For example, the large forest cover figure for Europe (a region with a long and

**Table 10.1**  Change in cover of forest and wooded land (in thousands of hectares) by world region

|  | 1995 | 2000 | 2005 | 1995–2005 | Change as % of 1995 |
|---|---|---|---|---|---|
| Africa | 699,361 | 655,613 | 635,412 | −63,949 | −9.14% |
| Asia | 574,487 | 566,562 | 571,577 | −2,910 | −0.51% |
| Europe | 989,320 | 998,091 | 1,001,394 | +12,074 | 1.22% |
| North and Central America | 710,790 | 707,514 | 705,849 | −4,941 | −0.70% |
| Oceania | 212,514 | 208,034 | 206,254 | −6,260 | −2.95% |
| South America | 890,818 | 852,796 | 831,540 | −59,278 | −6.65% |
| Total world | 4,077,290 | 3,988,610 | 3,952,026 | −125,264 | −3.07% |

*Source*: Adapted using data from the United Nations Food and Agriculture Organization's 2005 Global Forest Resources Assessment; www.fao.org/forestry/32033/en/

sustained history of forest loss) is increased enormously by the inclusion of the Russian Federation, with its vast Siberian forests (in continental Asia) that account for four-fifths of that number.

## The future of trees

Keeping this in mind, three things are still critically evident. First, forest cover continues to decline globally, with significant losses in Africa and South America especially. Second, however, and more subtly, declines in forest cover between 1995 and 2000 were greater than those between 2000 and 2005, suggesting that the net rate of deforestation (total deforestation offset by any regrowth or planting of trees) is declining, with possibly a cessation in the near future. Finally, some regions are actually experiencing reforestation, especially in the most recent period in Asia and Europe.

The return of trees to some areas is a complex matter (discussed at greater length below), but it can be accounted for by a number of processes. First, tree plantation often occurs on a massive scale. In the United States, for example, there are more than 17 million hectares of plantation forest, created and maintained by the government in conjunction with timber companies. That is a figure larger than the total forest cover of most countries in the world. So too, where older land uses are abandoned (like agricultural fields), forest cover often rebounds. Plantation forests are rarely ecologically similar to the original or older growth forests they come to replace. They are typically lower in diversity of tree species and are mostly even-aged, meaning the tree canopy lacks mixed structure and complexity. Plantations tend to support a lower range of native biodiversity and provide far fewer of the **ecosystem services** of native forests. In particular, this holds implications for **biodiversity**, since tree and forest protection and

**Ecosystem Services**  Benefits that an organic system creates through its function, including food resources, clean air or water, pollination, carbon sequestration, energy, and nutrient cycling, among many others

**Biodiversity**  The total variability and variety of life forms in a region, ecosystem, or around the world; typically used as a measure of the health of an environmental system

plantation is often not an end in itself, but rather a goal reflecting **preservation** values (see Chapter 5), which see the conservation of forests as a vehicle for the protection of myriad other species for which trees, forest soils, and complex forest growth form critical habitat. For many such species, plantations are a poor substitute for previously existing forests.

> **Preservation** The management of a resource or environment for protection and preservation, typically for its own sake, as in wilderness preservation (compare to conservation)

In all of these ways, more trees are not necessarily the same as more forests, since the latter represent complex ecologies from the interacting diversity of the former. Nevertheless, plantation of trees is an important component of the return of forests to the landscape.

Second, forests regrow on their own, though at rates that are difficult to measure and generalize. As noted above, secondary succession quickly brings pioneer species into cleared regions, which leads in time to a mixed tree canopy and increasingly dense forest, though possibly one that may differ significantly from the forests that came before. Even so, forest recovery dynamics demand acknowledgment as much as tree-cover losses.

All of this suggests that a long history of deforestation is quite real but that current trends are hard to generalize. Some areas of forest cover are definitely returning, however. This short history of trees may give us reason for cautious optimism.

Regrettably, other global forces are at work on the forests of the world. Chief among these is global climate change. While it is unknown what a 1–4°C warming will mean for all of the ecosystems of the earth, many forest systems are clearly at risk. The range of specific tree species will undoubtedly change, as will the composition of forests (van Mantgem et al. 2009). The forests we have come to know in the Amazon, the Appalachians, and Scandinavia will all change dramatically, if they do not disappear altogether.

Across the western part of North America, a massive increase in the mortality rates of trees is evident. This is likely due to water deficits resulting from recent prolonged droughts, which are predicted to continue in this region as the world gets warmer. Given the attrition of trees and the invasion of new species into forest areas, along with a rise in tree pathogens and parasites, the extent and health of forests in the future are not assured (Vandermeer and Perfecto 2005).

## Trees, people, and biodiversity

A final critical point is in order, however, one that again stresses the difference between trees – a diverse set of species and individuals – and forests, a systemic collection of many plants, animals, and soils. A significant proportion of the world's trees are not in forests. People's interaction with trees, especially in agriculture, has meant that trees are commonly a part of complex land-use mosaics maintained by people. Trees are typically mixed into crop fields throughout Africa, Asia, and Latin America, and tree crops themselves represent a significant agricultural product. These trees interact with their surroundings to produce valuable and complex ecosystems that differ both from forests and from those agricultural spaces free from tree cover.

Consider the example of coffee plantations. Though modern coffee production removes all of the trees from the landscape in order to maximize production, traditional methods of coffee production integrate coffee plants into forest canopy, leaving many trees present.

**Table 10.2** Comparison of the number of important insect species measured to be present in differing systems of coffee production in Costa Rica

|  | Traditional production | Modern agrochemical production |
| --- | --- | --- |
| Beetles in shade trees | 128 | 0 |
| Ants in shade trees | 30 | 0 |
| Wasps in shade trees | 103 | 0 |
| Ants on ground | 25 | 8 |
| Beetles on coffee bushes | 39 | 29 |
| Ants on coffee bushes | 14 | 8 |
| Wasps on coffee bushes | 34 | 30 |

*Source*: Adapted from Vandermeer and Perfecto (2005)

So-called "shade grown" coffee, with its mix of agriculture and tree cover, provides a range of ecosystem services and can actually sustain a great deal of native biodiversity, even while being used for commodity production. The data from Costa Rica shown in Table 10.2 demonstrates the dramatic difference between production systems that incorporate trees and those that do not.

Similarly, in the countryside of India, where intensive cultivation of cash crops is typical, the maintenance of productive tree-covered areas can actually contribute as much to conservation as native forest. In the growing of valuable arecanut (a nut used and consumed by 10 percent of the world's population in some form or another), for example, dozens of native bird species (including critically threatened hornbills; Ranganathan et al. 2008) are provided tree habitat. The diversity of important bird species is as high in densely populated agricultural areas as it is in protected conservation areas nearby. This emphasizes again the notion of **reconciliation ecology**, where even amidst productive human economic activities a range of ecosystem services are maintained and so biodiversity thrives. Trees have long been incorporated into the life worlds and production systems of human beings, even outside of "forests." This points the way forward to new ways of thinking about environment and society.

**Reconciliation Ecology**  A science of imagining, creating, and sustaining habitats, productive environments, and biodiversity in places used, traveled, and inhabited by human beings

## The puzzle of trees

This brief review of the history of trees reveals many things about the relationship of nature to society. First, it emphasizes that the history of human activity on the land, including the agricultural revolution, urbanization, and industrialization, is deeply entwined with trees, for better and for worse. Overall tree cover and the composition of forested and wooded areas have been transformed dramatically by human activity, although the complexity of disturbance and recovery dynamics is still not fully understood. More specifically, this review stresses several things about trees that make them a specific kind of puzzle:

**Box 10.1**  Shade Grown Coffee

Coffee is a crop for which there is a seemingly endless demand. Since the time of colonialism, when the plant was brought from its native Africa and Asia to Central and South America, it has been grown more and more intensively, meaning more coffee is produced on the same amount of land, year after year. To achieve these high levels of production, coffee is typically grown on vast plantations, where the land is cleared of forest, weeds and insects are vigorously controlled, and the full energy of the sun is used for growth.

The environmental costs of conventional coffee production are extremely high. Most notably, native forest is permanently destroyed in the creation of the plantation. In tropical and semi-tropical areas, this typically represents forest areas containing high floral, faunal, and insect diversity. It is hard to know exactly how much tropical forest and attendant biodiversity has been lost to coffee plantations over the past two hundred years, but the massive declines in forest cover in countries like Mexico in recent decades are closely associated with commodity production, especially coffee. Moreover, by managing this conventional "sun grown" coffee with intensive chemical inputs, non-target insect and plant species are typically killed and soils tend to degrade, losing nutrients and structure over time. This dependence on monocultural (single crop) coffee plantation also results in social and economic vulnerability. When prices for coffee collapsed in the late 1990s, many producers around the world became immediately destitute, with no resources at hand for recovery.

For all these reasons, there is growing interest in returning coffee production to its more traditional eco-logical and social roots. Originally, coffee was not grown in vast open areas. Instead, coffee bushes were seeded throughout existing forest with other crops, resulting in lower levels of production, but much higher diversity of crops, as well as wild flora or fauna. Such "shade grown" coffee could thrive amidst native vegetation, removing the either/or choice of forest versus coffee, or more generally: environment versus society. Shade grown coffee systems can vary enormously, from plantations that resemble intensive production but where coffee is simply intercropped with various trees, to systems so resembling native forest that it is difficult for visitors to tell by looking that farming is going on at all.

The problem with such a system is straightforward, however. Shade grown coffee, by definition, has lower yields than conventional plantation production. And during periods of low prices, as in the coffee price collapse of recent years, this leaves farmers even more economically vulnerable and indeed encourages them to clear forests for more intensive cultivation of different crops. To be sustained, shade grown coffee must, therefore, be supported somehow, either by consumers who are willing to pay a premium for more green or sustainable production, by governments interested in maintaining forest cover, or through cooperatives that can lower production and transportation costs for farmers. This makes shade grown coffee one of the more promising examples of reconciliation ecology, but raises questions about our willingness to collectively support sustainable enterprise.

- First, because of their universal symbolic value, trees have come to stand in for all environmental change, in a way that makes them a useful conservation target for people, but sometimes also sidetracks or obscures other environmental trends.
- Second, human influences have caused precipitous transformation in dominant tree compositions and overall decline of forest cover over many centuries. Trees are therefore excellent (albeit disturbing) indicators of human economic growth and expansion.
- At the same time, however, trees show an ability to recover from disturbances and shocks and forest regrowth is as important a part of environmental history as forest decline.

- Finally, trees have oftentimes been a basic part of human life systems and productive activities, leaving open the possibility that continued human economic activity need not lead to a zero-sum choice of people versus trees, despite the real threat of global climate change.

Trees therefore provide lessons for understanding where humans are positioned in the world, but they also present a *puzzle*: given the capacity for humanity to destroy forest cover and alter species compositions of forests, and the simultaneous ability of tree cover to recover over time, what explains areas where deforestation does or does not occur? What explains patterns of forest decline and recovery? And to the degree that protecting forests is an important part of sustaining the global environment, how might we best consider and include the fate of forests in our planning, actions, and environmental ethics? Differing perspectives provide differing insight into these questions, with a wide range of answers.

## Population and Markets: The Forest Transition Theory

Explaining why tree cover changes is no easy matter. People cut and plant trees for a range of reasons, and the health and survival of forests and woodlands are often tied to air quality, climate, and a host of other factors. Overall patterns hint at some interesting global trends.

Consider the fate of forests in Europe. Figure 10.2 shows the estimated coverage of forests in the region (here excluding forests in interior Russia and Siberia) over the long term (Richards 1990). Forest losses in Europe certainly predate the 1700s and major tree-cover declines likely occurred as early as the Bronze Age (ca. 1200 BCE). Nevertheless, dramatic deforestation is immediately evident during the early industrial period, in which urbanization and the rise of manufacturing led to rapid use of forest resources. This is also a period in which agriculture spread into many areas previously left out of the reach of the plow. There is a clear connection here between increased economic activity and decreased forest cover. This makes the recovery of forests, evident in the most recent historical period, somewhat confusing. After all, economic activity has not decreased in the past hundred years. To the contrary, cities, roads, and farms have continued to thrive and expand. Is it some combination of population trends and economic activity that has led to this outcome?

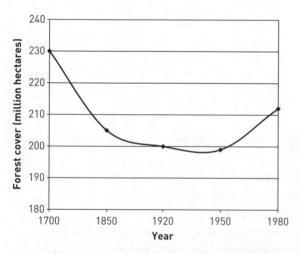

**Figure 10.2**  European forest cover from 1700 to 1980. The pattern of decline and regrowth follows a notably "U-shaped" curve, which has been a key argument in favor of the so-called forest transition theory. *Source*: Adapted from data from Richards (1990).

In Chapters 2 and 3, we learned that applying demographic and market-oriented thinking to environmental problems often produces counterintuitive insights. An approach that begins from combining demographic transitions and market logics, therefore, may suggest a hidden pattern beneath this apparently contradictory socio-environmental trend. According to this way of thinking, there is nothing inconsistent in this reversal. Consistent with the idea of **induced intensification** and the **market response model**, early socioeconomic activity might account for the initial decline of forest, while the later pace and intensity of ongoing growth might begin to provide opportunities for forest recovery. Initially, expanding populations and economic activity put pressure on forest resources and the destruction of trees frees up land for agriculture and raw materials for fuel and construction. Later, however, that ongoing pressure is manifested in new land uses. This is because intensification of agriculture, coming from increased population pressure but also from increasingly commercial farm production, allows more crops to be grown on less land (see Chapter 2), opening the possibility of unneeded or marginal land to transition back into forest. At the same time, the rise in urban employment takes more people out of rural areas, leaving lands fallow, eventually allowing **secondary succession**, and reforestation. In theory, as productivity increases through intensification, and as demographic transition results from later industrialization, the value of potential agricultural land underneath forests declines, and tree cover can rebound.

Recall also, from Chapter 3, the way human economies respond to scarcity. Typically, in keeping with the market response model, increasingly rare goods come to be coveted more highly, and so supply of these increases. Certainly the loss of forest cover in the highly developed European landscape has resulted in an increasing call for its restoration, which is arguably as much a kind of market response as it is an environmental urge. Tree plantation was visible throughout Europe during the nineteenth and twentieth centuries.

Extending this insight to form a more universal rule, a combined population and market-based approach to forest cover change predicts that, following a period of forest decline correlated with population and economic expansion, later economic activity and growth leads to forest return and recovery. This prediction is usually referred to as the **forest transition theory**, which holds that:

> over time, forest cover declines, but at some point a transition occurs, such that the decline halts and reverses and forest cover thereafter expands. This transition yields a U-shaped (or at least reverse J-shaped) curve for forest cover plotted against time … (Owusu 1998: 105–6)

The very economic growth that destroys forests is necessary for their restoration and return. From this way of thinking, it should be pointed out, high levels of expensive conservation efforts to protect forests must be seen as counterproductive. By delaying use and exploitation of forests, especially in economically poor but forest-rich nations, conservation is seen as retarding the use of crucial resources, which are essential to an economic transition that will yield less destructive economic forms and better environmental conditions down the road.

---

**Induced Intensification**   A thesis predicting that where agricultural populations grow, demands for food lead to technological innovations resulting in increased food production on the same amount of available land

**Market Response Model**   A model that predicts economic responses to scarcity of a resource will lead to increases in prices that will result either in decreased demand for that resource or increased supply, or both

**Secondary Succession**   The regrowth of vegetation and return of species to an area cleared or reduced by disturbance, as where a forest recovers its "climax vegetation" cover after a fire

**Forest Transition Theory**   A model that predicts a period of deforestation in a region during development, when the forest is a resource or land is cleared for agriculture, followed by a return of forest when the economy changes and population outmigrates and/or becomes conservation-oriented

Certainly this is consistent with the positions of key international development lending institutions around the world. The World Bank Group and the International Monetary Fund (IMF) were both highly supportive of large-scale forest use in the developing world in the 1980s and 1990s. The country of Ghana, for example, as part of its IMF-sponsored structural adjustment program, increased its revenue from forestry by a hundred million dollars by the early 1990s, generating large-scale economic activity and using the revenue to pay down a significant portion of its massive external debt (Owusu 1998). Assuming a more stable future economy in Ghana following the crisis (no small assumption), trees would be predicted to return in a country that is more urban, less agrarian, and decreasingly dependent on forestry and agriculture.

In theory, this more global prediction would mirror the process experienced in the growth and success of the European market economy (leaving aside for the moment the issue of debt repayment). And some evidence is indeed available that some forest recovery has occurred in developing countries around the world. Countries like Algeria (with a growing oil-revenue-based economy), India, Egypt, and China (with entirely plantation-based forest expansions), and Chile (with an increasingly diversified economy) all report forest cover increased in just the past few years after long periods of loss or decline.

## Limits of the U-curve model

Though attractive and convincing on the merits of the data available from the European historical case, the prediction of a U-curve recovery of trees through forest transition is by no means without problems. First, as noted previously, the ecology and composition of secondary forests is by no means similar to the primary forests they come to replace. These later forests tend to be lower in biomass, less diverse in the range of tree species they contain, and often less useful for local human populations, to say nothing of the fauna that depended on original forest cover for habitat. Plantation forests are even further ecologically removed from native tree cover, and are sometimes typified by a small range of fast-growing species. So even to the extent that forest recovery may be a product of economic transition, there is no guarantee that recovery does not still represent a significant and highly undesirable loss of original forests.

More troubling, a vast majority of nations around the world are experiencing deforestation and show little or no sign of immediate forest recovery. In the Ghanaian case described above, the country lost 1,300,000 hectares of old-growth hardwood forest between 1980 and 1990, with a mere 11 percent of its original closed-canopy moist forest remaining at the end of the millennium. Despite a massive revenue stream from forest exploitation in that country and a significant reduction of debt in the process, moreover, deforestation continues unabated, with very high 2 percent deforestation rates in both the periods 1995–2000 and 2000–5. From the point of view of both protecting native biodiversity and sustaining the still heavily forestry-dependent economy, if a U-curve recovery of trees is on the way in Ghana, it had probably better happen soon.

This latter problem, the unevenness of forest transition outcomes between countries, might be considered from a few different angles. In part, it may be necessary to conclude that forest transition is inevitably uneven and that the peculiar geographies and histories of some nations and forests make it more likely in some places than in others. There are significant differences between the forests of the Amazon, for example, where regrowth can be rapid and where agriculture is seasonally limited, and those of more arid Mexico, where labor migration and household resource demands are very different, and might continue to put stress on forest cover, even throughout economic transition (Stone 1974). We might conclude, therefore, that some countries are easier places for forests to recover than others.

It might also be argued, on the other hand, that this very unevenness is a symptom of a larger problem reflected by tree-cover decline. Specifically, it might be argued that the decline of forests in one region is *linked* to their recovery in another. In this view, the expansion of the global economy actually facilitates the extraction of forests from poorer regions (sites of deforestation) to serve as raw materials and commodities for wealthier economies (sites of reforestation). This explanation of forest-cover change would also scrutinize the seizure and control of long-held community forest ecosystems in poorer places and the circulation of value coming from forests in a global market. Such an argument is rooted in political economy.

## Political Economy: Accumulation and Deforestation

You will recall from Chapter 7 that the history of the expansion of economic growth is frequently viewed as inherently problematic for the environment. From a political economy point of view, decline in tree cover is foremost a problem of the expansion of development, here understood specifically as capitalist agriculture.

For example, small farm producers in a tropical country may traditionally operate by keeping their land in a range of crops, including some for the market and others for subsistence, leaving some proportion of the land in long fallow, and some portion of the area under trees and tree crops. Some local forests might also be held as community land or resources, a typical situation around the world.

Deforestation may begin when a large company, working either legitimately or illegally with the help of government officials, claims large areas of this communally held land, which is cleared for the most efficient mode of cash crop production: plantation agriculture. Those who have lost land or supplemental resources in this seizure (**primitive accumulation**) face some difficult choices. First, they may remain on the land and attempt to compete with larger producers, who are selling products at lower prices. This will necessarily entail putting more land under cultivation (and so cutting trees on cropland). It may also require that the producers increase their available cash on hand in order to pay for increased inputs into their land, in the form of fertilizers or pesticides. That new income will either demand an increase in cultivated land (again leading to tree clearing) or the sale of labor on nearby plantation farms, turning producers into workers.

> **Primitive Accumulation** In Marxist thought, the direct appropriation by capitalists of natural resources or goods from communities that historically tend to hold them collectively, as, for example, where the common lands of Britain were enclosed by wealthy elites and the state in the 1700s

For both the large plantation operator and the small producers left on the land, moreover, the extensification of production (expanding the area under crops) and the

**Table 10.3**  Some key tropical exports, their recent price fluctuations, and their leading export countries. Most major tropical commodities are grown in forest frontiers of countries experiencing tropical deforestation

| Tropical commodity | Recent low price (2008–9) | Leading exporters | Deforestation rates 2000 to 2005 |
|---|---|---|---|
| Robusta coffee | $1.50 per kilo | Colombia | –0.1 |
| | | Brazil | –0.6 |
| Cacao | $1.93 per kilo | Indonesia | –2.0 |
| | | Ivory Coast | +0.1 |
| | | Ghana | –2.0 |
| Bananas | $0.82 per kilo | Ecuador | –1.7 |
| | | Costa Rica | +0.1 |
| | | Philippines | –0.1 |
| | | Colombia | –0.1 |

intensification of production (putting more capital inputs like pesticides into the land to gain higher yields) have ecological impacts, both in the form of immediate forest cover loss, as well as declines in soil conditions favoring historical tree species coverage. Both retard the return or recovery of trees and both result in the increased availability of these cash crops (say, bananas) on the market, leading to a decline in prices. As prices fall further in a commodity crisis, the need to continue to extend farming into more forest land and to intensify production using modern, forest-averse techniques becomes greater.

While a full-scale decline of the agrarian economy could – in theory – lead to abandonment of the land and a return of forest cover (a kind of forest transition through economic collapse), the reverse scenario is more typical or likely. Producers will continue to expand and intensify production to make up for losses accrued in falling markets, leading to poorer prices, more forest clearance, and more intensive mining of soils. From a political economy perspective, therefore, deforestation is a symptom of inevitable periodic crises in capitalist agriculture.

Such bouts of sustained deforestation and their association with the booms and busts of commodity prices are a basic and repeated part of the ecological history of the poorer parts of the world, especially the forested tropics. As shown in Table 10.3, exports like coffee, bananas, and cacao are grown in forested tropical countries, typically at the forest edge, where deforestation advances.

## Deforestation as uneven development

From a political economy point of view, the second feature of this development history is that it has at its core a more insidious global process. By privatizing historically collectively owned forest lands, by turning forests into land for commodity production, and by using both of these transformations to increase the global circulation of capital, there is a ten-

**Table 10.4**  The predominant banana export companies operating globally, their share of the market, and their headquartered locations

| Banana company | Est. % of global market | Headquarters and financing |
| --- | --- | --- |
| Dole Food Co. (formerly Standard Fruit) | 25 | USA |
| Chiquita Brands International (formerly United Fruit Company) | 25 | USA |
| Fresh Del Monte Produce | 15 | USA (Chilean-based IAT Group w/capital held in the United Arab Emirates) |
| Exportadora Bananera Noboa (Bonita brand) | 9 | Ecuador (Grupo Noboa) |
| Fyffes | 7 | Ireland |

*Source*: BananaLink (http://www.bananalink.org.uk/content/view/61/21/lang,en/)

dency for the value from agricultural development in these forested areas to accumulate in faraway places. Specifically, the companies that own the coffee and banana plantations that displace forest cover, as well as the companies that buy, process, and resell these commodities, are typically headquartered in distant countries and financed by banks and firms far from the site of forest-cover decline. They are also highly concentrated and extremely few in number. In banana production and export, for example, five companies control the entire global trade in the commodity and are typically headquartered and financed in the United States, Europe, and elsewhere (Table 10.4).

The examples of tropical commodities like bananas and coffee are ones that most immediately come to mind when considering tropical deforestation, but they are by no means the only such crops. Soybeans present a perhaps even more spectacular example of the problem. This crop can be grown in a range of climates and is a dominant cash crop in North America and China among other locations. Yet the increasing amount of land cover given over to this crop in South America is astounding, with more than 20 million hectares of land under soybeans today in Brazil, for example, a country that grew no soybeans in 1960. The expansion of these cash crops, all processed by corporations in the United States and Europe, commonly comes at the continued expense of tree cover.

The luxury to allow and subsidize the regrowth of forests in places like Europe and North America, therefore, depends on economic growth and affluence, created from investments and profits made in deforesting tropical countries. In this sense, the forest transition in developed countries *depends upon* forest-cover loss somewhere else. The rise of forests in Europe is predicated on their continued decline elsewhere in the world. The political economy approach therefore may acknowledge a forest transition, but views it as one where forest cover moves from locations of exploitation to those of accumulation. The crisis of deforestation is not solved by economic growth in such an account, instead it is *simply moved around.*

**Second Contradiction of Capitalism**
In Marxist thought, this describes the tendency for capitalism to eventually undermine the environmental conditions for its own perpetuation, through degradation of natural resources or damage to the health of workers, etc., predicted to eventually lead to environmentalist and workers' movements to resist capitalism, leading to a new form of economy. Compare to the first contradiction of capitalism

**Social Reproduction**  That part of the economy, especially including household work, that depends on unremunerated labor, but without which the more formal cash economy would suffer and collapse

## Resisting deforestation

It is possible, following political economic thinking, that the declining yields and poor economic conditions of local workers and producers that follow in the wake of such extractive markets might lead to some form of organization on their part to demand amelioration of environmental destruction. Such political action is understood to emerge from the **second contradiction** inherent in capitalist agriculture: that expanding capitalist production leads inevitably to a decline in the environmental system's capacity to reproduce itself and the economy. In other words, the success of economic development both depends on a decline in forest cover and creates conditions that make its recovery less, rather than more, likely. The only avenue for forest recovery is through social and political action that develops from the failures and adverse effects of economic growth.

Such forest activism has been witnessed in places around the world. Consider the example of the "Chipko" movement which started in the Indian region of Uttaranchal in the early 1970s. Here, rather than allow government-sanctioned contractors to clear their forest land for profit, local villagers, especially women, wrapped themselves around trees (Chipko roughly translates as "stick to" in Hindi) marked for destruction. This paved the way for more widespread political action curbing contractor and state abuses of traditional community lands. This movement, it should be noted, is modeled on a nearly mythical prior event hundreds of years earlier, in which local villagers (largely women) from the Bishnoi caste in Rajasthan, known for their protection of the environment, attached themselves to trees and fell to the slaughter of ax-men who had come to clear their forest for the construction of a ruler's palace. In these cases, the solution to threatened deforestation notably derives from the activity of women. This reflects the tendency (noted in Chapter 7) for political economic activity and resistance to often be gendered, insofar as household activities differ between men and women, and the problems of **social reproduction** of households are commonly borne by women.

In another emblematic case from Brazil, government road-building efforts through the forest, tied to extractive logging interests, were opposed by the National Council of Rubber Tappers, a union formed by traditional, low-impact, forest users. The long war waged by the government on the rubber tappers and other local groups of small producers and indigenous communities led to wide-scale violence. In 1988 Chico Mendes, the leader of the rubber tappers, was murdered. In the wake of these events, extractive reserves were created throughout the Amazon to protect, at least in part, the rights of local residents and forest users.

From a political economy point of view, therefore, waiting for markets and economics to allow or encourage landscape transitions is folly. Instead, it would be argued, only direct political action in defense of local economies and livelihoods succeeds, as the cases of the Chipko movement, the rubber tappers, and many others suggest.

## Ethics, Justice, and Equity: Should Trees Have Standing?

**Ecocentrism** An environmental ethical stance that argues that ecological concerns should, over and above human priorities, be central to decisions about right and wrong action (compare to anthropocentrism)

As noted in Chapter 7, however, the insistence of a political economy approach on interpreting society's relationship to the environment as solely rooted in human economic interests, has some shortcomings. How might we explain the expanding environmental consciousness of the global public, including not only marginal communities but global elites? And to the degree that people in part hold **ecocentric** values (Chapter 5), how can these be extended to forest protection, management decisions, and more socially and ecologically desirable outcomes? How might we use ethical and legal tools *here and now* to rework society's relationship with trees?

This question has nowhere been more directly addressed than by legal scholar Christopher Stone (1974), who asks whether trees should have actual legal rights and standing in court. He argues that the history of Western enlightenment and common law legal traditions is one in which things initially without rights are progressively granted such rights over time. For example, under Western law for hundreds of years, children did not have rights. They could be legally punished, worked, or married off at the discretion of the patriarch. Much the same can be said of a number of people and entities that enjoy legal rights now, including women in traditional patriarchy, black people under slavery, or non-landholders under feudalism. It would seem absurd to us now that these people would not have the same legal rights as others (specifically, as adult, white, landholding men), but for long periods this was the case. Slowly, Stone argues, legal tradition expanded the number of people and things recognized as having rights. Most dramatically, in most contemporary legal systems around the world, corporations actually now hold the legal standing and rights of individuals (though arguably few of the responsibilities).

### What is it for something to have rights?

When Stone describes having "rights," he means it in the formal legal sense. This means, first, that if someone has a specific right (to vote, for example), some kind of public authoritative body (like a court) must be obligated to review any actions that might impinge on that person's right. Second, that person should have the ability to take legal actions (or for a guardian to take legal actions on behalf of that person) to defend these rights, and that a court must take injury to that person into account in their considerations. Generally, where something does not hold a legal right, moreover, it is typically considered the property of someone that does hold rights.

So as currently configured, trees definitely do not have rights. It is not in the interest of a tree to be cut down, for example, but no review is required for doing so as long as that action is taken by its owner. It might be the case that tree-cutting is reviewed by a court if it violates some kind of law (like the Endangered Species Act of the United States in the case where the tree was providing a place for a rare bird to live), but the review is not in defense of the right of the tree as an individual.

The same is true for natural beings and objects of all kinds. You cannot legally pollute a stream, for example, but the stream has no right to contest pollution of itself. Only people nearby who drink or use the water have that right. Moreover, if someone has to pay money damages for polluting a stream, they pay them to the people who live downstream, not to *the stream itself.*

## What would the rights of trees look like?

Stone is quick to point out that a legal right for a tree would not make it universally illegal to cut down trees. After all, he insists, people have rights but they cannot exercise them everywhere all the time. Sometimes the rights of individuals are "executed" and constrained. Trees need not be exempt from removal or any form of harm. And not all rights apply to all things or people. Corporations cannot plead the Fifth Amendment, for example. Trees need not have the right to vote, therefore, or any number of other specific and inappropriate rights.

What it would mean instead, Stone insists, is that certain specific rights of natural objects might be outlined, based on current law. Guardians, or representatives, would be legally established to have the obligation and opportunity to take grievances against those impinging on the specific right of that object directly to court for review. This is already done for children and for those deemed not fully capable to represent their own interests in court (a corporation is so represented, as is an invalid).

Some might argue that trees cannot communicate their needs or that lands already have guardians in the form of state agencies. On the first count, Stone points out, guardians make their best appraisals of the needs of those they legally represent all the time. Second, government guardians do not represent the interests of the beings in question, but instead represent the state itself. Countless examples of contradictions between the interests of these two things exist. In sum, according to Stone there is nothing legal, practical, or ethically problematic about the extension of rights to trees, fish, oceans, rivers, or anything else.

It should be pointed out that this is an approach totally unlike a market perspective on the question of forests. A market approach to ameliorating deforestation might encourage reforestation by assessing the costs of deforestation to people and perhaps providing economic incentives to discourage forest-cutting. In a rights-based scenario, on the other hand, the damages of deforestation *to trees themselves* would have to be weighed and evaluated by a court in the calculation of harm.

So too, it differs dramatically from a political economy approach to deforestation. Such an approach would stress the unjust actions of large corporate and state interests, for example, that cause value to flow away from local producers. It might also perhaps point to the decline in the productive value of the forest in question, and call for direct political action in defense of the rights of local people. A rights-of-nature approach, conversely, would force the acknowledgment of the *exploitation of trees themselves* and mobilize legal mechanisms to protect both local people and trees themselves against depredation.

In a fundamental way, this is the approach that most acknowledges that forests are constituted by trees, living things with interests upon whom deforestation acts. As counterintuitive as it is from **anthropocentric** ways of thinking, it is one that perhaps has the best view of the trees through the forest.

> **Anthropocentrism** An ethical standpoint that views humans as the central factor in considerations of right and wrong action in and toward nature (compare to ecocentrism)

## The Tree Puzzle

In this chapter we have learned that:

- The extent and composition of global tree cover has varied tremendously across global and human history.
- Humanity's complex and deeply emotional relationship to trees and forests has historically enabled the destruction of forests in the past but also a sometimes overly single-minded fixation on forest recovery.
- Human activities in recent centuries have clearly been linked to a significant overall decline in forest cover.
- Forest cover is actually increasing in some areas and data tracking forest-cover change are not foolproof.
- The dynamics of forest ecology raise questions about the role of human beings in protecting forests, especially when complicated by the likely effects of global climate change.
- Market approaches to the problem of deforestation stress the forest transition theory and the recovery of tree cover through economic growth and development.
- Political economy approaches to deforestation explain tree-cover decline as a result of the flow of value from forests, especially through commodity production.
- Political economy approaches stress that deforestation can only be opposed through direct political action by producers and affected local populations.
- Ethics-based approaches to the problem might employ an extension of legal rights to natural objects to address rampant destruction of forests and trees.

This brings us back to the grove of oaks in Berkeley. Despite the heroic efforts of the protesters and the legal efforts of activist groups, these trees no longer exist. Could a forest transition occur in North American cities (like Berkeley) through some form of valuation of trees? Did the protest here fail because it was not explicit enough and sufficiently directed towards the more fundamental political and economic structural constraints that directed priorities towards athletic facilities rather than trees? Was legal action insufficient here because the rights of trees themselves cannot be considered in a court under current legal precedent?

Whatever the answer to these questions, the Berkeley protesters did know and prove something fundamental. Trees are one of the most persistent puzzles in the history of environment and society interactions. The traces of humanity on earth are most evident in the dramatic decline and recovery of forests. Getting between an ax and a tree is an extremely powerful way of getting someone's attention.

## Questions for Review

1.  Describe a forest passing from disturbance through succession to climax, as theorized (and idealized) by Frederic Clements.
2.  What management implications arose from Clements' theory of succession? How has contemporary ecological theory altered these notions?
3.  What, generally, does the forest transition theory predict? Should its predictions be received as unqualified good news?
4.  What does it mean to say that forest clearance in the Global South is an extension of colonial relations with the Global North?
5.  Does Christopher Stone's proposition establish a truly ecocentric basis for forest conservation and management? How so or why not?

## Suggested Reading

Hecht, S., and A. Cockburn (1989). *The Fate of the Forest: Developers, Destroyers and Defenders of the Amazon*. London: Verso.

Shiva, V. (1988). *Staying Alive: Women, Ecology and Development*. London: Zed Books.

Tudge, C. (2005). *The Tree*. New York: Three Rivers Press.

Vandermeer, J., and I. Perfecto (2005). *Breakfast of Biodiversity: The Truth about Rainforest Destruction* (2nd edn). Oakland, CA: Food First.

Williams, M. (2006). *Deforesting the Earth*. Chicago, IL: University of Chicago Press.

## Exercise: Trees and Institutions

In this chapter, we have reviewed how the puzzle of trees might be addressed by population, markets, political economy, and ethics. Is there room for an analysis or set of solutions to deforestation that relies on institutions, collective action, and common property theory? How are forests and deforestation a collective action problem? What would an institutional solution to forest-cover loss look like? What are the limits of viewing trees as common property?

# 11

# Wolves

*Credit*: Ron Hilton/Shutterstock.

## Chapter Menu

## Keywords

- Anthropocentrism
- Apex predators
- Background extinction rate
- Biodiversity
- Conservation biology
- Ecocentrism
- Ethics/ethical
- Extinction crisis
- Masculinity
- Maximum sustainable yield
- Natural resource management
- NEPA
- Niche
- Rewilding
- Stakeholders
- Sustainability
- Trophic levels

### January 12, 1995, Yellowstone National Park

The Roosevelt Arch is one of the most famous and recognizable icons of the National Park Service (Figure 11.1). Erected in 1903, early in Theodore Roosevelt's presidency, the arch marks the northern (and oldest) entrance to the world's first national park. In 1872, the US Congress established Yellowstone National Park, as the founding document stated, "for the benefit and enjoyment of the people." In the first 123 years of its existence, over a hundred million visitors have enjoyed its world-renowned geysers, waterfalls, and wildlife. Most visitors come to the park during the busy summer months. Winter mornings in Yellowstone, on the other hand, tend to be relatively quiet. The morning of January 12, 1995, however, was anything but. Hundreds of people lined the road near the Roosevelt Arch. Local and national press crews vied for the best camera angles, and scrambled to interview anyone who looked official. What was all the fuss about? What could draw all these people outside to brave the freezing weather?

The crowds were there to witness the return of one of Yellowstone's most famous residents. A converted horse trailer was the centerpiece of the celebrity motorcade. Within the trailer were eight wolves, each in a small cage. Their destination was the Lamar Valley, about 40 miles by road inside the park. Wolves had been eradicated from the region decades earlier, but now the federal government was taking the bold step of reintroducing them to the park. The motorcade passed under the arch to snapping camera shutters and a cheering crowd. Inside the park, the motorcade stopped so that Bruce Babbitt, Secretary of the US Department of the Interior (the agency that runs the Park Service), and Mollie Beattie, Director of the US Fish and Wildlife Service (the agency in charge of managing endangered species), could greet the crowd and make short speeches. Babbitt called it a "day of redemption and a day of hope" (Fischer 1995: 161). The mood was triumphant, but the good vibes were quickly dampened, at least for a few hours (Figure 11.1).

**Figure 11.1**   The wolf motorcade passes under the Roosevelt Arch as it enters Yellowstone National Park on January 12, 1995. Environmentalists, the press, and schoolchildren bussed in on field trips look on. *Credit*: National Park Service.

The night before, the Wyoming Farm Bureau (a group representing agricultural interests, including livestock raisers) had filed an emergency appeal in federal court to halt the reintroduction. Standard practice is for judges to issue "stays" (temporary halting of actions) to give themselves time to review the appeal. The judge initially called a 48-hour stay on the reintroduction, meaning that the wolves could not be released into the park for two days. Agents scrambled for a backup plan, as it would be dangerous to the wolves to leave them drugged and penned for that much time. The judge, as it turns out, did not need two days to decide whether to grant the appeal; early that same evening the stay was lifted. At about 10:30 p.m., the wolves were released from their pens into the Yellowstone snow. Environmentalists cheered. Yellowstone was, as many emotionally said, finally "whole" again. Ranchers and many other locals, however, greeted the event with mortification.

How and why, then, did it ever happen that wolves were intentionally eliminated from nearly their entire natural range, *even in national parks*? Moreover, what could explain both the frenzy that accompanied their return to Yellowstone and the deep feelings of the animal's opponents?

## A Short History of Wolves

The wolf (or the "gray wolf," more technically), *Canis lupus*, is the largest member of the Canidae, the dog family (Figure 11.2). Other canids include foxes, coyotes, jackals, African wild dogs, and domestic dogs. Indeed, domestic dogs are so genetically similar to wolves that they are now considered by most to be a subspecies of the wolf, *Canis lupus familiaris*; it is likely that all breeds of domestic dogs are biological descendants of wolves. Many subspecies of the gray wolf have been identified and

**Figure 11.2**   The gray wolf, *Canis lupus*. *Credit*: Al Parker Photography/Shutterstock.

include the critically endangered red wolf (*Canis lupus rufus*) of the Southeastern United States, the Mexican wolf (*Canis lupus baileyi*), and the Iranian wolf (*Canis lupus pallipes*).

The gray wolf evolved into something very close to its present form several hundred thousand years ago. Likely evolving in North America, through successive ice ages (during which a land/ice bridge opened up between North America and Eurasia) the wolf spread throughout most of North America and Eurasia. A very adaptable species, wolves occupy desert, grassland, forest, and tundra ecosystems. Adults range in size from around 45 pounds (the smallest subspecies is the Arabian wolf) to over 100 pounds (which would be a large northern gray wolf). Coat color varies widely, even within local populations, ranging from white-blond to every conceivable gray and brown to slate black. Most wolves live and hunt in packs and feed primarily on ungulates – large, geographically widespread, hoofed animals which include wild species such as elk, deer, antelope, and bison as well as domestic species such as cattle, sheep, and goats. Indeed, the ubiquity of domesticated ungulate species is one of the primary reasons that the wolf has for so long been an object of concern and fear for agricultural human societies.

Despite conflict with humans, wolves still exist over much of their historic range (Figure 11.3 and Table 11.1). Eurasian wolves live as far east as Eastern Siberia (Russian Federation), Mongolia, and parts of China (and at least two subspecies of wolves inhabited Japan until

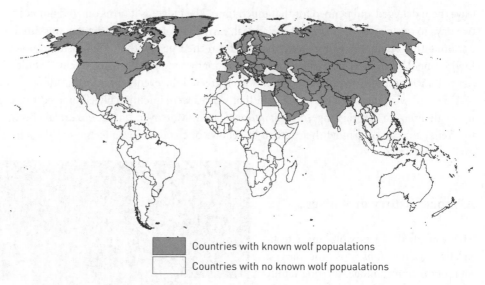

Countries with known wolf popualations

Countries with no known wolf popualations

**Figure 11.3**  World map of countries with known wolf populations. *Sources*: 1. Mech and Boitani (2003), pp. 322–3. Chicago: University of Chicago. 2. International Wolf Center: http://www.wolf.org/.

**Table 11.1**  Countries worldwide with wolf populations of 2,000 or greater

| Country | Wolf population |
|---|---|
| Canada | 52,000–60,000 |
| Kazakhstan | 30,000 |
| Russia | 25,000–30,000 |
| Mongolia | 10,000–20,000 |
| China | 12,500 |
| United States | 9,000 |
| Kyrgyzstan | 4,000 |
| Tajikistan | 3,000 |
| Belarus | 2,000–2,500 |
| Spain | 2,000 |
| Ukraine | 2,000 |
| Uzbekistan | 2,000 |

*Source*: Mech and Boitani (2003)

around the turn of the twentieth century). The largest Eurasian populations exist across Siberia southwest through the Central Asian Republics. Though critically endangered across the southern portion of their range, wild wolves still persist from Bangladesh across the Indian subcontinent, west through the Arabian Peninsula (where very little is known about their distribution and numbers), even into Africa, where a small population of perhaps 50 individuals survives in Egypt. Perhaps surprisingly, wolves still roam most of continental Europe, but they have been extinct on the British Isles for several hundred years. In North America, both Alaska and Canada have substantial populations, even though they were eliminated from nearly half of their North American range after the arrival of European settlement.

## The social success of wolves

Wolves live in packs, extended families of about five to eight individuals. Larger packs have been witnessed but are unusual. Usually, only one female per year per pack gets pregnant after copulating with one male member of the pack. Litters commonly consist of six to eight pups, but litters as small as one and as large as 14 have been noted by field biologists. Wolf pups – blind and deaf and weighing about one pound – are born in dens that shelter them from predators and the elements. Pups are raised by the entire pack, but with the most attention coming from their biological parents. They are nursed exclusively for the first couple of weeks, after which pack members feed them regurgitated meat. Within a few months, pups are eating raw meat and their social activities have broadened beyond play. Within the first year, most are hunting with the pack. Young wolves will live as sub-ordinate members within the pack for between one and four and a half years.

The social structure of the wolf pack is intimately tied to every facet of the pack's survival and reproduction. At least three hierarchies exist in every wolf pack: one for the females, one for the males, and a third non-gender-specific hierarchy that varies from season to season. The lead male and the lead female – often called the "alpha" wolves – are usually, but not always, the parents of the year's pups. Socially structured activities include raising young, play, territory defense, and hunting. Sound familiar? There exists a large literature on the parallels between wolf and human sociality.

The social structure of wolves has (like ours) proven to be evolutionarily successful. Eventually, young wolves "disperse," meaning they leave their packs. Lone wolves that find a mate and available territory will mate and, if successful, establish a new pack. In Isle Royale National Park (an island in Lake Superior) a pair of young wolves colonized some unclaimed territory and successfully raised seven pups *in the first year*! While this 450 percent increase was unusually high, wolves have very high reproductive rates. Highest rates occur when recolonizing vacant territory (e.g., Yellowstone in the first few years after reintroduction) or in years following a significant reduction in population size (whether due to human action or by natural mortality factors such as disease). Annual population increases of 30–50 percent are common, meaning it is not unusual for a wolf population to double its numbers within two years (Terborgh et al. 1999).

## The ecological role of the wolf

Ecosystems – geographical areas inhabited by assemblages of plants and animals interacting with their physical environment – contain multiple **trophic levels** which comprise the local food webs. Over-simply, these map out a food chain in which photosynthesizing plants are consumed by herbivores, which are in turn consumed by

**Trophic Levels** Parallel levels of energy assimilation and transfer within ecological food webs; in terrestrial ecosystems, photosynthetic plants form the base trophic level, followed "up" the web by herbivores and successive levels of carnivores

predators. Wolves exist as **apex predators** (also commonly called "top carnivores") in that they have no natural, regular predators. Most ecosystems will have multiple apex predators. For example, in the Northern Rocky Mountains wolves, grizzly bears, lynx, wolverines and golden eagles are all apex predators.

The **biodiversity** of a given ecosystem – the number of different species – is determined by a number of abiotic (non-biological) and biotic (biological) factors. Abiotic factors include availability of water, wildfires, storms, harsh winters, and the like. Biotic factors include limiting factors such as competition, disease, parasitism, and predation, as well as availability of food and habitat. Another way of classifying processes that affect species distribution and abundance within ecosystems is by dividing biotic factors into "bottom-up" and "top-down" processes (Mech and Boitani 2003: 41). A bottom-up process would be one sparked at the bottom (or at least a low trophic level) in a food chain. If, for example, the distribution of a native plant species in an ecosystem is dramatically reduced (perhaps by an introduced non-native species) and if this reduction causes a decline in an herbivorous species that depends on these plants, this might lead to a decline in predators, who rely on these herbivores as their prey. Such controls on predator populations are *bottom-up*, in that it is the limit of the resource base that impacts the consumer and predator population. As such, bottom-up effects can change the species makeup, biodiversity, and relative stability of an ecosystem.

Species distribution and abundance within ecosystems is also regulated by *top-down* processes. *Absence* of predators, for example, has effects on the distribution, numbers, and the behavior of prey species. If an apex predator is removed from an ecosystem, as wolves have been in countless ecosystems worldwide, their prey will increase in numbers (at least temporarily) as fewer succumb to predation. But the effects go beyond just the reduction or elimination of direct predation. Prey species freed from predation pressure spend less time hiding and burrowing down, therefore freeing up more time and energy for reproducing.

So what is wrong, you might ask, with more deer or more elk in an ecosystem in which wolves have been eliminated? Like bottom-up processes, top-down processes do not just affect the next trophic level down the food web, but rather cascade down through multiple levels. This is where we begin to see the significance of the ecological role (what ecologists call the **niche**) of the wolf within an ecosystem. In our current example, if wolves are eliminated and deer and elk (their primary prey in some ecosystems) increase their numbers, this has an effect on the next trophic level down. As deer and elk, both herbivores, increase in number, the local vegetation is affected.

Yellowstone National Park, site of a relatively recent elimination of wolves, followed several decades later by their reintroduction, provides a real-world laboratory of such top-down effects. Biologist Douglas Smith referred to Yellowstone after reintroduction as "the best chance we've ever had to figure out the difference between a wilderness system with wolves versus one without wolves" (Smith and Ferguson 2005: 118).

So what changes did the wolves cause after they were reintroduced in 1995? One of the most prominent is the rapid recovery of willows. With elk (who feed on willows) more constantly on the move since wolf reintroduction, willows are not fed on as intensively. The renewal of willows, moreover, has led to the reemergence of beavers, since beavers use willows for dams and lodges. These dams, once restored, lead to a proliferation of reptiles and amphibians. Increases in eagles, magpies, and ravens have been noticed due to the wolf-kill carcasses left scattered about their feeding areas as well, with carcasses finally providing food for invertebrate species like beetles and decomposing in the soil, becoming food for detritus feeders and eventually becoming soil nutrients. The bottom line is simple: the Yellowstone ecosystem is a different place with wolves. So how was it that an enormous national park like Yellowstone came to be bereft of wolves?

## Three centuries of slaughter: Wolf eradication in the United States

When European Americans settled the New World, wolves roamed across nearly the entire continent. Only parts of present-day California and possibly portions of the Great Basin were without wolves. Their numbers were as impressive as their extent, with probably at least 350,000 wolves inhabiting the American West alone (Barclay 2002). By 1958, they had been extirpated from nearly all of the contiguous United States and Mexico (see Figure 11.4). So what, exactly, can explain the demise of such an intelligent, adaptable animal that existed in such abundance over such a large area? For over three centuries, wolves were subject to a systematic, and ultimately successful, extermination campaign.

Wolf bounties – moneys paid by the government to citizens who kill wolves – are one of the oldest forms of tax expenditures in American history. Wolf bounties in the American

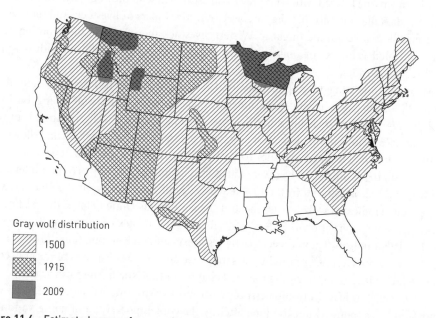

Gray wolf distribution
- 1500
- 1915
- 2009

**Figure 11.4** Estimated range of gray wolves in the contiguous United States, 1500, 1915, and 2009. *Sources*: Map is original, adapted from the data from the following sources: Current (2009) distribution: USFWS: www.fws.gov/midwest/Wolf/archives/2006pr_dl/2006pr_dlsum.htm and Sightline.org: www.sightline.org/maps/maps/Wildlife-Wolf-CS06m/?searchterm=None; 1915 Distribution: Young and Goldman (1944), p. 58; 1500 Distribution: Mech (1970), p. 31.

colonies go back at least as far as 1630, five years before the first public school was opened in Boston. The share of revenue dedicated to wolf eradication is impressive. In Chester County, Pennsylvania, for example, 28 percent of tax revenue was spent paying out wolf bounties. Dozens of other counties and municipalities across the colonies spent 10 percent or more of their budget on wolf bounties into the eighteenth century (Fischer 1995). Killing methods ranged from individuals shooting wolves on sight, to organized group hunts, to "wolf pits": holes dug in the ground over which baits were placed onto a collapsible false floor. The campaign was successful. By 1700 wolves were extinct throughout most of New England and by 1800 they were absent east of the Appalachian Mountains. In the nineteenth century, Midwestern states followed the colonial model, establishing bounties and quickly and efficiently eradicating their wolf populations (with the exception being the wolves that survived near the Canadian border in the North Woods of Minnesota, Wisconsin, and Michigan).

After the Civil War, wolf eradication in the Great Plains coincided with the well-known history of the incarceration of Native Americans on reservations. Conquering the Indians required destroying the bison, the primary subsistence base for many indigenous nations. During the first few decades after the Civil War, "buffalo hunters" killed literally tens of millions of bison. The systematic elimination of wolves soon followed. The last three decades of the nineteenth century indeed represents the period where the largest numbers of wolves were killed in the shortest period of time.

As cattlemen moved west, their livestock served as the primary feedstock for the remnant wolf populations. After all, the ecosystem had been transformed, in a span of a couple of decades, from one littered with wild prey to one scarce in wild prey but rich with livestock. Wolves, adaptable animals that they are, were effective killers of livestock, and represented the single largest economic threat to Western livestock men. Despite the shooting, trapping, and increasing use of poisons, straggler populations and crafty individual wolves persisted for decades.

Many of these straggler populations lived on federal lands. In the first decade of the twentieth century, the federal government officially entered the wolf-killing business. Wolves preyed on livestock, but additionally, game managers felt wolf predation kept wild game species below optimal harvest levels. The last confirmed wolves in Yellowstone National Park were trapped and shot by park rangers in 1926 (Foreman 2004).

All in all, thousands of wolves were killed in the United States by private hunters, cattlemen, and federal officers in the first few decades of the century. In the 1950s, having successfully eradicated wolves from most of the contiguous United States, the eradication campaign went international. Federal officials traveled to Mexico, bringing "Compound 1080," a lethal poison that had been partly responsible for the elimination of some of the last wolves in the West. Within eight years the Mexican wolf (a subspecies of the gray wolf) had been eliminated, representing the final chapter in this North American saga.

Fortunately (at least from the perspective of most environmentalists), the epic of wolf eradication was followed not too long after by the contemporary era of conservation. By the 1940s, nearing the end of their systematic elimination, wolves were slowly gaining advocates within the scientific and conservation communities. Prominent ecologists were penning journal articles, and scientific and popular books were fitting the wolf within ecology. Wolves began to appear not as things that threatened the stability of nature, but instead as part of larger ecological processes and networks. By the end of the 1960s, the modern environmental movement was in full swing, and efforts to conserve or reestablish historically persecuted species were at the forefront of activists' concerns (the slogan "Save the Whales!!" was practically synonymous with environmentalism in its early era). Key pieces of landmark legislation would give legal teeth to this popular trend. CITES, the Convention on International Trade in Endangered Species, was drafted by members of the World Conservation Union in the 1960s and signed in 1973 (see the CITES website at www.cites.org). As its name implies, CITES bans or highly regulates the international trade in plant and animal species either currently threatened with extinction, or whose unregulated international trade might lead to endangerment. The wolves of India, Pakistan, Nepal, and Bhutan are so-called "Appendix I" species, meaning they are given the strictest level of protection under CITES. All other wolves are "Appendix II" species, meaning the international trade in these animals or their parts (e.g., wolf pelts) is monitored and strictly regulated.

In 1973, the United States passed the Endangered Species Act (ESA), one of the most far-reaching and ambitious national environmental laws ever established (Box 5.1, "The Endangered Species Act"). All existing populations of wolves outside of Alaska were included in the original list of endangered species covered by the ESA. Powerful environmental groups like Defenders of Wildlife and the National Wildlife Federation pushed for aggressive programs to reestablish their presence throughout large parts of their former range (websites at: www.defenders.org and www.nwf.org), an effort that ends in our opening story: Thursday, January 12, 1995, when eight wolves were released into the snow of Yellowstone National Park.

## The puzzle of wolves

This brief history of wolves provides insight into many facets of the relationship between nature and society. For one, it highlights that the relationship of humans to both wild and domesticated animals is quite complicated. Dogs, after all, are the same biological species as wolves, yet our relationships with each are different. Beyond this, these relationships change over time and across cultures. Wolves represent how particular "pieces" of non-human nature can be powerful symbols of broader aspects of the human–nature relationship. Moreover, the history of eradication and conservation of wolves shows how the very fate of the species is tightly tied to its cultural symbolic value. Finally, wolves serve as a humbling reminder of the difficulties of trying to live in the world without destroying its wildness.

Specifically, wolves are puzzles in that they force us to ask the following questions:

- Exploring our sometimes divisive relationships to one another – centered on wolves – how do we assert environmental priorities and govern our behavior in a way that is fair and democratic?
- Critically assessing our present relationship with wolves, where can we make room for wolves in a human-dominated world, and *how*?

## Box 11.1   Predator-Friendly Beef and Wool

In the opening chapter, we discussed our preference for forms of **reconciliation ecology**, whereby new alliances forge innovative and more sustainable models of nature–society relationships. One laudable recent example of such an alliance can be found in the Rocky Mountain Northwest, where livestock ranchers and conservationists are allying to simultaneously conserve native predators and maintain ranching as an economically viable livelihood. That this alliance even came about is notable, as ranchers and environmentalists have historically been rather adversarial parties. The source of a considerable amount of the tension between these two groups was the traditional "shoot on sight" mentality of many ranchers toward predators. Even though ranchers are far from solely responsible for the demise of predator populations worldwide, they nonetheless were vilified by many environmentalists, who felt that concern for the conservation of native predators should override economic hardships of livestock producers.

Ranchers and environmentalists in some places, however, have moved beyond their differences to work toward a shared goal, the preservation of open space. The common enemy is exurban development. In places near rapidly growing cities and towns like Fort Collins, Colorado and Livingston, Montana, enormous portions of open range are being swallowed up by subdivisions, roads, and shopping malls. Many environmentalists quickly realized that their best potential allies in the fight against this land-use transformation are livestock ranchers, the people who were working and living on the very lands being transformed into McMansions and parking lots. Environmentalist–rancher alliances have resulted in a substantial network of "conservation easements" in the region. In a typical easement, the rancher-landowner is granted some sort of tax break or other financial incentive in return for a guarantee that they will not sell and subdivide their land. The result is a win–win for both ranchers and environmentalists. For ranchers, the financial incentives help them remain economically viable as working livestock producers, a livelihood that has seen profit margins decline substantially over the past several decades. For environmentalists, the guaranteed preservation of open range (even with cattle) is far more desirable than the habitat loss and fragmentation that result from exurban development.

Building on the productive relationships developed in the conservation easement movement, some ranchers and environmentalists have taken their alliance even farther. One of the newer and more innovative alliances is the "predator friendly" certification program. The Bozeman, Montana-based conservation organization Keystone Conservation will certify beef and wool with their "predator friendly" stamp of approval if ranchers manage their herds using non-lethal methods. This might mean guard dogs instead of shotguns or poison, and it also might mean losing more livestock to predation. But ideally, the producers will be able to offset the higher costs of operation by selling to a customer base willing to pay a premium to know that their beef or wool was raised in a predator-friendly manner. Currently, the scale of the predator-friendly certification program is quite small, but it serves as a promising example of the creative efforts being made to find ways for humans to live on and work the land in more ecologically sustainable ways.

### Reference

Keystone Conservation website: www.keystoneconservation.us/, accessed March 20, 2009.

- And assessing our historical relationship with wolves, what can explain both our persecution and subsequent celebration of this species?

Wolves, therefore, present something of a window into humanity. When we look at wolves, we can see how our own science, industry, and emotion become entangled. If we can work toward solving the puzzle of wolves, by implication, we may come a bit closer to forging a more sustainable relationship with non-human nature.

## Ethics: Rewilding the Northeast

In Chapter 5 we established environmental problems as **ethical** dilemmas, problems of right versus wrong human actions toward nature. An ethics perspective is commonly applied to questions regarding the treatment of animals, both wild and domestic. Increasingly, however, ecological ethics look beyond just the treatment of individual animals to questions of right action toward species, entire ecosystems, and future generations of humans and non-humans. An ecological ethics perspective can have a good deal to say about solving the puzzle of wolf conservation.

**Ethics/Ethical** The branch of philosophy dealing with morality, or, questions of right and wrong human action in the world

**Sustainable/Sustainability** The conservation of land and resources so as to secure their availability to future generations

**Ecocentrism** An environmental ethical stance that argues that ecological concerns should, over and above human priorities, be central to decisions about right and wrong action (compare to anthropocentrism)

**Anthropocentrism** An ethical standpoint that views humans as the central factor in considerations of right and wrong action in and toward nature (compare to ecocentrism)

### Wanted: An ecocentric ethic of sustainability

An ethics of **sustainability** informs us that we have a moral obligation to sustain the quality and productivity of the environment for future generations. With this guidepost, we can critically assess current actions and ask, "Can such and such an action be continued more or less indefinitely? *Is it sustainable?*" If so, great! If not, then the *right* thing to do is to alter our actions so that they are sustainable. **Ecocentrism**, for its proponents, serves as a corrective to **anthropocentrism**. Seeing humans as part of nature, ecocentric ethics looks beyond narrow, human priorities to judge right and wrong actions within a broader, ecological context.

Even the deepest of ecologists, however, cannot really conform to a *solely* ecocentric ethic. We *must*, it seems, allow anthropocentric priorities to guide our environmental policy and planning, at least in part. For example, we need to divert some portion of the fresh water available in every ecosystem for use by humans. We should, in other words, simultaneously work to sustain the environmental conditions in which human communities can meet their needs.

There are some issues, however, that cannot be adequately judged through a strictly anthropocentric calculus. If you are a proponent of reintroducing wolves to New England or Scotland, for example, it is difficult to make your case arguing along solely anthropocentric lines. Granted, wolf reintroduction in these regions might spark an ecotourism boom that could help revitalize rural communities, as it has for a handful of communities around Yellowstone. This is a thin line to walk, however, as it is difficult to predict the long-term economic impact that reintroduced wolves would actually have. Besides, what if economic projections show marginal or even no economic benefit from reintroduced wolves? Is that reason *not* to proceed? Most wolf supporters would answer with an emphatic "no," and as such, many environmentalists argue that issues like species reintroductions should be guided by ecocentric concerns.

**Rewilding**   The restoration of natural ecological functioning and evolutionary processes to ecosystems; rewilding often requires the reintroduction or restoration of large predators to ecosystems

**Conservation Biology**   A branch of scientific biology dedicated to exploring and maintaining biodiversity and plant and animal species

**Extinction Crisis**   The current era of anthropogenically induced plant and animal extinction, estimated to be between one thousand and ten thousand times the historical average, or background extinction rate

**Background Extinction Rate**   Usually given in numbers of plant and animal species per year, the estimated average rate of extinction over long-term, geologic time, not counting mass extinction events

This ecocentric spirit is most evident in the "**rewilding**" movement, which argues exactly along these lines, based in the philosophy of deep ecology and the science of **conservation biology** (Foreman 2004). Deep ecology arms rewilding advocates with the philosophical tools to argue for an ethics of sustainability (*what to preserve, and why*) and conservation biology provides the scientific roadmap to get there (*what it will take to make it happen*).

## Rewilding, Part I: The ethical dimension

Every year, we lose a significant portion of the earth's biodiversity. The magnitude of the **extinction crisis** is quite alarming. It is estimated that the current extinction rate is somewhere between one thousand and ten thousand times the historical average or **background extinction rate**. There have been five historical major extinction events in the past half-billion years, the most famous of these also being the most recent. About 65 million years ago, dinosaurs and (less famously) many mollusks became extinct. As you likely learned in childhood, many scientists believe that this extinction event was kick-started by an asteroid impact. Today's extinction rate is so high that it ranks on the same level. The difference between those that came before and the current crisis is that today's extinction crisis is an anthropogenic (human-caused) event.

Anthropogenic causes of the current extinction crisis include: fragmentation of habitats, loss of ecological processes, invasions of exotic species, air and water pollution, and climate change (Wilson 1993). Concomitant with the extinction crisis, it is argued, is the impoverishment of genetic diversity upon which adaptation and evolution rest. The broader significance of the extinction crisis, therefore, lies in its potential irreversibility and longevity. Take tropical deforestation, another justifiably high-profile current environmental issue, for comparison (see Chapter 10). Even a tropical rainforest can rejuvenate itself given a century or two. In other words, when we chop down a rainforest, we are "undoing" a century or two's worth of ecological work. Compare this, by contrast, to the possible elimination of 50 percent (or more) of all living species on the planet, as the eminent biologist E. O. Wilson (2002) believes could happen *within this century*. Add on to that a massive slowdown of evolution and we have (some argue) something far more worrisome than a razed forest or depleted fishery. Rather, what we have is the undoing of tens of millions of years of the planet's biological heritage.

Considered in this light, it is no surprise that many holding a consciously ecocentric worldview place this issue – reversing the extinction crisis and renewing evolutionary processes – atop their list of environmental priorities. So what does this all have to do with wolves? Many conservation biologists argue that we should rewild the landscape to enable the rejuvenation of evolutionary processes. The reintroduction of apex predators to portions of their former ranges is central to the project of rewilding.

## Rewilding, Part II: How to get from here to there

Conservation biology is the scientific study of the biodiversity crisis, as well as a largely activist discipline, meaning that conservation biologists (generally, though some still resist) actively work toward promoting changes that they believe would help stem the biodiversity crisis. Many North American conservation biologists have been actively promoting the rewilding of the continent. Rewilding advocates argue that "vast expanses of self-willed terrain [should] be protected and recovered" (Barlow 1999: 54). These advocates argue that evolution cannot be sustained (or rejuvenated) without the important top-down ecological functions of large predators (as discussed above).

In North America, supporters of rewilding quickly realized that if their broad goal of a continental (or even global) scale renewal of biodiversity and evolution was to become a reality, areas outside of official wilderness designation required special attention (Sessions 2001). Indeed, there are many areas outside of formal wilderness where one might have "free nature," as Arne Naess, the founder of deep ecology, refers to it. Zones of free nature would be areas of sparse enough human habitation and intact enough natural ecosystems where the full complement of natural ecological-evolutionary processes could be restored and maintained (Nie 2003). Wolves would have to be reintroduced and managed on areas beyond government land, where people live, ranch, farm, and play.

## Wary of the wild: Deep ecology and democracy

Which leaves us with "only" one small problem: How, exactly, do we manage these wolves once they are reintroduced to places where people live? Deep ecology may offer up eco-centrism as an appealing corrective to anthropocentric ethics and their attendant ecological destruction. Beyond giving us rather broad guideposts, however, deep ecology provides little help as to *how to make this happen all the while maintaining democratic forms of decision-making.* Some critics have even gone so far as to see deep ecology as leading to an *inherently authoritarian (or "eco-authoritarian") politics.* The potential of individual "self-realization" notwithstanding, when it comes down to managing real natures in real places, most proponents of deep ecology invariably fall back on *scientific* ecology as the final word for decision-making. Science, however, (ecology or otherwise) is an elite form of knowledge production, inaccessible to the uninitiated. Granting scientific ecologists sole responsibility for all management considerations regarding reintroduced wolves would be giving them literal authoritarian power; that is, decision-making control not subject to the checks and balances that constitute a democracy. This is at least a good part of the reason why proponents of, for example, "radical democracy" (a social movement dedicated first and foremost to maintaining decision-making power at all strata of society) are often wary of calls for science (solely) guiding politics.

Happily, deciding "between" ecology and democracy is a false dilemma. It is possible to have our wild nature and our democracy, too. One of the more promising forms of such win–win ecologies arose out of this very puzzle, wolf management, in the North Woods of Minnesota.

## Institutions: Stakeholder Management

Wolves are unusual, though not unique, as *resources*. If we define resources, typically, as "things useful to humans," the question arises: *For whom are wolves useful? How so, and why?* For certain, they are not (at least not legally) useful in the typical "consumptive" sense. That is, we do not remove them from "nature" for some or another use (such as we might pick blueberries or capture tropical fish to put in a home aquarium). Wolves' *value* is harder to pin down. In line with the previous section, wolves have value in helping to restore ecosystem functioning. Wolves also have symbolic value. Their presence makes many of us feel better about the world, that at least some corners of the world are more wild than tame. (On the other hand, one might more cynically look at the multi-million-dollar budget of Defenders of Wildlife, whose symbol is the wolf, and say "yeah, symbolic value: wolves sell t-shirts.")

The question of the value of wolves can also, however, be turned on its head. The presence of wolves, for some, is objectively more bad than good. Livestock owners are the most obvious example. Wolves kill livestock. With wolves around, these people stand to lose money in an already economically marginal livelihood. There are also those who perceive the presence of wolves to mean lost game for hunters and lost clients for paid hunting guides.

There are, in sum, people who have different reasons to *positively or negatively* value wolves. This is part of what makes wolf management so vexing. It is not a matter of assessing how to divvy up a commonly valuable resource, like a patch of forest or an ocean fishery. Yet wolves are still, in a way, common property resources. In Chapter 4, we reviewed the management of commons as problems often successfully solved through *institutions*, that is, "systems ... leading to orderly and constrained use of natural resources." The puzzle of wolf management is no different. If wolves are to persist (at least without an authoritarian de-peopling of the landscape), environmentalists must *constrain* their "use" of the animals (i.e., they cannot be everywhere; their populations cannot expand indefinitely). Hunters and livestock owners, just the same, must constrain their traditional treatment of the wolf (i.e., shoot on sight). As such, people are experimenting with new institutional structures.

Of course, wolf management is very complex; it is not a simple two-group affair, neatly divided into proponents and opponents, each unified and speaking with a single voice. It is, rather, a diverse continuum of groups, each containing diverse voices and viewpoints. The best hope for wolves, some say, is to get these so-called "**stakeholders**" together, and have them craft rules and responsibilities collectively.

**Stakeholders**  Individuals or groups with a vested interest in the outcome of disputed actions

**Natural Resource Management**  Both the academic discipline and professional field dedicated to the management of environmental conditions, goods, or services for social goals, which may range between instrumental human utility to ecological sustainability

### Public participation in resource management

**Natural resource management** in the United States has been, for most of the modern era, a highly centralized, "expert" affair. Up through the 1960s, professionally trained range managers, wildlife managers, and foresters "scientifically" managed their resource stocks for **maximum sustainable yield**, but with little or no input from the lay public. By the 1960s many resource professionals were thinking *ecologically*, integrating assessments of ecological connections and eco-system health into their attempts to maximize resource yields. But even as management emphases evolved, the public (including the growing body of environmental activists) remained, for the most part, left out of the decision-making process.

> **Maximum Sustainable Yield**  The largest seasonal or annual amount of any particular natural resource (e.g., timber, fish) that can be harvested indefinitely
>
> **NEPA**  The National Environmental Policy Act of 1970 commits the US government to protecting and improving the natural environment; after NEPA, the federal government is required to write environmental impact statements (EIS) for government actions that have significant environmental impact

The seeds of a more participatory management structure were sown in 1970 with the passage of the National Environmental Policy Act (**NEPA**), which among other innovations required "environmental impact statements" (EIS) for any government action. Further, these statements required the inclusion of *public input and participation* in the process. The era of public involvement in environmental management had begun (Andrews 1999).

Many endangered species conservation plans, wolf reintroduction in Yellowstone included, must fulfill the full scope of the EIS process, including public participation. The arrival of wolves to Yellowstone suggests, however, that although reintroduction may have heralded a shift in management priorities (from eradication to conservation), it hardly marked a new era in management *style*. Recall the opening story of the chapter: Bruce Babbitt and Mollie Beattie – two high-ranking federal officials – literally carrying the cage of the first wolf to be released from the truck and into the snow. For some, this was a powerful (and negative) symbolic gesture: wolf reintroduction was a top-down, federally imposed action. As such, for many residents of Montana and Idaho, wolf reintroduction was perceived as business-as-usual: East Coast priorities imposed in their backyards. This was so much the case that some environmentalists feel reintroduction may have done as much harm as good, further polarizing opponents against proponents of environmental protection in the region (Fischer 1995). It was not until a few years later that a different model, one that moved from public *input* to a more truly *participatory* structure, would be tested in the highly volatile context of wolf conservation.

## Stakeholders in Minnesota wolf conservation

In 1998, the state of Minnesota began implementing the first *stakeholder* model wolf man-agement plan in the United States. Home to about 2,500 wolves, Minnesota was given more leeway than other states by the federal government because it had the largest and most stable population of wolves.

The process began with 12 informational meetings held throughout the state, in which attendees were informed of the state's plan, and additionally could provide comments and recommendations on the process. After the 12 meetings, a wolf management "roundtable" was convened. The roundtable included representatives from environmental groups (e.g., the Sierra Club), agricultural interests (e.g., the Minnesota Farm Bureau), wolf groups (e.g., HOWL – Help Our Wolves Live), hunting and trapping groups (e.g., Minnesota Deer Hunters Association), and tribal interests (there are 11 Native American reservations in Minnesota). These were joined by regional and scientific representatives.

The roundtable had a specific mandate to develop a management plan for the state's wolves, formed entirely through *consensus*; *every* member of the roundtable had to back the plan or it was back to the drawing board. After several meetings, including a marathon 10-hour final session, the roundtable voted unanimously on a management plan. Highlights of the plan included:

- The state's wolf population would be managed to expand its numbers and range.
- Landowners could kill wolves caught depredating on livestock or pets.
- No wolf hunting or trapping seasons would be considered for at least five years.
- Livestock and pet owners would be compensated for losses.
- The state would encourage non-lethal methods of livestock protection.
- A stiff fine of $2,500 for illegal wolf kills would be implemented.

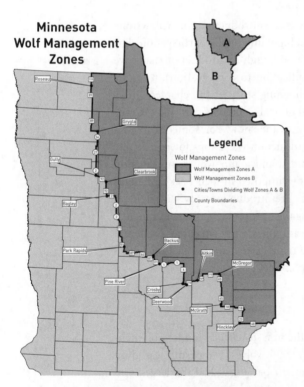

**Figure 11.5** Wolf Management Zones in Minnesota. In Zone "A," wolves are managed by the plan crafted by the stakeholder roundtable. This zone contains 90 percent of existing wolf range and 98 percent of the state's wolf population. Zone "B" is designated an "agricultural zone," where landowners and managers have considerably more flexibility in the management option, including the right to shoot wolves suspected of threatening property or livestock. *Source*: From online state management plan: http://wildlife.utah.gov/wolf/pdf/mn-wolf-plan-01.pdf, p. 45.

After submitting its plan to the state department of natural resources, most roundtable members felt they had reached a reasonable consensus and most Minnesotans held a positive impression of the process. It became quickly apparent, however, that roundtable members had worked (tirelessly, and for months) under the false impression that their plan would be implemented as written. This proved to be far from the case. As it turned out, the Minnesota state legislature continued to hold final authority, and they initially proceeded by largely overlooking the group plan altogether and presenting their own, which included extensive hunting and trapping. Only after a public outcry did a revised plan emerge, representing something of a compromise. The revised plan implemented the stakeholder plan on 90 percent of the wolves' range (Figure 11.5), making the remaining 10 percent into an "agricultural zone" in which landowners have much more flexibility in killing wolves (Williams 2000).

## Evaluating the results

As would have to be expected, there is mixed opinion on the merits of the roundtable and the resulting management plan. Supporters of the final outcome highlight that the roundtable group was able to forge a consensus plan that was deemed biologically sound by state scientists. Supporters also point to the significance of the *process* of the roundtable, in that it built bridges and opened lines of communication that will make future wolf (and other) management issues more easily resolvable. Critics still abound, however. Some in the environmental community felt that the final plan was too much of a compromise and did not represent a true commitment to wolf recovery. Most pro-wolf critics of the plan most vociferously opposed the splitting of the state into two management zones.

Which is the fairest assessment? Certainly, the process was not perfect, and the stakeholder team should have been aware of their limited power from the start. But wolf protection advocates – even those most adamantly opposed to compromise – may in the end have gotten more than they realize. According to veteran wolf biologist David Mech, one nearly sure route to failure in wolf recovery is to *overprotect* them, allowing them to become "pests" to livestock owners. Indeed, this is exactly what happened in Poland: *three times* wolves were reintroduced with full protection and no leeway for control, and all three times they were all shot by landowners (Emel 1998). So, in the case of Minnesota, by including livestock owners, hunters, and trappers in the decision-making process, perhaps the agri-*cultural* attitude toward wolves in Minnesota will change from a "shoot on sight" mentality to one where living with wolves is viewed as the best goal for all concerned. If, via the new institution of stakeholder management, Minnesota turned that corner, then its wolves probably stand a much better chance of long-term recovery than they would have otherwise. But changing human views of wolves, for either environmentalists or for ranchers, is no easy affair, because of the powerful *social constructions* of the wolf amongst differing communities.

## Social Construction: Of Wolves and ~~Men~~ Masculinity

In Chapter 8 we introduced the social construction of nature. Viewing nature not as some preexisting entity but rather as a social construction – a product and process arising from human images, stereotypes, and cultural norms – can be a productive exercise in a number of ways. In the puzzle of wolves, a social constructivist analysis can help shed light on the connections between dominant representations of the wolf and our treatment of and attitudes toward these animals.

### Man as righteous hunter, wolf as evil hunter

The fact that European Americans exterminated the wolf across so much of its native range is tragic, but perhaps not particularly surprising. After all, early American settlers carried with them a culture that feared wilderness and the wild creatures that inhabited the forests and "empty" lands beyond cultivation (see Chapter 8). Moreover, wolves were a *practical* problem for agrarian settlers. Wolves may not have killed people, even if Western wolf mythology might lead one to believe so (think "Little Red Riding Hood"), but they did (and *still do*) kill livestock. But can these two factors – a disdain toward the wild and an agrarian mode of production – really explain the fanaticism of the killing? Can these factors explain the gruesomeness and carelessness of some of the techniques used to kill wolves? Does this explain the use of poisons so deadly that they affected literally thousands of

**Figure 11.6** An early-twentieth-century government wolf trapper, chaining to a tree the one wolf pup whose life was temporarily spared after its littermates were killed. Any adult pack members who returned to the den in response to the pup's cries were shot on sight. After the returning adults were killed, the "spared" pup would be shot. *Credit*: Unknown photographer.

**Masculinity** The socially agreed upon characteristics of behavior associated with men in any society; these may vary significantly between cultures, locations, and periods of history

non-targeted species – including many overtly desirable –just to rid the land of every wolf? What about the public torturing and/or burning of trapped wolves? Or killing all but one of a litter of wolf pups, and then tying the pup by its leg to a tree so that its cry would bring back its mother (to be shot) (Figure 11.6)? As the geographer Jody Emel ponders:

*What is this all about?* It is not solely about protecting livestock, because … the slaughter went on long after the economic threat ended. It continues to this day when almost no one in the continental United States has seen a wolf. (Emel 1998: 201, emphasis added)

An examination of constructions of **masculinity** during this era can help us more fully understand its causes and learn to avoid repeating history. First clues come from the ways in which ideals of masculinity were historically perpetuated, particularly through one of the masculine stereotypes of the day: the gentleman hunter. The gentleman hunter was the pinnacle of manhood in turn-of-the-century America. He embodied all the qualities of the idealized rugged, independent American. He was a survivor, wise in the ways of nature, thriving in solitude that would strike fear in lesser men. Yet he was selective and restrained, only ever killing as much as he could use.

The wolf, too, is a hunter. Such a good hunter, perhaps, that for many men, there may have been some envy at work. Moreover, wolf-as-hunter fell short of cultural standards of "the hunt." First off, wolves hunt in packs. This appeared cowardly when compared to the idealized solitary huntsmen. Wolves also sometimes kill more than they need, leaving "wasted" flesh strewn about kill sites. Of course, most predators do this, a practice that opens up ecological niches for scavengers, which are integral to terrestrial and marine ecosystems everywhere. But this is not how wolf kills were viewed. In contrast to the *humane* gentleman hunter, which treated his game with *mercy*, the wolf was viewed as savage or merciless. This attitude is typified by noted turn-of-the-century natural historian William Hornaday: "There is no depth of meanness, treachery, or cruelty to which [wolves] do not cheerfully descend" (Hornaday 1904: 36). If there was to be any hope of bringing light to the wilderness (see Figure 8.3), these creatures had to be done away with.

Clearly, the man for the job was the wolves' opposite: a righteous, gentleman hunter, bringing order to the wilderness. Wolf hunters were *heroic*. Of course, it is difficult if not impossible to assess exactly how much causality to place on the contrasting constructions of the hellish wolf and the righteous, heroic hunter. It is also difficult to untangle all the seemingly double standards at work in American society at that time. For example, how did society reconcile these constructions of the evil wolf while at the same time witnessing (*subsidizing* even) the merciless slaughter of bison on the Great Plains? And even as we can never hope to conclusively answer these questions, we can recognize the power of constructions in guiding our actions and our notions of acceptable versus unacceptable actions. For centuries, negative constructions of wolves fueled an extreme hatred toward them, and this hatred was manifest in a terrible slaughter that went nearly unquestioned by society as a whole.

## Wolves save the wilderness, but for whom?

A critical constructivist analysis of wolves should not, however, stop at the turn-of-the-century anti-wolf attitudes; it must be brought to bear on their opposite. Wolves are no less powerfully constructed today than they were a century ago. Wolves have become, in North America at least, nearly synonymous with "wilderness," but as reviewed in Chapter 8, wilderness itself is far from an innocent social construction. While wilderness is (currently) typically constructed as land outside human control, actual wilderness areas are, rather, marked by formal sets of rules that dictate their use. The ecological "use" of wilderness (as reviewed in the "Ethics" section above) is a relatively recent justification for wilderness preservation. Even as late as the 1960s, the primary official justification for wilderness was recreation. This is codified in the Wilderness Act of 1964, the legislation that formalized the US system of designated wilderness areas. As went the typical language of the day, "man" needed a place for solitude, a respite from modern urban civilization.

As such, the allowed uses of wilderness are non-motorized, low-impact activities like canoeing, backpacking, and fishing. These are all activities that are quite literally exclusive, in regards to both gender and class. The wilderness was, and to a good degree still very much is, the domain of middle-class men. It takes time and money to, say, float for six days on the Salmon River through the Frank Church/River of No Return Wilderness in Idaho. Gendered exclusivity exists on many levels as well. For example, how many working-class women with children at home have the freedom to head off for a week (or an afternoon for that matter!) of quiet recreation in the wild? Debates over wilderness areas, as such, are not so much arguments over using versus not-using a piece of land, as much as debates over who gets to use them and in what manner.

But the era of protecting wilderness for the respite of the white man has (justifiably) waned. Today, justifications for land preservation rest more commonly on ecological grounds. Enter the wolf. Today, the presence of wolves *makes* (read: constructs) a "rewilded" land, a *wilderness*. Regardless of formal "Wilderness" designation, getting wolves onto the land – *constructing wilderness* – favors some land uses (and some land *users*) over others. Wolves in northern Vermont could, for example, give weight to opponents of snowmobilers, as wolves tend to avoid areas of high snowmobile use. It might make it easier if they can argue that affluent bed-and-breakfast patrons are visiting the area to listen to wolves howling and not snowmobiles humming.

Wolf reintroduction would also spur a local growth-industry in conservation biology research. By arguing for wolf reintroduction, scientists are arguing for personal and somewhat exclusive access to the land (now constructed as the "ecosystem"). Granted, it would be unfair and cynical to argue that this is *the* motivation behind their advocacy. But it is undeniable that scientists have something to gain, personally and professionally, from successful reintroduction. It is also undeniable that, like earlier dominant uses of the wilderness, the backwoods "muddy boots" fieldwork of large carnivore science is an overwhelmingly white and well-educated affair. Perhaps the reticence toward ambitious rewilding efforts like wolf reintroduction held by many rural residents in Minnesota, Vermont, and elsewhere is not fueled by some archaic disdain toward wolves. For many, it is undoubtedly driven more by the recognition that wolves *do* mean wilderness, and that while a local wilderness might make the perfect getaway or laboratory for some, it will also mean exclusion and hardship for others.

## The Wolf Puzzle

In this chapter we have learned that:

- Wolves are an evolutionarily successful, adaptable species.
- Humans have a long history of intimate interactions with wolves, dating back thousands of years to their domestication, through to the more recent era of systematic programs of eradication, to the even more recent trend of wolf conservation.
- Humans are the primary mortality factor for wolves worldwide.
- Novel institutional models of wolf conservation offer more democratic and possibly also more effective methods of ensuring the species' long-term health and survival.
- Ecocentric ethics suggests that we need to think well beyond the bounds of human usefulness if we are to allow significant spaces for wolves to thrive.
- Conservation biology suggests that the fate of the wolf may be linked to the broader fate of global biodiversity and evolutionary processes.
- Social constructions of the wolf have changed through time, and these constructions are inextricably tied to societal norms and stereotypes.
- Social constructions of the wolf have effects; they are tied to the way in which we treat these magnificent animals (note: the irony of this statement is not lost on us; we realize that wolves-as-magnificent-animals itself is a construction, reflecting our culture, and having real effects).
- Social constructions of wolves are also tied to how we treat each other.

This brings us back to Yellowstone. As of the time of writing, the Yellowstone wolves are doing fine. Indeed, as of April 2009, the secretary of the interior was moving to delist the wolf from the federal endangered species list. If delisted, their management (as in Minnesota) will be in the hands of states, all of which traditionally have politics influenced by the livestock industry. The next decade or two of wolf conservation *around* (if not so much *in*) Yellowstone will be a telling reflection of 1) shifting social-environmental priorities, 2) competing socio-environmental discourses, and 3) increasing enviro-institutional experimentation.

Globally, calls for wolf recovery will increase. What will be the fate of the wolves of India? Of Egypt? Of Denmark? Will wolves be reintroduced to the Highlands of Scotland, as some have proposed? The answers will, no doubt, vary from place to place, from population to population. Hopefully, we can learn from the past and sort through the complex spaces of our mutual interaction. Wolves are, to be sure, one of the more difficult, persistent, and intriguing puzzles of nature and society.

## Questions for Review

1.  List a species from each trophic level (producer through apex predator) in a food web somewhere near your home. (The food web can be terrestrial or aquatic.)
2.  How has the reintroduction of wolves affected the ecosystem in and around Yellowstone National Park?
3.  Write an ethics-based argument in favor of *or* against reintroducing wolves into the northeastern United States.
4.  Contrast the Minnesota wolf "roundtable" to traditional "expert-based" wolf management.
5.  Explain the following statement: The presence of wolves helps construct a wilderness.

## Suggested Reading

Lopez, B. (1978). *Of Wolves and Men*. New York: Scribner.
Mowat, F. (1963). *Never Cry Wolf*. Toronto: McClelland and Stewart.
Steinhart, P. (1995). *The Company of Wolves*. New York: Alfred A. Knopf.
Walker, B. L. (2005). *The Lost Wolves of Japan*. Seattle, WA: University of Washington Press.

## Exercise

In this chapter, we have reviewed how the puzzle of wolves might be addressed by ethics, institutions, and social construction. Explain how you might understand this problem using a population-centered framework (as described in Chapter 2). How might human population growth affect existing populations and habitats of wolves and pose challenges for wolf conservation? What are the limits of a population-based approach to wolves?

# 12

# Tuna

*Credit*: Richard Levine/Alamy.

## Chapter Menu

## Keywords

- Animal rights
- Bycatch
- Conditions of production
- Consumer boycott
- Dolphin safe tuna
- Exclusive economic zones (EEZs)
- Fordism
- Green consumption
- Longliners
- Maximum sustainable yield
- Moral extensionism
- Post-Fordism
- Purse-seine fishing
- Second contradiction of capitalism
- Social construction
- Sustainability
- Transnational corporations
- Utopian

# Blood Tuna

In the 2006 film *Blood Diamond*, starring Leonardo DiCaprio and Djimon Hounsou, unsuspecting viewers received a jolt. This was not your standard international action film, with a fantastical plot, far removed from the comfortable lives of viewers in the West. Rather, for many viewers, the diamond jewelry they wore implicated them directly in the violence they witnessed on screen. *Blood Diamond* brought to light (albeit in rather over-the-top, Hollywood fashion) the violence and human bloodshed that are part and parcel of the international trade in West African diamonds. Shocked viewers realized that, through their diamond purchases, they may have unwittingly funded a pointless and ruthless war, with all of its attendant killing, rape, smuggling, torture, and illegal incarcerations. Exposure of this violence resulted in some real changes in global diamond markets. Many consumers were now demanding – and were soon able to purchase – certified "cruelty free" or "conflict free" diamonds. The anti "blood diamond" movement even spawned a fashion trend that might have been unthinkable a couple of decades earlier, *haute* synthetic diamonds. Diamonds, of course, are not the only natural resource whose troubling history gets hidden within commodities.

Eighteen years earlier, another film had produced similar reactions and results. The 1988 film, however, was no Hollywood production. It was compiled footage from low-budget handheld movies filmed by Sam LaBudde aboard the Panamanian tuna boat *Maria Luisa*.

> The beaks of Sam LaBudde's first dolphins strained against the net that had formed a canopy above them. Their flukes churned the ocean white. They thronged at the surface, desperate to force slack in the net sufficient to free their blowholes for breath. Their shrieks and squeals began high in the hearing range of humans and climbed inaudible scales above. LaBudde wanted to scream himself. (Brower 1989: 37)

LaBudde was a spy of sorts. A biologist by training, he had gotten a job as a galley cook aboard the *Maria Luisa*, but with the hidden intent of filming the slaughter of dolphins. Dolphins, for reasons still unknown to biologists, school above tuna in some parts of the world's oceans. In one part of the Pacific Ocean, for about 30 years, fleets of tuna boats had been targeting the easily spotted dolphins to locate the tuna that schooled beneath them. The tuna were destined for the lucrative US canned tuna market. In three decades over six million dolphins were tangled in the massive tuna nets and killed, most drowning before they could be released (Gosliner 1999). LaBudde got what he came for. The film – which included graphic images of dead and dying dolphins – was shown across the United States and then around the world. Shocked filmgoers cringed to think that this seemingly innocuous commodity – canned tuna, staple of school lunches everywhere – could have left such a trail of blood in its wake. As they later would for diamonds, consumers responded in force to this saga of "blood tuna," a commodity far more humble and far more common.

Indeed, few wild-caught food products are as ubiquitous as tuna. *Annual* catches for some species of tuna are in the *billions* of pounds! Yet, like most animal products, tuna arrives in our kitchens with little visible evidence of its natural history and ecology, or the human labor that got it from sea to table. What does a tuna even look like? (Not a whole lot like "Charlie Tuna," the mascot of the StarKist brand). How are they caught? Where are they caught? By whom? What about the effects of tuna fishing on tuna populations, and on ocean ecosystems more generally? In this chapter, we will examine these questions through the lenses of markets, political economy, and ethics. By closely examining the puzzle of tuna, we learn a great deal about the problem facing the world's oceans, about the massive impact of food consumption on ecosystems, and about the awkward relationship between ethics and economics.

## A Short History of Tuna

"Tuna" is an informal designation referring to a suite of similar fish species, all belonging to the family Scombridae, which also includes bonitos and mackerels. Most tuna are members of the genus *Thunnus*, but a few, including the commercially important skipjack tuna (*Katsuwonus pelamis*), are not. All tuna share a similar morphology. They are largest in the middle and smoothly tapered toward both front and back. They have a powerful and distinctive "crescent-shaped … *heterocercal* (equal-lobed) tail" (Ellis 2008: 22). Their powerful tails, streamlined bodies, substantial musculature, and unique physiology combine to make tuna some of the fastest swimmers in the ocean (tuna have been clocked swimming at nearly 70 miles per hour). Shared with only a handful of species of sharks and predatory fish, the unusual physiological adaptation that gives them much of their power is endothermy, a type of warm-bloodedness. To simplify a complex and multifaceted biological adaptation, "one advantage [to tuna] of being warm-blooded is that warmth increases the delivery of oxygen to … the muscles …; more oxygen, more power" (Whynott 1995: 33). They are not just fast. They have stamina, too. Tuna travel on some of the longest migrations of any animal, many with seasonal migrations of four thousand or more miles. Not only is this undeniably impressive, but it has socioeconomic implications as well. No one, nor any country, anywhere, has sole control over any stock of tuna. Because of their far-ranging nature, only truly global efforts will prove successful for the long-term conservation of these fish (Figure 12.1).

Tuna are also, of course, incredibly valuable fish that are prepared in many different ways. Tuna is eaten raw (as "sashimi," which means thinly sliced raw seafood). Tuna filets have become recently popular cooked as "steaks" on outdoor grills. Tuna is popular in soups and chowders in many different cuisines worldwide. And, of course, as many American schoolchildren come to know it, as "tuna fish," tuna comes in a can. The most commercially important species of tuna are bluefin, yellowfin, bigeye, albacore, and skipjack. Each has different markets, with bigeye tuna eaten mostly fresh in Japan and Korea, for example, and albacore consumed as canned "white" tuna in the United States (Ellis 2008).

In the sections that follow, we will discuss the bluefin and yellowfin. These varieties are emblematic, first because they are both objects of familiar consumption to us, but also because they represent a crisis of overfishing in the world's oceans (especially bluefin), and because they raise issues of collateral damage to other species in their harvesting (especially dolphins in yellowfin fisheries).

## Bluefin tuna: From horse mackerel to ranched sushi

The bluefin tuna is divided into three species, the northern bluefin (*Thunnus thynnus*), the southern bluefin (*Thunnus maccoyii*), and the Pacific bluefin (*Thunnus orientalis*). The northern bluefin is the largest tuna, with individuals commonly weighing over 1,000 pounds.

The oldest known bluefin fisheries (a "fishery" is simply a harvested fish stock) are in the Mediterranean Sea. The earliest may have been that of the Phoenicians, living on the eastern Mediterranean coast around modern-day Lebanon, who for thousands of years used ingenious methods of trapping, herding, and eventually harpooning the corralled bluefin. These fish represented some of the most

**Figure 12.1**  The sleek, powerful bluefin tuna (*Thunnus thynnus*). The bluefin are the largest of the tuna (the largest one ever caught weighed 1,496 pounds). Prior to the 1960s, bluefin were considered "trash fish" and fished primarily for sport and sometimes sold for pet food. Giant bluefin are now, pound for pound, the most valuable fish in the ocean. A single 444 pound bluefin fetched a record $173,600 ($391/pound) at auction in Tokyo. *Credit*: Brian J. Skerry/National Geographic/Getty Images.

important catches to coastal cultures (including the Ancient Greeks) and as such are no small players in the development of Western civilization (Ellis 2008). For most of the modern era, however, bluefin were not a commercially valuable catch. When the fish was known commonly in North America as the "horse mackerel," its flesh was considered unfit for human consumption and was sold as an ingredient in commercial dog and cat food. It has since become a popular sport fish (Ellis 2008: 84).

The fate of the bluefin changed dramatically in the 1960s, when "maguro" (sushi from large bluefin) started to be highly prized as the highest of high-end sashimi. Seafood has been central to the Japanese diet for millennia; sashimi in various forms since at least the seventeenth century. However, the taste for maguro (especially "toro," the marbled red belly meat of the bluefin) has arisen only very recently. It was not until the 1960s, when Japan built a fleet of refrigerated tuna freighters that could remain at sea for weeks, that the boats could catch, freeze, and deliver to domestic Japanese markets from the mostly more distant giant bluefin fisheries. The popularity of maguro, as soon as it was widely available, increased faster than the fish could be caught. Quickly, non-Japanese fishers – from the United States and Canada to the Mediterranean – jumped into the market. It is not surprising that so many fishers would want to get into this seemingly insatiable market, as a single giant bluefin can sell for over a hundred dollars a pound, and the fish can weigh several hundred (even over a thousand) pounds. The demand for giant bluefin has increased

**Purse-Seine Fishing**  An effective fishing method for species that school near the surface; a large net is encircled around the targeted catch, after which the bottom of the net is drawn tight like the strings of a purse, thus confining the catch in the net

**Longliners**  An industrial fishing method deploying lines baited with hundreds or thousands of hooks; longlines are usually several miles long and often result in significant bycatch

**Sustainable/Sustainability**  The conservation of land and resources so as to secure their availability to future generations

even further as maguro and toro have become internationally haute cuisine. Indeed, in 2008, the most expensive restaurant in New York City was a sushi bar where a meal can run to more than $1,000 (Gosliner 1999: 152). The money to be made has unsurprisingly resulted in increasingly heavy harvest. These fisheries are some of the most overfished in the world and have driven the bluefin to biological endangerment.

One of the latest ways that fishers have tried to maximize their bluefin production is by "ranching" them. In the South Pacific and the Mediterranean, entire schools of young bluefin are caught in "**purse seine**" nets; so-called because after the net encircles the school, the bottom of the net is drawn together like the strings of a purse. The enormous nets are then towed to a facility where the tuna are raised (*fattened*, really, in a manner not dissimilar to feedlot-fattened cattle) for a few months in pens before being harvested and exported, mostly to Japan.

Distressingly, tuna ranching has substantially decreased wild bluefin tuna stocks, thus furthering their endangerment. The added pressure bluefin tuna ranching puts on wild stocks is due primarily to two factors. One, it is almost completely unregulated. Tuna caught for ranching do not count against legal catch quotas, even though they are being permanently removed from wild populations. But ranched tuna do not fall under aquaculture regulations either, because they are not reared in captivity. The second key factor is the way ranching is affecting the age structure of wild populations. Prior to ranching, the heaviest fishing pressure was on the largest tuna. Now that younger tuna are valuable, more younger bluefin are being removed from wild stocks. Bluefin populations are losing the very fish that will be needed to replenish their numbers in the future. Considering the development of technologies like increasingly huge purse-seine freighters, **longliners**, and tuna ranching, coupled with the seemingly insatiable growth in demand for sashimi, bluefin fisheries worldwide are being harvested well past **sustainable** levels.

Bluefin tuna, like dozens of other wild-caught ocean fish species, are overfished; meaning simply that commercial fishing operations target these species and catch more than are able to naturally regenerate. Overfishing is socially and ecologically undesirable. Socially, overfishing results in considerable hardship in fishing-dependent communities. In many cases, "artisanal" (small-scale) fishers suffer first and most. Commonly, fish stocks only become severely depleted after international fleets concentrate activity in a particular fishery for a number of years. As these boats are owned and operated by corporations (or individuals) with considerable capital investment, these large-scale operations are able to relocate to another fishery if their current target fishery is overfished and falls into decline. This is not, however, an option for small-scale artisanal fishers, who are tied to their local fisheries by economic constraints (and other factors). Many well-known fish species that are readily available at supermarkets are actually severely overfished species (Table 12.1). Overfishing, however, is only one of a host of problems besetting marine ecosystems. Another problem directly related to modern large-scale fishing operations – and directly

## Box 12.1    Longliners

Along with bottom trawling, drift nets, and purse seine nets, longlines are one of the intensive, modern fishing methods causing widespread overfishing of the oceans and tremendous bycatch. As their name suggests, longlines are long fishing lines, consisting of a main line, usually over several miles and sometimes up to 60 miles long, and hundreds of hooked secondary lines branching off the main line. A single longline will have several hundred to over a thousand hooks. Most longliners (boats deploying longlines) target large, high-trophic-level pelagic (free swimming) fish like sharks, tunas, and swordfish. A smaller number of longlines are set on demersal (bottom-dwelling) species like the endangered Patagonian toothfish and halibut. (Most longliners targeting the Patagonian toothfish, it should be noted, are doing so illegally.)

Longlining, as you might guess, is a rather indiscriminate fishing method. Swordfish, for example, have suffered significant population declines since most commercial swordfishers replaced pole and line and harpoon with longlines beginning in the 1960s. Not only do longliners simply catch many more adult swordfish than were caught prior, but significantly larger numbers of non-targeted juvenile swordfish die from longlining as well. Harpooning is, of course, the most selective way to fish for swordfish. As a commercial fisherman, there would simply be no reason to set a harpoon on a small swordfish that you would not be allowed to sell. And even with pole and line, the survival rate of hooked and released juvenile fish is relatively high. Not so with longlines. The hooked juveniles stay on the line for so much longer (often a day or more)

that even those that may be still alive when brought to the boat die soon after release. The effect on the fishery has been devastating. At the turn of the twentieth century, the average swordfish landed was between 300 and 400 pounds. Today, the average is around 88 pounds. Juvenile swordfish, however, are just one example of the problematic bycatch associated with longliners.

Longline bycatch of sea turtles and seabirds such as albatrosses is just as controversial. Sea turtles worldwide represent some of the most critically endangered animals on the planet. In 2001, the US National Marine Fisheries Service pegged longlines as the primary threat to two endangered turtle species, the Atlantic loggerhead and the leatherback, and subsequently ordered 2.6 million square nautical miles of the North Atlantic closed to commercial longlines to protect these species from extinction. Albatrosses, large seabirds, will often follow longliners and go after baits nearly as soon as they reach the water, after which hooked birds are dragged underwater and die. Legal and illegal longliners are responsible for the death of over 100,000 albatrosses each year, many of which belong to endangered species.

### References

Chambers, J. (2001). "Going, going, gone." *Big Game Fishing Journal*, Jan–Feb.

Ellis, R. (2003). *The Empty Ocean: Plundering the World's Marine Life*. Washington, DC: Island Press/ Shearwater Books.

relevant to discussions of tuna – is **bycatch**, non-targeted organisms incidentally caught by commercial fishing operations.

> **Bycatch**  Non-targeted organisms incidentally caught by commercial fishing operations, including many fish species, but also a large number of birds, marine mammals, and sea turtles

### The Eastern Tropical Pacific yellowfin tuna fishery

Yellowfin tuna are smaller than bluefin but are still quite large, some exceeding 400 pounds. Throughout most of the twentieth century, most yellowfin tuna caught for US markets

**Table 12.1** Overfished species "red-listed" by the Blue Ocean Institute. Red-listed species are the most ecologically problematic commercially available fish species

| Fish species | Fishing method | Interesting facts |
| --- | --- | --- |
| Chilean sea bass (real name: Patagonian toothfish) | Primarily bottom longlines | Longline bycatch includes large numbers of endangered albatrosses |
| Atlantic halibut | Bottom longlines; pole and line | Overfished for centuries; may become extinct in the wild |
| Orange roughy | Bottom trawling | Can live to be over a hundred years old; common bycatch includes several threatened shark species |
| Grouper | Various (e.g., longlines, traps, pole and line, spearfishing) | 98.5% of grouper sold in the US come from Mexico, which has no management plan for conserving grouper stocks |
| Atlantic cod | Bottom trawling; pole and line | Officially closed as a commercial fishery since 1992, the North Atlantic cod population shows no signs of rebounding |

*Source*: Blue Ocean Institute: www.blueocean.org/

was canned as "light" tuna, but in other parts of the world it is also eaten raw and cooked in various ways. Yellowfin also has become a popular less-expensive alternate to bluefin for sashimi, and is popular sold fresh as tuna "steaks" for grilling.

One of the most commercially valuable and historically most heavily fished tuna fisheries in the world is the yellowfin tuna fishery in the Eastern Tropical Pacific (ETP) Ocean. The ETP stretches south from southern California along the coast of South America, covering some 18 million square kilometers. The valuable yellowfin fishery sits at the southern end of the ETP, reaching as far as Chile. This fishery was first commercially exploited when improvements in refrigeration technologies in the 1930s allowed new boats in the US tuna fleet to stay out at sea longer and venture farther offshore to more distant tropical waters. Since typically used cotton nets degraded quickly in the warmer southern waters, the boats targeting southern yellowfin switched to pole and line fishing (Gosliner 1999).

Pole and line fishing was considerably more labor-intensive than purse-seine fishing, but remained profitable for a few years. In the 1950s, however, as the US market was flooded with cheap, imported tuna, the pole and line yellowfin fishery became unprofitable, and was saved only by two key technological innovations, which together portended serious ecological consequences. One was nylon nets that did not break down in the warmer waters. The other was a more powerful power block that could haul in once-unimaginably large nets. The reconversion of the fleet to purse-seine fishing (as described above) was rapid. Indeed, within a few years nearly all boats in the southern ETP were using large, nylon purse-seine nets to catch schools of yellowfin, making the fishery more productive and profitable than ever (Joseph 1994).

Unlike skipjack, which (as their name suggests) "skip" along the surface and are therefore easily spotted from boats or spotter planes, yellowfin swim at depth, which makes them difficult to spot. For reasons still unknown to ecologists, however, ETP yellowfin (and, strangely, not yellowfin elsewhere) often swim below schools of dolphins, allowing fishermen to target schools of tuna by searching first for easily spotted dolphins, which swim at the surface and breach often. By 1960, this had become the primary method for finding yellowfin. Once a dolphin school was spotted, a fleet of speedboats would "corral" the dolphins to concentrate them. After this was accomplished, a boat would haul a huge (sometimes a mile long or longer) purse-seine net around the school of dolphins, simultaneously trapping yellowfin beneath.

In the early years of the purse-seine dolphin method, little effort was made to reduce dolphin bycatch mortality. A few dolphins would breach over the net and escape, but most would be hauled aboard in the nets. A few would be thrown back alive, but most were thrown back dead, having drowned after entanglement. The numbers of dolphin deaths associated with this fishery during the 1960s are staggering (Figure 12.2), and it is no surprise that, when these figures were made available to the public in 1968 through the media, there was an immediate and substantial public outcry and political response (Joseph 1994). This was rather unfortunate timing for the tuna fishing industry. The environmental movement was just at this time gaining prominence and confidence and – along with anti-nuclear activism ("*No Nukes!*") – marine mammal conservation

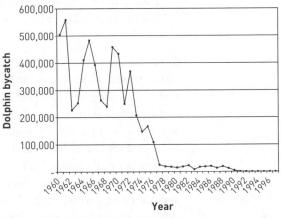

**Figure 12.2** Dolphin mortality at the hands of United States fishing vessels using purse-seine tuna technology in the Eastern Tropical Pacific, 1960–97. *Source*: Adapted from data from Gosliner (1999), p. 124.

was one of the highest-profile issues of the day ("*Save the Whales!*"). The popular television show *Flipper*, about a helpful bottlenose dolphin living with the family of a Chief Warden of an imaginary marine reserve (effectively recreating the popular dog program *Lassie*, but under the sea!), also aired between 1964 and 1967, and provided a popular representation of dolphins in the United States. Even long before LaBudde's exposé, therefore, concern for dolphins was on the rise. The "tuna–dolphin problem" (like the problem of overfishing of bluefin today) became the order of the day, creating a human/environment puzzle with global implications.

## The puzzle of tuna

This overview of the fish we call "tuna" sheds light on some of the seemingly intractable problems bound within the rapidly changing relationship between people and the world's oceans. The review shows us how changes over time in society – including evolving tastes, technological developments, emerging ideas of right and wrong, and a globalizing economy – have all radically changed the way we harvest and eat tuna. And, of course, these changes have also, by orders of magnitude, increased the sheer tonnage of fish taken from our oceans. Specifically, our review suggests that tuna are a kind of puzzle when we acknowledge that:

- Unlike humans (but like the rest of non-human nature), tuna respect no political boundaries. Yet their fate is inexorably tied to geopolitical decisions.
- Tuna are valuable; strangely, almost unexplainably valuable. The demand for tuna seems to have no ceiling, yet wild stocks of tuna are an undeniably finite resource.
- Their seeming ubiquity (for example, on store shelves) masks the declines of populations in the wild.
- Their puzzling relationships with other sea creatures (like dolphins) make it very difficult to heavily harvest tuna without impacting other species as well.

Tuna therefore provide a window through which we can see the effects on nature of our tastes, trends, technologies, and cultures. As such, tuna present us with an important puzzle: Can the global community make agreements that will preserve wild stocks of these freewheeling but incredibly valuable creatures? Considering that marine flesh is the largest single source of protein for human beings worldwide, it would not seem an overstatement to say that the fate of people and the fate of tuna may indeed be inseparable. By specifically viewing the ETP tuna–dolphin controversy through the lenses of markets and commodities, political economy, and ethics and ecocentrism, we see that different perspectives paint very different pictures of the history of this debate. Furthermore, they point us toward different types of solutions, in this fishery and beyond.

## Markets and Commodities: Eco-Labels to the Rescue?

We began the discussion of markets and commodities in Chapter 4 by asking the strange question: "*Can using more stuff lead to the availability of more stuff?*" Can this same logic apply to overexploited marine resources? Stated more baldly: Can the free market save the very resources it has depleted? Intuitively, it would seem like it would not, yet a slight tweaking of the question cuts to the heart of one of the most ambitious and widespread recent conservation efforts. So, in this case, perhaps we might ask: *Can consuming the* right *fish save our oceans?*

This is the bet on which many scientists, government officials, environmentalists, and seafood corporations have recently put their money. Despite centuries of overexploitation of ocean resources; despite decades of governmental regulation at local, national, and supranational scales; despite the best efforts of environmental organizations at highlighting the plight of the seas: despite all this, the oceans continue to be increasingly overharvested to this day. Our best chance at reversing this trend, some say, is by harnessing the forces of the market through informed and enlightened **green consumption**. If society wants sustainable ocean resources, the mandate is simple: we must buy the right things.

> **Green Consumption**  Purchasing of products that are purportedly environmentally friendlier or less harmful than their alternatives; a model of environmental protection that relies on consumer choices to change the behavior of firms or industries rather than regulation
>
> **Dolphin Safe Tuna**  Tuna caught without killing dolphins as bycatch

In this section, we review the campaign for "**dolphin safe**" labeling of canned tuna. For decades, thousands of dolphins were killed as "bycatch" in tuna fishing operations. Initially, the "dolphin safe" labeling program gave consumers a choice to use their power in the market to try to force producers to adopt more environmentally desirable fishing methods. The label was so popular that Congress passed a law that standardized the use of the "dolphin safe" label and, even more significantly, banned the importation of non-dolphin safe tuna.

## Attempts at solutions through legislation

Largely in response to the public outcry over dolphin deaths at the hands of Eastern Tropical Pacific yellowfin purse seiners, in 1972 the US Congress passed the Marine Mammal Protection Act (MMPA). The MMPA, most broadly, prohibited the killing, selling, importing, and exporting of marine mammals or marine mammal parts within the United States. The bill included a handful of exceptions to the moratorium, however, including confusing and even conflicting exceptions for ETP tuna. In one section dolphins in the tuna fishery are to be managed for "optimum sustained population," allowing for dolphin kills as long as they do not reduce their population numbers below biologically acceptable thresholds (Gosliner 1999: 1). The tuna fishers are also, however, ordered within the Act to reduce dolphin mortality over time to a target of zero mortality. From a fisheries management standpoint, these are conflicting directives. The legislation proved vague and confusing enough, and contained enough loopholes, that dolphin mortality continued in high numbers for years after passage of the MMPA (Figure 12.2).

Amendments to the MMPA in 1976 led to a substantial decline in dolphin mortality. One was the implementation of the "backdown procedure." The backdown procedure is simply the backing up of the boat when the net is about half pulled in. This causes the end of the net loop to dip below the water's surface. Inflatable boats inside the net can "herd" the dolphins toward the submerged part of the net, where they can swim over and escape. The other was the addition of a "Medina panel" to the net (named after its inventor, Joseph Medina). The "panel" was simply a much finer section of mesh at the back end of the net, where the majority of dolphins were ensnared. With the panel in place, the dolphins are nudged away from the net until they either clear the net via breaching or wait and swim over during the backdown procedure (Figure 12.3). Even with these changes, and a subsequent requirement to place government observers on purse seiners to document dolphin mortality, the original goal of movement toward zero mortality remained elusive. Indeed, 1984 amendments to the MMPA allowed the US fleet (which was by this time no longer the only country catching ETP tuna with purse-seine fishing) to kill 20,500 dolphins annually, a testament to the persistence of the problem (Joseph 1994).

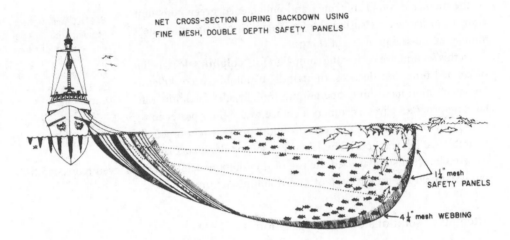

**Figure 12.3** The "Medina panel," one of a few technological and operational changes made by purse-seine tuna fishers in attempts to reduce dolphin mortality. The Medina panel greatly reduced, but did not eliminate, dolphin bycatch in this fishery. *Source*: United Nations Food and Agriculture Organization Fisheries and Aquaculture Department www.fao.org.

## Consumer activists to the rescue

**Consumer Boycott**   A method of protest that aims to pressure corporations into changing their practices by urging people to forgo purchasing products associated with the targeted corporations

Many environmentalists remained outraged that a dozen years after the MMPA, dolphin mortality was still considered acceptable. The regulatory route was proving unsatisfactory. Some other sort of action was needed. In the late 1980s, the environmental non-governmental organization Earth Island Institute (EII) organized a **consumer boycott** of canned tuna, urging consumers to forgo purchasing canned tuna until producers could show that their product did not result in the deaths of dolphins. The boycott was substantial, yet it was not until 1988 when the Sam LaBudde film – as told in the opening story of this chapter – gave the campaign the shot in the arm it needed. After the nationwide showing of the film, the boycott grew in strength, tuna sales declined, and environmentalists launched a massive letter-writing campaign to tuna producers and politicians. A coalition of environmental organizations (led by EII) started promoting the idea of labeling tuna caught without targeting dolphins as "dolphin safe."

In 1990, the campaign celebrated its breakthrough success. The three largest US tuna brands (StarKist, Chicken of the Sea, and Bumblebee) agreed to purchase only "dolphin safe" tuna (Newsweek 1990). These canners make up about 90 percent of the US market, so this was clearly a major coup for the environmentalists. The federal government passed a law that required canners who used the "dolphin safe" label to adhere to the standards as defined by the EII consortium. Informed US consumers understood, then, that any tuna purchased with the "dolphin safe" label did not involve the pursuit or capture of dolphins. Congress also passed a trade embargo against the importation of tuna from any countries that could not prove their tuna fishing practices were "comparable" to US regulations

(Bonanno and Constance 1996: 6). This effectively shut out Mexico, Venezuela, and Panama (the countries with the largest ETP tuna fleets still targeting dolphins) from their largest export market.

## The label stays intact

Coalitions of Latin American states have since filed lawsuits against the United States on charges of discriminatory protective measures in violation of free trade agreements, but have repeatedly failed to gain access to US markets for their purse-seine tuna fishers. Internal pressure has sometimes also been brought to bear. Twice, in fact – once under the Clinton administration and again a decade later under the Bush administration (and each time *with the backing of the presidential administration*) – attempts were made to water down the "dolphin safe" standards, in support of free trade. Both times the existing standards were upheld by courts. Despite domestic and international political pressure, therefore, the original "dolphin safe" standards as defined by EII have so far remained intact.

From many angles, the campaign for dolphin conservation in the ETP appears to represent an unqualified success of market forces put to work in the cause of conservation. Dolphin mortality has declined steadily since the activist campaign was launched, down close to *99 percent* from its peak in the 1960s. Moreover, it was pressure from activists to sell only "dolphin safe" tuna that undeniably forced the hand of major tuna processors. US consumers will see some or another dolphin label on any can of tuna they purchase (US law allows canners to put whatever label logo they choose to market their tuna as "dolphin safe," as long as the product abides by federal regulations). Most, no doubt, go away feeling good about their purchase, safe in the expectation that their sandwich is not soaked in the blood of dolphins. Similar labeling campaigns have since sprung up in many European countries, Australia, and New Zealand, though not all "dolphin safe" labels worldwide, it should be noted, adhere to the same standards.

But the success also could lead one to question the overall impact of the practice. We might wonder, for example: How many consumers see "dolphin safe" and understand its definition? How many, perhaps, equate "dolphin safe" with "eco-friendly" or "ecologically sustainable"? This would be a problematic assumption (as will be discussed in the environmental ethics section that follows), insofar as tuna harvesting overall remains a serious pressure on ocean ecosystems in terms of over-extraction, pollution, and a host of other effects.

More generally, can we assume that green certified products in general are ecologically preferable to uncertified products? Probably not. As green labels proliferate, is it really possible for the average consumer to track the legitimacy and regulations associated with every "green" certification? Undoubtedly not. One thing, however, is certain. The rise of green certification marks a change in power over who controls consumer products, and how it is done. With the aid of a political economy perspective, we can assess this transformation of power. In the following section, we will do just that for "dolphin safe" tuna and "green" seafood more broadly.

## Political Economy: Re-regulating Fishery Economies

In Chapter 7, we stated that it is impossible to understand our relationship with nature without bringing the economy into the picture. Moreover, we showed how and why the economy is never separate from politics (hence: "political economy"). The previous section, which focused on a "markets and commodities" perspective, shared a similar focus on the economy, but with a different set of assumptions. Political economy does not, for example, operate under the assumption that a "free market" is the one system of social arrangements to which all future alternatives would best conform. Unlike a perspective that seeks out solely market-based solutions when problems arise, political economy takes a converse approach, analyzing the ways in which capitalist production often *causes* these very problems; how capitalist production, no matter how many patches are applied, can never escape its own crises and contradictions.

So let's not be so quick to assume …

**Second Contradiction of Capitalism**
In Marxist thought, this describes the tendency for capitalism to eventually undermine the environmental conditions for its own perpetuation, through degradation of natural resources or damage to the health of workers, etc., predicted to eventually lead to environmentalist and workers' movements to resist capitalism, leading to a new form of economy. Compare to the first contradiction of capitalism

**Conditions of Production**   In political economic (and Marxist) thought, the material or environmental conditions required for a specific economy to function, which may include things as varied as water for use in an industrial process to the health of workers to do the labor

In the case of the ETP yellowfin tuna fishery, we can, for example, see the **second contradiction of capitalism** at work. As soon as tuna fishers started targeting schools of dolphins, they were degrading the ecosystem on which they relied for their livelihood. Whether continued setting on dolphins for several more decades would have ultimately sent ETP dolphins – and possibly their schooling mates, the large yellowfin tuna – toward ecological endangerment we will (thankfully) likely never find out. The fishers did not abandon the ETP due to overfishing, as has happened in many other fisheries.

The North Atlantic cod provides a notorious example: Once perhaps the most valuable fishery in the world, the North Atlantic cod was overfished to the point where the government of Canada permanently closed the commercial fishery in 1992. Bluefin tuna fisheries, as noted above, are on the verge of collapse.

Even though yellowfin tuna have not been overfished per se, the means of production (boats and nylon purse-seine nets) that expanded over the decades through cutthroat competition caused ecological degradation that disrupted (and ultimately derailed) the production process. The contradiction, therefore, is still clear: the tuna industry was unable to continually increase production without befouling its **conditions of production**.

It could be argued that, in the absence of the rise of global capitalism, the changes in fishing methods (from pole and line to purse-seine mega-nets) would never have taken place. The transition was, after all, made to increase harvests at a cheaper cost of production. Hundreds of fishers were laid off as the yellowfin fishery transitioned from a labor-intensive to a capital-intensive means of production, to the tune of hundreds of thousands of jobs, at the global scale. During the pole and line ETP tuna fishing era (pre-1959), the boat owners, who were sometimes, but not always, the captains of their vessels, may have been the owners of "firms," but the vast majority of those employed in the fishery were *mere* workers, paid for their labor, onto which a premium was added as profit for the owners of the operations.

To refer to the fishers as "mere" workers is not to denigrate their livelihoods or their labor, but to highlight their expendability. Had the workers owned and controlled the boats, refrigeration, and processing equipment – and been able to maintain their domestic market by, for example, keeping cheaper imported tuna from coming into the country – there would have been no incentive for them to have moved to larger nets, and eventually to a wholesale reliance upon dolphins to land their catch. Put another way, they would have had no reason to put themselves out of work. They would have, ideally, worked to maintain both their livelihoods and the ecological conditions necessary to sustain them. Granted, we cannot go back in time and hand the boats over to the fishermen. Neither can we go back and force the state to enact the protective measures that would have been necessary to protect even a worker-owned, labor-intensive fishing sector from foreign competition. Nor, to be sure, can we know that even a government-protected worker-owned fishing cooperative would have necessarily had the knowledge and foresight to work in such an ecologically harmonious manner within the fishery.

Speculative analyses like this are somewhat **utopian** and as such are not perfectly reliable gauges against which to measure real-world outcomes. Nevertheless, a political economy perspective does shed light on the simultaneous pressures that capitalist accumulation exerts on both society and nature. Moreover, utopian or otherwise, such a perspective provides glimpses of how alternative social relations might have concrete ecological advantages.

Another strength of a political economy approach to nature–society puzzles is the insight it provides in explaining the way resources are exploited and regulated. Why are tuna managed the way they are? How have global changes in tuna production and consumption influenced regulation of the seas? Below we use a political economy perspective to show that, in the case of Pacific yellowfin tuna, regulation of the oceans often reflects international struggles over control of fisheries resources, that global upheavals in the food economy have led to a de-centered multinational system of tuna production, and that this new system has led to new and highly questionable forms of "green" self-regulation.

> **Utopia/Utopian**   Imaginary, idealized social conditions arising from socio-political systems that facilitate cooperation over competition
>
> **Maximum Sustainable Yield**   The largest seasonal or annual amount of any particular natural resource (e.g., timber, fish) that can be harvested indefinitely

## Geopolitics of tuna

Beginning in 1949, US and Latin American tuna fishing nations, in an effort to maintain the ETP yellowfin fishery, agreed on annual tuna harvest quotas based on projections of **maximum sustainable yield**. This catch limit for the fishery set a total number of fish that could be taken overall, but was reached on a first-come, first-served basis; whoever could catch the most fish the fastest was the winner in the fishery. Once the quota was reached, no more tuna could be removed. This encouraged a reckless "race to harvest" with bigger boats and bigger nets (Bonanno and Constance 1996: 131).

**Exclusive Economic Zones (EEZs)** Usually extending 200 nautical miles off the coasts of sovereign states, EEZs are sea-zones within which states claim ownership over fishery and mineral resources

**Fordism** Relations of production dominant in many industrialized countries in the first several decades of the twentieth century; marked by large, vertically integrated corporations, high wages and rates of consumption, and considerable state power

As a result, by the 1970s the ETP fishery had become more complex despite, and perhaps because of, regulation, with many countries vying for a bigger share of the ETP tuna fishery. In that decade, Mexico and four other Latin American countries with Pacific coasts extended their "**Exclusive economic zones**" (**EEZs**) – their sea-zone ownership jurisdictions – out 200 miles from their coasts, following the lead of the United States, which had been the first nation to do this in 1973. The Mexican government began seizing US boats for fishing without permission in its EEZs. As a result of a range of other factors, moreover, including El Niño years causing tuna number declines, boats switching from US to Mexican flagging, and US boats leaving the ETP for the Western Pacific, the US ETP fleet shrank dramatically during this period.

The Mexican tuna fleet failed to capitalize on its new dominance in the fishery, however. Instead, global oversupply depressed tuna prices. US markets, most problematically for Mexico, became off-limits due to the dolphin embargo. Since 1949, the United States had exploited the quotas system with its superior technological and investment base (every year, they won the "race to harvest"). Now, they held developing nations at bay by forcing them to conform to US regulations. These crises and struggles over the ETP fishery do not simply reflect the impacts of markets, therefore, but also of the waxing and waning of US regional power (Bonanno and Constance 1996; Constance and Bonanno 1999). In other words, environmental regulations are often the product of geopolitical struggles to control natural resource production and profits as much as they are institutions aimed at sustainability. At the same time, changes in systems of production may also lead to new forms of regulation.

## From a Fordist to a Post-Fordist fishery

The early era of US dominance in fisheries from the 1950s to the 1980s typified what political economists call "**Fordism**." Named after Henry Ford and the early Ford Motor Company production model, Fordism refers to relations of production that combine high wages, mass production, and mass consumption. Fordism is typified by vertically integrated domestic corporations (where firms own or control the production chain from bottom to top) and forms of state power that favor tariffs against low-cost foreign competition, sustaining continued profitability for large firms.

The tuna industry from the 1950s to the early 1980s fits the Fordist model quite well. The large tuna consuming countries (namely, the United States and Japan) were also the largest tuna catching and processing countries. US tuna boats sold their tuna to US canners, who sold to US customers. Some tuna boats were even owned by tuna processing firms, typifying the vertically integrated Fordist corporation. The state maintained the profitability of the industry by securing and maintaining US dominance in the ETP (as described above). Even through the 1970s, the US government foiled environmentalist opposition and foreign competition through MMPA exemptions and trade embargoes.

The state could only forestall broader trends from affecting the US fleet for so long, however. As Fordism morphed into new relations of production referred to as "**Post-Fordism**," the locus of power shifted away from the state and to **transnational corporations (TNCs)**. The 1980s saw a restructuring of the global tuna industry, typified by the dismantling of the vertically integrated US tuna firms. The trend was kick-started by a combination of factors, including the rise of increasingly cheap, factory farmed (Chapter 4) beef and poultry, which led to a decrease in consumer demand for tuna. The decrease in demand contributed to a growing crisis of overproduction (see Chapter 7), and continued price declines. As a result, the old system broke apart, with formerly vertically integrated corporations divesting significant portions of their operations, selling their purse-seine boats, and purchasing tuna on contract, seeking out the lowest-priced suitable fish. Processing for the "big three" US brands moved from California to Puerto Rico, taking advantage of lower-cost labor. Finally, the large US tuna brands were all absorbed by Asian TNCs. Much of even the canning and processing is now secured through subcontracting. Thus, the tuna "company" is now merely a "brand" owned by a large TNC that searches the globe for the least expensive materials and labor to assemble its products. This transition to a Post-Fordist global tuna industry is now complete (Klein 2007), with implications for regulation of the world's fisheries.

> **Post-Fordism**  Arising in the last decades of the twentieth century, the current relations of production in most industrialized countries; marked by decentralized, specialized, and often subcontracted production, the prominence of transnational corporations, and diminished state power
>
> **Transnational Corporations (TNC)**  Corporations operating facilities in more than one country; also commonly called multinational corporations (MNCs)

## Post-Fordist regulation: The Marine Stewardship Council

With a fully decentralized, global tuna production system, traditional modes of regulation (based within nation states) become less tenable, leaving space for new ones to emerge. In 1997, the World Wildlife Fund and the agro-food mega-transnational company Unilever established the Marine Stewardship Council (MSC). The MSC became a wholly independent, self-funded and governed organization two years later. The mission of the MSC is to provide market-based incentives to corporations for the production of "sustainably" harvested seafood, to ameliorate overfishing of the world's oceans. If a fishery meets the MSC guidelines, its associated seafood products can gain the MSC seal of approval (Figure 12.4). In a similar spirit

**Figure 12.4**  A label from a can of Alaska salmon bearing the Marine Stewardship Council stamp of approval. (MSC label blown up to show detail.) This was one of only three MSC-certified products we could find at our local chain supermarket; all were canned Alaska salmon. None of the fresh seafood bore any standardized green stamps, but there was a sign at the fish counter telling us the supermarket was committed to selling sustainably harvested seafood products.

to the "dolphin safe" label, consumers can be part of the solution by sending their seafood dollars to firms that are "doing the right thing." Though an attractive model for green consumption, there are reasons that consumers might want to be wary of granting this trend a full embrace.

From a social justice perspective, the MSC is already receiving troubling reviews from small fishers throughout the Global South (as defined previously to include vulnerable and typically postcolonial nations dependent on primary production and exports), who are shut out from MSC certification due to not having the proper connections and capital investment to even begin the certification process. So while large companies are moving into the waters previously fished by poorer nations, fishers from those areas are locked out of markets in wealthy countries by apparently "green" labels, pushing towards further consolidation and loss of jobs, markets, and opportunities in the poorest parts of the world. In this way, producers from poorer nations are, nearly undeniably, bearing a disproportionate share of the "shock" that accompanies the "opening up" of the world to free markets (Klein 2007). More disturbingly, green labels appear to (somewhat ironically) favor large-scale, corporate, and conglomerated firms over small craft producers.

Moreover, it is difficult to imagine that groups like the MSC, structurally wedded as they are to free market solutions and TNCs, will do much to support any competing, non-market-based, alternative forms of regulation. For grassroots environmentalists, parallel concerns arise. Will groups like the MSC, allied as they are with powerhouses like the World Wildlife Fund, be able to successfully resist proposals for environmental management, especially those that restrict corporate activities? Proponents of democracy, informed by this political economy perspective, might evaluate new environmental regulations and regulatory bodies in terms of the degree to which they add more voices or more balance to governance and accountability. New forms of post-Fordist regulation like the MSC, increasingly corporatized and supranational in scale, are even more inaccessible to citizens (except as consumers!) than the Fordist state regulations they replace. Where green certification systems work to enhance democratic structures and ecological sustainability, they might be supported. They should not, however, be accepted uncritically. From this point of view, we would be well served to check what lies behind the labels. This is especially true if we begin to consider the complex ethical tradeoffs that differing forms of regulations entail, for individual species, and the ecosystem more generally.

## Ethics and Ecocentrism: The Social Construction of Charismatic Species

*"Tuna is a fun food. If it's associated with the harassment and killing of a noble creature like the dolphin, that's not right. Hopefully, in a few years, the safest place for tuna will be swimming underneath a dolphin."* (Ted Smyth, Vice President of Heinz Foods, which at the time owned the StarKist tuna brand (United States Fish and Wildlife Service 2008))

Why were so many people so moved to action by the images in the LaBudde film? Why has the picture of tuna boat deckhands slinging a dead dolphin overboard

(Figure 12.5) become such an icon of environmental wrongdoing? Thousands of people were moved to act on what became one of the most effective consumer boycotts in history. Yet, during the same period of time, countless other ocean fisheries were increasingly ecologically jeopardized due to overfishing. Is there some irony in the fact that tuna – some of the most unsustainably harvested ocean fish – have become nearly conflated with dolphin conservation? One way to view the success of the dolphin safe tuna campaign and the simultaneous failure of ocean fisheries conservation (tuna included) is as a victory of **animal rights** over ecological ethics (see Chapter 5).

**Figure 12.5**   Still image from the Sam LaBudde video. Caption from the Earth Island Institute website: "Dead dolphin being dragged off the deck of a tuna boat, from an undercover video." *Credit*: Samuel La Budde/Earth Island Institute.

### Rights for "noble creatures": The case against dolphin-setting

When speaking of "conservation" or "ecology," most discussions of right and wrong actions deal with the plight of *species*. The United States, for example, passed the Endangered Species Act (ESA) in 1973. The intent of the ESA is to stop plant and animal species from going extinct. The ESA has a complicated and mixed history, with a few spectacular success stories and a number of notable failures. Let us take the American alligator, one of the ESA's success stories, as an example. The alligator lives in swamplands of the United States, from North Carolina south and west to eastern Texas. By the 1950s, alligator numbers had dropped to all-time lows throughout almost all of its native range due to habitat loss and market hunting (Ellis 2008). The alligator was listed as an endangered species upon enactment of the ESA in 1973. As such, alligator hunting was banned and habitat conservation measures were enacted. In 1987, the US Fish and Wildlife Service (the agency in charge of managing endangered species) declared the alligator fully recovered and it was removed from the endangered species list. Since delisting, every state with alligator populations has enacted a hunting season on alligators. As long as alligator populations are managed so that they are not again threatened with endangerment, the states can allow hunting of the animal, and have more leeway regarding the loss of alligator habitat as well.

> **Animal Rights**   An ethical position and social movement that states that non-human animals, particularly intelligent mammals, should be granted rights as ethical subjects on par or at least similar to human beings
>
> **Moral Extensionism**   An ethical principle stating that humans should extend their sphere of moral concern beyond the human realm; most commonly, it is argued that intelligent or sentient animals are worthy ethical subjects

Endangered species protection, as such, is an expression of *ecological* ethics: Through federal legislation society has deemed it *wrong* to allow species to go extinct due to human action. However, we (as a society) never said that killing alligators is wrong. Stated another way, society laid down no ethical mandate concerning the past or future treatment of *individual* alligators. Such measures, as such, are expressions of ecological ethics but *not* statements regarding animal rights. However, questions of environmental ethics, Singer argues, need not stop at the level of species.

As reviewed in Chapter 5, Peter Singer (1975) argues that we should extend our moral concern to all creatures that have interests. He is careful to make clear that such a call for **moral extensionism** is not meant to imply that all beings are *equal* and therefore holding *equal rights*. Different animals, rather, have different types and degrees of interests. Consideration of this can inform discussions of right and wrong actions toward individual animals. The success of the dolphin safe campaign is a testament to the ubiquity of this sentiment. Dolphins are intelligent, highly evolved animals who can most definitely *suffer*. As such, they fit highly into Singer's (and no doubt any animal rights activist's) hierarchy of interests. It is easy, therefore, to make a case that it is *wrong* to drown hundreds of thousands of dolphins a year in tuna nets. For that matter, the argument can be (and was) made that it is wrong to kill *any* dolphins in tuna nets. It is no wonder that so many people were so appalled by the images from the LaBudde film – images of obviously sentient creatures meeting rather horrible fates.

It is clear that there is, and has been for some time, strong popular support for outright *moratoriums* against the killing of any and all marine mammals. Witness the widespread international opposition to Japan's maintenance of its "scientific whaling" program, even when only minke whales, a biologically abundant species, were being killed. The public outrage against ETP dolphin mortality and Japan's whaling program shows that these are cases, primarily, for animal rights.

**Figure 12.6**   Does a tuna have rights? Mutilated tuna rest here on pallets at a seafood wholesaler in Tokyo, Japan. The image accompanied a "tree-hugger.com" article about overfished tuna stocks. No mention was made of the tunas' rights or their dignity being violated, either before or after catch. *Credit*: Copyright (2008) P. Y. Yee, Singapore.

**Social Construction**   Any category, condition, or thing that exists or is understood to have certain characteristics because people socially agree that it does

Tuna, on the other hand, do not sit highly in such a hierarchy of interests. Even for most activists, tuna are not perceived as holding interests comparable or even close to dolphins. We simply do not empathize or identify with tuna like we do with dolphin. Imagine the outcry if mutilated tuna (Figure 12.6) were dolphins. Granted, this treatment of tuna is anathema for some vegans and radical animal rights activists. For most people, however, this is simply a picture of fish for sale. Moreover, the dolphin safe campaign did little or nothing to ameliorate any of the other human impacts on the habitat of tuna and dolphins, or the overall condition of the seas. The success of the dolphin safe campaign can be chalked up more as a victory for animal rights than for ecological ethics, therefore. It might also be said that it is the **social construction** (see Chapter 8) of dolphins as charismatic species with interests, relative to tuna or any other sea creature, that makes this distinction, and its confusing ethics, possible.

## Can a rights victory produce an ecological defeat?

Reading the dolphin–tuna controversy as one involving two sometimes conflicting types of environmental ethics, animal rights and ecological ethics, we gain clues as to why ocean conservation lags so far behind marine mammal conservation. One answer, perhaps, is this: Marine mammal conservation has always been fueled primarily by popular sentiment for animal rights. Fish, on the other hand, are not viewed with such benevolence. (Indeed, bestselling author Zane Grey described yellowfin tuna as "stupid pigs" in his 1925 high-seas adventure book *Tales of Fishing Virgin Seas* (Safina 2001: 185).) Fish conservation law, on the other hand, is guided by a management perspective aiming for maximum sustainable yield. The imperative to always maximize fish harvests invariably overrides considerations of the maintenance of ocean ecology (even as catch projections are informed by scientific studies). This is not to say that society lacks an ecological ethic of the sea. Rather, the dominant ethic is that it is *right* to manage the seas for maximum harvest (or, perhaps, maximum *profit*). This ethic does not result in conservative (or *precautionary*) approaches to fisheries management, and, as such, could be argued to be only a slight improvement over the earlier era of unregulated "tragedy of the commons" overexploitation of the oceans. This would support the common, broader argument made by many ecological ethicists that, until society develops a truly ecological ethic, the increasingly unsustainable exploitation of nature will continue.

What implications does this have for the oceans? Consider this: tuna sales in the United States are at all-time highs. Moreover, *every* can of tuna sold in the United States carries the dolphin safe label. So what happened to all the boats that were setting on dolphin in the ETP when the "dolphin safe" campaign drastically reduced demand for their product? One thing is for sure. They did *not* simply revert to pole and line fishing. While this would have reduced stress on the yellowfin, eliminated stress on the dolphins, and resulted in new jobs for fishers, it was not a cost-effective option for boat owners and thus not even considered a realistic option (see the section on political economy above).

The purse seiners, rather, did one of two things. Some boats remained in the ETP, but switched from setting on dolphins to setting their purse seines around logs and other floating objects, under which (like dolphin schools) yellowfin tend to aggregate. The problem is, yellowfin schools under dolphin tend to be much "cleaner" catches than those under logs; meaning that nets set on logs "can entail hundreds of times the bycatch that netting around dolphins does" (Bloomberg Newswire 2008: 436). Considering that 30–40 percent of all bycatch is dead before it is thrown back, and much of what is thrown back alive dies soon from the associated stress, increasing bycatch by two or three orders of magnitude could be viewed as a substantial ecological tradeoff for the lives of individual dolphins.

Still more ETP purse seiners, however, simply moved to the Western Tropical Pacific (WTP), where dolphin bycatch is not a problem. The resulting, massive increase in WTP yellowfin harvests has resulted in stocks that are now either fully fished or overexploited. Indeed, some Pacific yellowfin stocks are so severely depleted that eight Pacific Island nations, concerned about collapse of the fishery, have called for an international moratorium on tuna fishing in an area the size of Alaska (Back et al. 1995).

Many consumers see the "dolphin safe" label and assume it means they are purchasing an eco-friendly product. Consumers have been given the *green* light to purchase canned tuna and feel good about doing so. And, granted, thousands upon thousands of dolphins *were* spared thanks to the campaign. However, the main *ecological* problems intensified: bycatch increased and overfishing just moved west. Is there something *wrong* with that?

## The Tuna Puzzle

In this chapter, we have learned that:

- Tuna exist in almost unimaginably large numbers in the sea, yet nearly everywhere they exist, they are being unsustainably harvested.
- Wild tuna stocks have suffered from multiple new fishing technologies (purse seiners, longliners, tuna ranching) as each technology has allowed us to pull more fish out of the sea.
- Tuna fetch a high price; as usual, when there is big money to be made, ecological considerations can get put on the backburner.
- In the Eastern Tropical Pacific Ocean, schools of large yellowfin tuna swim below easily spotted schools of dolphins so for many years, dolphins died so that consumers could get cheap canned tuna.
- Green consumer campaigns have successfully re-regulated tuna production through "dolphin safe" labeling, suggesting that through markets, consumer advocacy can move corporations to "do the right thing."
- Political economy approaches, conversely, urge us to view market-based solutions with a critical eye, closely assessing the altered sites and scales of power that accompany each market-based success story, even "green" ones.
- Ethics approaches show us that, at least in the ocean, our ecological ethics has yet to catch up with our support for animal rights.

Tuna are only one small node in the complex network of relationships that are the world's oceans. Large-scale overfishing of many other species, massive marine pollution, and numerous other impacts are radically transforming the seas (to say nothing of global warming), well beyond the tangle of dolphins in nets. Nevertheless, tuna is perhaps one of the clearest avenues to help us understand the power our individual consumption choices exert on global ecosystems, while simultaneously underlining the deeply structured way the seas are incorporated into global trade. To the degree that reforms of tuna fisheries have occurred, there may be precedent for new ways of dealing with the global crisis facing the oceans. To the opposite degree, however, that massive efforts to restrain extraction have resulted in only nominal ecological changes, the case of tuna points to the long swim ahead.

## Questions for Review

1. How valuable was bluefin tuna meat a century ago? How about today? How has this change affected their harvest?
2. Describe the following three methods of commercial fishing: pole and line, longlining, and purse-seine fishing. Over the past few decades, which methods have declined and which have grown? (And why?)

3. Who launched a consumer boycott to save ETP dolphins? Was it successful? Explain.
4. From a fishery management perspective, what is different about marine fisheries located within 200 miles of coasts compared to open ocean fisheries located farther offshore?
5. Of tuna and dolphin, which are more common recipients of moral extensionism? Why is this so?

## Suggested Reading

Grescoe, T. (2008). *Bottom-Feeder: How to Eat Ethically in a World of Vanishing Seafood*. New York: Bloomsbury USA.

Mansfield, B. (2004). "Neoliberalism in the oceans: 'Rationalization,' property rights, and the commons question." *Geoforum* 35: 313–26.

Rogers, R. A. (1995). *The Oceans are Emptying: Fish Wars and Sustainability*. Cheektowaga, NY: Black Rose.

## Exercise: Eco-Labeling and Certification

Tuna are only one of countless products labeled for "safe," "green," or "sustainable" consumption. But who are the organizations that oversee the certification of these products? What procedures do they use? Where competing labels exist, how do they differ? Find an eco-labeled product and try your best to answer the following questions: 1) What does the label assure? 2) Have the characteristics of the product assured by the label improved or changed environmental practices of the companies making the product? What do the words on the label mean (for example: "natural"). What does the label not assure? 3) Who oversees that label; are they "third parties" (people or groups apart from the company)? 4) What procedures does the product undergo to claim the label and how is that confirmed? 5) Are there competing labels that assure the same thing, provided by differing groups/companies? How do they differ?

Now consider the following: how much time and labor was it to confirm that the label in question is reliable and does what it says? Can we confirm their claims? What would it take to do so for all such products you consume, assuming you wanted to? How much *trust* must be placed in labels? Is it well placed?

# 13

# Bottled Water

*Credit*: Jason Keith Heydorn/Shutterstock.

## Keywords

- Commodity
- Common property
- Desalinization
- IPAT
- Life cycle analysis
- Overproduction
- Primitive accumulation
- Risk assessment
- Risk communication
- Risk perception

## Chapter Menu

## A Tale of Two Bottles

Luisa Guzman climbs the steps of a makeshift street on the outskirts of Tijuana, Mexico carrying a five-gallon bottle of water, a more than 40-pound load. The winding alley is steep and runs with sewage, part of a labyrinth of recently made roads that snake along the hilly flanks of one of the fastest-growing cities in the world. Luisa carries her heavy load through the streets of this *colonia* – an informal, unplanned, squatter settlement – towards her home on the high hills. This encumbered walk is a daily task in her household, where these bottles are the sole source of drinking water for her family.

The underfunded Comisión Estatal de Servicios Públicos de Tijuana, an agency charged with supplying water to this city of five million people, is incapable of connecting most of these settlements to the municipal water supply. Perhaps 25 percent of Tijuana residents live "off the grid," therefore, with no running water coming from their taps. As a result, city residents are incredibly creative in securing access to water. Luisa harvests water from her roof and stores it in large plastic barrels. Her neighbors nearby, with equal inventiveness, have indirect access to water by illegally tapping nearby municipal pipes. All of these households use these water sources for critical household purposes like bathing, and then reuse the bath water for multiple loads of laundry, finally discharging the remainder to their gardens. In this way, Tijuana residents, and people like them throughout Mexico, are among the most efficient users of water in the world.

But drinking water poses an exceptionally difficult problem, so most households depend exclusively on bottled water. These five-gallon plastic jugs are sold off of trucks throughout the city by water vendors. At 10 pesos for five gallons (around US $1), this is a modest but critical part of Luisa's household budget, and it is a standard expense not only for *colonia* dwellers throughout the city, but wealthier households as well. The sources of the bottles vary, but in most cases, they are filled directly from the municipal supply, making the bottled water business a lucrative trade for those who sell the state-provided commodity to citizens like Luisa who have not been linked to basic services.

A mere few miles north across the US border in the city of San Diego, Carlos Perez turns his Prius, off of Harbor Drive, into the parking lot of a 7-11 convenience store facing the sea. Entering the store, he walks to the back refrigerator, pulls a 16-ounce bottle of Dasani water from the case, and brings it to the register. He pays $1.49 (approximately $60 for five gallons at this rate) and drives home.

Like almost all San Diegans, Carlos is connected to a municipal water and sewer supply that provides tap water around the clock, every day of the week. Nevertheless, he will consume roughly 30 gallons of bottled water in a year, near the national average, spending hundreds of dollars annually for a product provided by his tax money for free. Certainly, he likes the taste of the water he buys in the store, and he feels safer with water that has undergone "reverse osmosis." Even so, like Luisa's water, Carlos' is nothing more than treated local tap water, indeed, the very same Colorado River water that supplies Tijuana. But unlike bottled water in Tijuana, that sold in San Diego is wrapped with a slick blue label by the Coca-Cola Company and sold at a price 60 times higher than its Mexican counterpart, and hundreds of times more expensive than its tap water source, which costs fractions of a penny per gallon.

How did a ubiquitous, free-flowing, fundamental building block of all life on earth come to be a commodity that is captured, marketed, and distributed around the world? To what degree does the marketing of water reflect a real state of scarcity, a plausible response to a health risk, or merely the clever gimmick of a marketing executive? And with what different implications for Luisa and Carlos, for Mexico, the United States, and the rest of the world?

## A Short History of Bottled Water

Presumably, water has been bottled since the advent of bottles, dating to the time of the first clay vessels, thousands of years ago. Given the uneven distribution of water over space (stream sources and wells are unevenly available) and time (rainfall and snowmelt are seasonal and often unpredictable across most of the world), storing transportable water is absolutely nothing new.

Yet water as a marketed commodity, specifically bottled for resale and consumption at distant sites, is a relatively recent phenomenon. The earliest markets for bottled water were rooted in the perceived health benefits of "spa waters," sourced in hot springs throughout Europe, like the famous waters of Bath Spa in England and Vichy in France. Perceived to have medical value, these were captured and sold around the world as early as the 1700s. Later advances in chemistry led to a decline in the sale of these waters, as many of their purported benefits were debunked in the 1800s. The passage of the Pure Food and Drug Act of 1907 in the United States, followed by similar mandates around the world, spelled the end of widespread medicinal use of *spring* water, along with a range of products whose benefits were fraudulently exaggerated (i.e., "snake oil"). The persistence of a perception that bottled *mineral* waters are healthful and scarce, on the other hand, helped sustain bottled water drinking throughout the nineteenth century, all the way to the present day (Beverage Marketing Corporation 2009).

By the beginning of the twentieth century, "artesian" spring waters and other water from natural sources entered a new market phase, especially throughout North America. Many brands familiar today (Poland Springs, for example) were first established in the middle of the 1800s. These waters were still typically associated with spring waters simultaneously operating as therapeutic retreats and baths. Such bottled products were therefore somewhat elite, certainly expensive, and not in any way typically consumed by average American, British, and European people who were just then beginning to receive reliable modern municipal tap water (Figure 13.1).

**Common Property** A good or resource (e.g., bandwidth, pasture, oceans) whose characteristics make it difficult to fully enclose and partition, making it possible for non-owners to enjoy resource benefits and owners to sustain costs from the actions of others, typically necessitating some form of creative institutional management

In other parts of the world, bottled water sales during this period were entirely unheard of, and would likely have seemed bizarre. For the average agrarian producer or rural-dweller in India, Mexico, Egypt, or China in 1900, pure water sources were available in the form of local wells or tanks, and a significant labor input (typically by women) would have been dedicated to drawing freely available water from these common areas (Figure 13.2). Even today, in much of the world, water is considered a free good and part of the **common property** of communities.

Later in the twentieth century, bottled water became more common and broadly distributed in industrialized countries, especially in commercial settings. The ubiquitous "water cooler" was a commonplace office artifact by the 1950s. Home delivery of bottled water occurred, but for only a small portion of the population. This would change radically, however, in the closing years of the twentieth century. Beginning in the 1970s, elite international brands of waters, including Evian and Perrier, began to make their way into consumer markets. Though sustaining a small "niche" consumer market, these products did not make a significant dent in typical consumption patterns. In the 1990s, however, an incredible acceleration of bottled water consumption began around the world, with continued acceleration until the present.

**Figure 13.1**  The Poland Spring "Spring House" in 1910. Like other naturally occurring springs, these water sources had become elite destinations in the 1800s. Markets for their waters were associated with luxury, health, and opulence. They were not a typical household expense, however, for working people who were only then beginning to receive regular water delivery through modern plumbing. *Credit*: Photograph provided courtesy of the Poland Spring Preservation Society.

## The current global state of the bottled water market

By 2009, an enormous range of bottled water products had become available to consumers. These products are very different, and treated somewhat differently under national and international laws. Specifically, bottled waters include:

- *Spring water and "artesian" spring water*. This is water drawn from a single underground water source. Spring water can flow naturally or be forced to the surface by pumping. In the case of "artesian" water, this merely represents groundwater from a confined aquifer where the water source flows to the surface.
- *Mineral water*. To be considered real mineral water, the water must contain 250 parts per million of minerals, and

**Figure 13.2**  Women draw water from a communal well in Rajasthan, India. Though groundwater declines threaten these rural water supplies, villagers depend on freely available communal water sources for survival, and upon the significant women's labor require to draw, carry, and dispense water every day. *Credit*: Trevor Birkenholtz.

they must occur naturally and be derived from the water source. The actual minerals typically vary enormously, but might include calcium, sodium, magnesium, and fluoride, among others, thereby accounting for the distinct taste.

- *Purified water.* By far the most common form of bottled water, this is water that may come from surface water sources (like rivers and streams) but often comes from a municipal source. The water is treated through filtration, and perhaps through "reverse osmosis" (see below) or other related techniques. This is not much more than expensive tap water and represents the vast majority of the bottled water market.
- *Fortified products.* This includes a range of new bottled waters with additional nutritional additives, like vitamins or electrolytes.

Between 2002 and 2007, global consumption of bottled water of all kinds grew from 34 billion gallons annually to 49 billion, representing a 7.6 percent annual increase over the period. The leading consumer nations of bottled water are diverse and represent tremendous differences in the types of markets for this commodity. In overall consumption, the United States leads the world, consuming nearly nine billion total gallons of bottled water annually, followed by Mexico at 5.8 billion gallons. In terms of per capita use, the world leader is the United Arab Emirates (UAE), followed by Mexico, a number of European countries, and the United States. The average person in Mexico consumed 68 gallons of bottled water in 2007. In the United States, each person consumed almost 30 gallons in that year. People around the world are clearly drinking more bottled water every year (Landi 2008) (Figure 13.3).

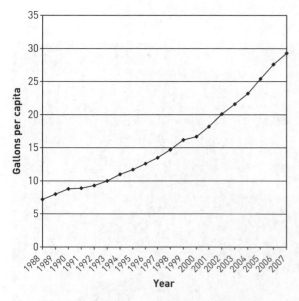

**Figure 13.3** US per capita consumption of bottled water from 1988 to 2007. The steady increase in US bottled water consumption shows no signs of abatement. Consumption has quadrupled in the past 20 years. *Source*: Adapted from Landi (2008).

This represents a massive growing economy. In 2007, sales of bottled water in the United States amounted to more than 11 billion dollars. This puts bottled water in the current number two sales position for all beverages marketed in the United States, behind only carbonated soft drinks. The price of these branded waters varies enormously, with globally traded waters like "Fiji" and some authentic geographically-specific spring waters holding the more expensive end of the market. Most markets are dominated by domestic mid-priced products, like "Dasani" (owned by Coca-Cola) and "Aquafina" (owned by Pepsi), which account for a lion's share of sales. Notably, these mid-market domestic products are sourced directly from the municipal water supplies in the cities where they are sold.

In part, the rise of bottled water consumption, in both poorer and wealthier countries, arises from a perception (both justified and unjustified, as we will see below) that traditional and municipal water sources are compromised. Recent outbreaks of water-borne disease (as in cryptosporidiosis, see Box 13.1), nitrogen alerts, and other crisis events have certainly contributed to consumer concern over water quality. This is tied to a less coherent, but no less persuasive, feeling amongst consumers in wealthy countries that bottled water is "natural," a social construction that associates bottled water with healthy and environmentally friendly living. And to be sure, relative to corn-syrup-laden soft drinks that otherwise dominate the market, bottled water is probably an improvement in this regard.

In countries like Mexico, India, and Egypt, on the other hand, the water system has been dramatically transformed in recent years, driving demand for bottled water in both rural and urban areas. In part, this demand has been a result of significant urbanization where, as in our introductory story, the capacity of municipal services to keep up with demand has been outstripped by growth. On top of this, however, has been a change in the way water is owned and managed around the world. In many countries water services have been privatized, often as a result of poor state management, and the distribution of water has come to be governed through water markets. In such cases, the purchase of water, in bottles or directly from tanker trucks, is increasingly a fact of daily life.

## Environmental impacts of bottled water

Contradicting the warm "green" feelings often associated with bottled water (and its typical labels depicting pristine mountain streams), bottled water has unquestionable environmental costs and externalities. Consider first the package itself. Water bottles are made from plastic (PET or polyethylene), a product that is itself derived from petroleum; for every kg of bottle, roughly two kilos of petroleum are required. Every bottle consumed propels carbon into the atmosphere. Second, consider the energy and material impacts of simply moving water around in this way. While it is true that most bottled water comes from regional sources (often the city tap water in the town where it is sold), it must nevertheless be bottled, trucked, or shipped from place to place, requiring the consumption of petroleum and the expulsion of more carbon dioxide, adding further environmental costs to a product that is actually very close at hand for most consumers. Finally, the life cycle of the typical water bottle ends in a landfill, with the associated costs and impacts of garbage management more generally. These include the costs and environmental impacts of hauling, as well as the land demands for garbage disposal. More than 30 million tons of plastic entered the municipal solid waste stream in 2007, less than 7 percent of which was recovered (Päster 2009). Certainly water bottles are recyclable. But only a small proportion of bottled water containers ever are.

The total footprint for a bottle of water depends enormously on brand and location so a full **life cycle analysis** of the product remains elusive. So too, differing assumptions can lead to dramatically different estimates. Still, one well-informed calculation for the most exotic of all bottled waters – Fiji brand – suggests a staggering total. The bottle itself likely requires more water to produce than it contains. In addition, fuel demands are estimated at 160 grams of fossil fuels to produce the bottle in China, 2 grams to deliver it to Fiji, and 81 grams to get it to

**Life Cycle Analysis** The rigorous analysis of the environmental impacts of a product, service, or object from its point of manufacture all the way to its disposal as waste; also known as cradle-to-grave assessment

a store in Europe or America (Hunt 2004). Weighed against the near negligible environmental costs of a glass of tap water, bottled water's environmental footprint is dramatically higher.

## The puzzle of bottled water

This brief review of the history of bottled water suggests several things about the relationship between environment and society. First, it is a relatively recent phenomenon, meaning that it is a product that can only be explained with reference to specifically current changes in human life. Second, it is a case of a product that, despite being wrapped in an image of good health and clean living, has negative environmental implications.

This review stresses several things about bottled water that make it a specific kind of *puzzle*:

- Bottled water has been associated historically with rarified health benefits and elite patterns of consumption, making people's desire for the product more than a merely rational choice.
- Increases in bottled water consumption in industrialized nations exploded in the last decade as a result of a combination of perceived health risks from public supply and a marketed "lifestyle" choice associated with affluence.
- In "underdeveloped" or poorer nations, on the other hand, bottled water use is on the increase as traditional community supplies have become scarcer, more polluted and increasingly privatized.
- In both industrialized and poorer nations, water has become more expensive, and consuming bottled water has increased human environmental impacts associated with its production, transportation, and landfilling.
- All of this points to the strange case in which – in a few short years – a freely available, ecologically innocuous, and historically communal resource has become an environmentally detrimental privatized commodity.

Bottled water therefore provides lessons for understanding the evolution of human resource use over time under very different development conditions, but it also presents a *puzzle*: given its rapid turn towards a commodity, does bottled water represent the solution to a problem of water scarcity, a rational response to risks associated with declining water quality or, rather, an artificial market cleverly engineered to sustain profits? As with other objects explored in this book, differing perspectives provide differing insights into these questions, again with highly contradictory answers.

## Population: Bottling for Scarcity?

It is commonly observed that water on the earth is both ubiquitous and terrifically scarce. A quick look at the global water budget confirms that there is plenty of water, though not in the places and forms most needed by people. Of the earth's roughly one and a half billion cubic kilometers of water, approximately 97 percent is the saline water of the oceans. Much of the rest is in glaciers and groundwater, leaving 0.01 percent of the total water in lakes, rivers, and circulating in the atmosphere. In this sense, water is absolutely scarce. In quickly growing underdeveloped countries, moreover, roughly a quarter of the population goes without access to safe drinking water. This is a significant problem given that water-borne disease accounted for three and a half million deaths worldwide in 1998 alone (Rogers 1996).

With population continuing to grow over the past 20 years at a rapid pace (see Chapter 2), and the total amount of available water essentially fixed, one might reach the conclusion that bottled water is a solution to the increasing scarcity imposed by dwindling available resources. As locally available sources of water become more difficult to access, are compromised by human activities, or are drained completely, we can imagine a thirsty planet turning to middlemen who transport water from locations of greater availability, marking up the price to cover costs.

Looked at another way, however, the scarcity of water is really quite relative; a great deal of freshwater exists on earth (approximately 55,000 cubic kilometers), though it is not equally distributed or efficiently utilized. The total amount of freshwater available in sub-Saharan Africa, for example, is around 6,000 cubic meters per person. People living in the region utilize barely 2 percent of that for agriculture, industry, and domestic purposes combined. In Canada and the United States, on the other hand, people use at least a quarter of the water available. Table 13.1 describes the usage of renewable freshwater resources around the world (meaning the use of surface water and groundwater), as well as the percentage of water use in different sectors and the average quantity used domestically every year.

Notable here is the relative degree of use in differing sectors, across differing regions. While Europe uses about a third of its water for agriculture, Asia dedicates a full four-fifths to that usage. Also outstanding are the staggering differences in domestic usage between

**Table 13.1**  Freshwater usage around the world. The total average withdrawals of freshwater per person vary enormously around the world, as do the proportions of those dedicated to domestic use

| World region | Per capita annual withdrawals ($m^3$) | Agricultural withdrawals (%) | Industrial withdrawals (%) | Domestic withdrawals (%) | Domestic withdrawals ($m^3$) |
|---|---|---|---|---|---|
| North America | 1,663 | 38 | 48 | 14 | 233 |
| Oceania | 900 | 72 | 10 | 18 | 162 |
| Central America & Caribbean | 603 | 75 | 6 | 18 | 109 |
| South America | 474 | 68 | 12 | 19 | 90 |
| Europe | 581 | 33 | 52 | 15 | 87 |
| Middle East/North Africa | 807 | 86 | 6 | 8 | 65 |
| Asia | 631 | 81 | 12 | 7 | 44 |
| Sub-Saharan Africa | 173 | 88 | 4 | 9 | 16 |

*Source*: Adapted from World Resources Institute, Earth Trends Data Tables

**Table 13.2**  Typical interior water usage in the developed world. Faucet usage is minimal proportion of water use in the average home

| Use | Percent of total |
| --- | --- |
| Toilets | 28.4 |
| Washing machines | 21.2 |
| Showers | 21.2 |
| Faucets | 11.7 |
| Bath | 8.9 |
| Toilet leakage | 5.5 |
| Dishwasher | 3.1 |

*Source*: Rogers (1996)

regions, with North Americans each using almost 15 times as much water as their sub-Saharan African counterparts. If bottled water usage is rising in some underdeveloped countries – like Mexico, for example – it is at least in part because available water is not directed to domestic use (owing to poor infrastructural development). It may also be a product of demands and potential waste in non-domestic sectors, especially agriculture.

Broken down further, the average American's domestic water usage is also revealing. As is clear from Table 13.2, much domestic water usage is dedicated to water-demanding toilets and other conveniences. So too, these are all highly amenable to technological improvements and conservation measures (5 percent of household use comes from toilet leakage!). If water is a product of scarcity in the United States, then it is a scarcity that results largely from inefficiencies.

This matters in terms of water scarcity for consumption in a number of ways. In places where vast quantities of water are dedicated to agriculture, for example, the technological efficiency of irrigation becomes the main determinant of water available for other uses. If, as in Israel, irrigation uses a highly efficient drip system, then water left over for household use will be high. So too, where household uses are more efficient, as where low flow toilets replace water-demanding and leaky ones, the total water freed for any given population increases dramatically.

## Who drinks bottled water?

When we compare these usage estimates against the countries where bottled water has a large and growing market, moreover, the picture becomes even more complicated. Table 13.3 shows a selection of the countries who lead the world in bottled water consumption, including the top three countries, United Arab Emirates (UAE), Mexico, and Italy. Starting with the case of the UAE, we see an extremely high annual rate of bottled water consumption. This country (really a federation of princely states) is not a heavily populated one, indeed it is tiny, but it consists of wealthy oil states located in a desert with almost no freshwater resources at all. Extensive consumption of bottled water is unsurprising here.

**Table 13.3** Selected leading per capita consumer nations of bottled water, 2007

| Rank | Countries | Bottled water consumption per capita 2007 (liters)[1] | Actual renewable water resources per capita (m$^3$)[2] | Population density 2002 (people per km$^2$)[2] | Per capita GDP 2008 est. (US $)[3] |
|------|-----------|------|------|------|------|
| 1 | UAE | 259.7 | 49 | 32 | 40,000 |
| 2 | Mexico | 204.8 | 4,357 | 52 | 14,200 |
| 3 | Italy | 201.7 | 3,336 | 191 | 31,000 |
| 5 | France | 135.5 | 3,371 | 108 | 32,700 |
| 9 | USA | 110.9 | 10,333 | 30 | 47,000 |
| 15 | Croatia | 92.0 | 23,890 | 82 | 16,100 |
| 18 | Thailand | 89.3 | 6,459 | 125 | 8,500 |

[1] *Source*: Adapted from Beverage Marketing Corporation (2009)
[2] *Source*: World Resources Institute, Earth Trends Data Tables
[3] *Source*: CIA World Factbook

Hydrology and climate tell us little, on the other hand, about the situation in Mexico, where freshwater resources abound. Here, water access is especially problematic in rapidly urbanizing areas with little infrastructural capacity to meet citizen needs. Under conditions of considerable poverty and large-scale informal urban settlement (as described in the opening of this chapter), the adoption of bottled water is also unsurprising, though for dramatically different reasons. Much the same might be said of Croatia and Thailand.

For Italy, France, and countries like them (including the United States), neither of these explanations appears as compelling. Here are nations with modest overall populations, well-developed infrastructures, and relatively abundant natural water sources. And yet they sustain significant bottled water markets. As we shall see, some proportion of this consumption is driven by the perception in these places that, while public water may be abundant, its quality may be questionable. Where this is not the case, it is also evident that bottled water is a mark of a certain kind of lifestyle in these countries, perceived to be desirable.

While increasing the number of people on earth clearly marginally increases the stress on the global water system in an abstract way, therefore, the immense variability in global conditions means that the wealth of countries and their infrastructural technology certainly play the key role in determining water use, as per the **IPAT** formula laid out in Chapter 2. In this formulation, you will recall, the *impact* of people on the environment is the result of a combination of factors, taken to include the overall quantity of human *population*, but exacerbated or mediated by the *affluence* of that population and the available *technology* they use to live. This is certainly true in a number of ways. As noted above, water use technologies profoundly influence the proportion of water available for household use and for human consumption.

**IPAT** A theoretical formula holding that human Impact is a function of the total Population, its overall Affluence, and its Technology; this provides an alternative formulation to a simple assumption that population alone is proportional to impact

Similarly, water treatment infrastructure – and its notable absence in many underdeveloped settings – heavily impacts the real need for a treated supply of bottled water. Even when viewed in this light, however, the critical variables do not appear to be the number of human consumers of water, but instead the range of technologies and money resources that control the flow of water. Put this way, drinking water scarcity driven by human population stress alone does not explain the explosion of bottled water consumption around the world in the past decade.

Still, some sort of relative scarcity is at work, through highly variable sets of local conditions and human expectations. One can conclude that bottled water is best viewed as a response to scarcity, therefore, but clearly of at least three different kinds: hydrological scarcity, techno-economic scarcity, and perceptual scarcity. In the first case (hydrological scarcity), some unique combinations of climate, affluence, and human population, as in the oil-rich Gulf States most notably, present a hydrological regime where bottled water is an almost inevitable response. Such places are also the ones most likely to turn to other capital and energy demanding water solutions, including **desalinization** of sea water.

> **Desalinization**  A technology that removes salts and other minerals from water, especially sea water; prohibitively expensive in most contexts, current techniques are highly energy demanding

In the second case (techno-economic), we find conditions of underdevelopment exacerbated by swiftly urbanizing populations moving into burgeoning municipalities. Combined with slipshod water distribution and treatment facilities in these cities, a form of scarcity emerges where we might predict the occurrence of large-scale bottled water markets. Potable water here is scarce, even though natural water supplies are plentiful.

In the final case, even where water and water treatment are widespread, a perception that bottled water is safer or healthier may account for a kind of scarcity (perceptual scarcity), where affluent consumers turn to an alternative for a widely available staple resource, like water. In countries where urban populations are growing by leaps and bounds, where income is highly stratified and infrastructure is unevenly developed, and where climate and hydrology are highly variable (as is the case in Mexico), one might predict all three forms of scarcity.

## Risk: Health and Safety in a Bottle?

Leaving aside the problems of underdeveloped safe water supplies in the poorer parts of the world (to which we will return shortly!), it is possible to ask whether consumers in developed countries are making a kind of risk assessment when they choose to drink bottled water. In this case, when relatively wealthy consumers with plenty of available tap water turn to Dasani and Evian, questions arise about the actual condition of public water supplies, and people's perception of water quality. Ignoring for a moment issues surrounding the taste of bottled and tap water (see the next section for more on this), it is likely that a significant proportion of bottled water drinkers purchase the product for health and safety reasons. Though survey results on this question vary, many consumers – as many as half – report either that bottled water is a "healthier option" or that tap water is "riskier" than bottled water (Napier and Kodner 2008).

This makes bottled water an extremely interesting puzzle to which we might apply insights from the field of risk analysis and risk perception (Chapter 6). As you will recall, these approaches to environment and society issues examine the world in terms of hazards, those features that may or may not be dangerous to people. These fields address such hazards by examining the role of risk – a probability of harm – not only in terms of how risky certain decisions are, but also why people assign levels of risk to actions and behaviors that may be far higher or lower than experts would suggest.

For water, we can reasonably ask what, if any, actual risks are associated with tap water – a **risk assessment**. If the actual risks of tap water are not comparable to the ones people imagine, we might also consider the problem of **risk perception**, that not-always-rational, socially and culturally influenced, cognitive process where people come to evaluate the risk of a product or behavior. To the degree that people perceive the risks of their tap water in a way that may be misleading, we might finally consider techniques in **risk communication**, which seek to convey technical information (about trace elements in water, for example) in an intuitive way, and so lead to better informed decisions.

**Risk Assessment**  The rigorous application of logic and information to determine the risk – possibility of an undesirable outcome – associated with particular decisions; used to so reach more optimal and rational outcomes

**Risk Perception**  A phenomenon, and related field of study, describing the tendency of people to evaluate the hazardousness of a situation or decision in not-always-rational terms, depending on individual biases, culture, or human tendencies

**Risk Communication**  A field of study dedicated to understanding the optimal way to present and convey risk-related information to aid people in reaching optimal and rational outcomes

### Risk assessment: Is bottled water "healthy" or "less risky"?

The earliest bottled waters were spa waters touted for their health benefits, deriving mostly from ideas about their mineral content. In theory many minerals commonly occurring in spring water are good for people, including calcium and magnesium, for example. Research on these waters, however, is highly inconclusive, because the contents of each source vary enormously. This is also true of tap water around the world, making any health claim for choosing bottled water over tap water generally untenable. Given that many bottled waters *are* tap water, this comparison becomes all the more fruitless. What's in the bottle is probably no better for you than what comes out of your garden hose (Landi 2008).

Of course, compared to other available beverages, water is always a pretty good option. In the United States, for example, where bottled water still lags behind soft drink sales and remains on par with beer, the beverage profile of the average consumer might certainly benefit from more water.

The bottled water industry – who defends this position vociferously – has a strong argument here. Over the past few decades the growth in bottled water consumption appears to have come in part at the expense of soft drink consumption. In a country like the United States, adult onset diabetes is on the rise, typically caused by high-sugar diets over the course of a lifetime. A switch from cola and root beer to bottled water would probably be beneficial for many people, given rising healthcare costs. This might also benefit society as a whole, since many healthcare costs are borne collectively through social programs and insurance costs.

On the other hand, most regular bottled water drinkers appear to be people who already consume organic foods and "health" foods, raising questions about the degree to which the proliferation of water bottles has really changed the healthiness of national consumption overall (Doria 2006; Napier and Kodner 2008).

The converse question, whether bottled water is safer than tap water, is also amenable to risk assessment. Certainly, in places where the water supply is poorer overall, and during periods of obvious declared water emergencies (like a cryptosporidium outbreak), bottled water is undeniably less risky than tap water.

---

## Box 13.1    Cryptosporidium Outbreaks

Cryptosporidiosis is a disease caused by a tiny protozoan parasite, which results in acute watery diarrhea that, in immune depressed or weakened people, can result in more deadly forms of the illness. The disease spreads by infecting a host, reproducing quickly (causing serious sickness), and being expelled from the host as a microscopic, thick-walled spore (or oocyst) through excretion. Water (or food) becomes contaminated by exposure to the feces of an infected person or animal, and so enter new hosts, starting the cycle again.

The public health challenge of the disease is exacerbated by two features of this marvelously adapted disease. First, the oocysts can persist in the environment for months, providing a long opportunity for a latent contamination to turn into a disease outbreak. Second, these spores are highly resistant to typically reliable forms of disinfectant, including chlorine bleach. Since chlorination is a typically relied-upon mode of water treatment, many water supplies with older treatment systems are vulnerable to the disease. Worse still, the disease is essentially untreatable. The worst symptoms of the disease, especially including dehydration or more severe forms of the illness in immuno-compromised people (such as AIDS patients), can be treated with intravenous fluids and antibiotics. Typically, however, the sickness is forced to run its course.

In recent years there have been a number of "crypto" outbreaks across North America and the United Kingdom. Most notably in Milwaukee during the spring of 1993, the public water supply became fully compromised by cryptosporidium. As a result, over the next few weeks, 403,000 people became ill, of whom 4,400 were hospitalized. A study of the outbreak later revealed that several factors made the outbreak especially severe. First, discovery that the source of widespread illness was in fact "crypto" was delayed by lack of public knowledge of the problem and the specialized nature of testing for the disease. The original source of the disease was never pinpointed, but was likely waste from cattle, slaughterhouse facilities, or humans. The treatment system in at least one of Milwaukee's facilities was inadequate to remove the disease from the water, moreover. Older plant designs and lower-than-necessary (though technically legal) treatment standards were also identified (MacKenzie et al. 1994). Water quality standards have been updated in years since.

The disease does not tend to cause serious illness in normally healthy people and the total number and frequency of outbreaks is so low as to make this a somewhat rare hazard. The politics of risk in this case center on the fact that we may be experiencing a range of unacknowledged (and unknown) low-impact disease outbreaks on a regular basis, raising questions about national and local water quality standards. So too, these several recent outbreaks, and the press they have received, have unquestionably affected people's *perception* of risk of their water supply. Needless to say, bottled water companies have seized the marketing opportunity to make a great deal of these cryptosporidium outbreaks in public water, even though many companies depend largely on these same municipal supplies themselves.

### Reference

MacKenzie, W. R., N. J. Hoxie, M. E. Proctor, et al. (1994). "A massive outbreak in Milwaukee of cryptosporidium infection transmitted through the public water supply." *New England Journal of Medicine* 331(3): 161–7.

But further investigation of bottled water raises real questions about whether or not it is really any safer than other daily sources, especially municipal treated water. Most notable is a revolutionary 1999 report from the National Resources Defense Council (NRDC), which tested bottled water extensively in the United States and examined the regulations in that country at the time. This study concluded that about a third of all bottled waters were simply packaged municipal water, and that of those bottled waters that were effectively tap water, many were not treated any further (Napier and Kodner 2008).

Of course, municipal tap water is already heavily treated. This is because the water that comes to your tap is derived from a huge range of sources, including lakes, rivers, and groundwater, all of which are exposed to a lot of different pollutants, ranging from live bacteria to viruses and metals. It is understandable that municipal water is strictly regulated (in the United States by the Safe Drinking Water Act) and treated extensively. Typically, water is pumped from the source through screens and put in tanks for flocculation and coagulation, where chemicals (like alum) are added to help fine suspended matter drop out. This usually destroys 99 percent of viruses or other contaminants. The resulting water is then filtered through sand and gravel. The water is fluoridated at this point (fluoride is added), and sometimes chlorinated. Ozonation might follow, where ozone gas is introduced to eliminate any possible remaining microbes. In some cases, ultraviolet (UV) radiation might also be applied. Tap water and any bottled water derived from tap water are pretty safe as a result.

The NRDC results concluded, however, not only that bottled waters were no safer than tap waters but in some cases they were *less safe*. US tap water, the study emphasizes, must meet rigorous standards set by the US Environmental Protection Agency, whereas bottled water is regulated by the Food and Drug Administration, which often provides exemptions and loopholes for bottled waters. More recent research has raised further concerns and some bottled spring water sources, for example, may have trace minerals (like arsenic) with negative health implications (Doria 2006).

The water industry is cautious in this regard. The International Bottled Water Association, an organization that includes perhaps 80 percent of the industry, has voluntarily adopted guidelines that member companies follow. This includes an insistence that bottled water advertising should not explicitly compare their product with the safety of tap water and municipal supplies (Napier and Kodner 2008).

Further research has raised questions about the dangers that lurk beyond the water and in the bottle itself. While more urban legends have proliferated on the internet about these purported risks than have careful and rigorous scientific assessments, a body of research on the question is growing. First, it has been suggested that leaving water in polycarbonate plastic bottles (in which bottled water typically comes) over a period of time may increase the presence of undesirable chemicals, like bisphenol A, which has an estrogen-like effect. Research has also raised the question of whether a bottle of water, once opened or refilled, might become home to various bacteria. These potentially include pathogenic organisms that might be especially dangerous to children or older people. Whatever the ultimate results of these studies, they will certainly not demonstrate plastic bottled water to be safer than tap water (Anadu and Harding 2000).

Working under the understanding, therefore, that bottled water is not generally healthier than tap water and by no means any safer, a straightforward rational risk analysis appears not to be at the core of growing consumer water purchases in countries like the United States and United Kingdom, where tap water is ubiquitous and highly regulated. We would do better, therefore, to think of this as a problem of *risk perception*. In this regard, what do people think about bottled water, and why do they think it?

## Risk perception and the limits of risk communication in water quality

Research suggests that a proportion of the population consistently thinks tap water represents some kind of health and safety risk. Bottled water users are similarly most likely to cite health and safety as their leading reason for not drinking from the tap. Indeed, some proportion of bottled water drinkers (as much as a quarter) typically believes that bottled water is entirely free of contaminants (Napier and Kodner 2008).

These views are not surprising if we consider some of the general trends discovered in risk perception science. As you will recall from Chapter 6, people tend to have an inflated distrust of phenomena over which they do not have direct control, relative to things they do control. Drinking water quality, and the complex systems of treatment public water undergoes, definitely is not under the direct control of most people. It is also the case that given the invisibility of water quality risks (you cannot see cryptosporidium), there are few objective ways for the average person to directly know the quality of their water. As a result, people tend to base their judgments on taste or color. Poor taste in public water, however, is not a good indicator of water quality or safety, as all water sources have idiosyncratic and local flavors. As a result, people consistently and irrationally overestimate the hazards of public tap water, and for fairly predictable reasons (Anadu and Harding 2000). On the other hand, this fear is not always entirely irrational. It has also been shown that people who live in towns with public water contamination histories are more likely to worry about their water than those whose water supplies have no record of problems (Johnson 2002) – we would hope so!

Still, most tap water is regulated and safe, and a lot of bottled water is merely tap water. Can biased perceptions be overcome through clearer information about our water supplies? According to risk analysts, there is plenty of room for improvement in the way people think about this problem. Improved risk communication can overcome some of these irrationalities, at least in theory. Specifically, decision scientists assert that, when it comes to water, the public is increasingly interested in side-by-side comparisons of their tap water to that of other towns and utilities, as well as to bottled water. These comparisons, it is believed, will help allay consumer fears about perfectly good tap water, which have in the past caused them to mistakenly report their own water supply as worse than that of other utilities or of bottled water.

Early experiments in this area are promising. Side-by-side "blindfold" comparisons have been conducted in communities where municipal tap water quality is in fact comparable to bottled water quality. Presented with the study results that show that they are unable to consistently distinguish between the two sources, people have become far less likely to under-rate their water supply. Put simply, people can learn to overcome their tendency to dislike their tap water. This new perception, however, has not been demonstrated to actually affect people's behavior, especially their purchase of bottled water.

People who have seen such side-by-side comparisons of their water with bottled water have come away complaining less about their municipal supply, but they still continue to buy bottled water that is no healthier or safer (Anadu and Harding 2000).

This final finding suggests the intractability of the bottled water market to improved risk communication. It hints at the limits of a risk/hazards approach to explaining people's actions. Despite the optimistic hope that better information leads to more rational behaviors, something more deeply rooted may be going on. The view from a political economy perspective, conversely, points to more stubborn forces propelling the bottling of water, beyond individual or group perceptions. Instead, political economy might suggest, bottled water is part of a larger insidious privatization of public goods that both steals from poorer people and imposes itself as a false commodity on wealthier ones.

## Political Economy: Manufacturing Demand on an Enclosed Commons

For thousands of years, drinking water has been a classic common property good, typically drawn from wells and rivers and managed through informal community rules (or institutions, as per Chapter 4). Indeed, in many parts of the world, people's access to water remains free, though important restrictions protect community water supplies against over-exploitation or pollution. Consider, as only one example, that the Koran specifies *Chafa* – the "right to thirst" – which dictates the universal right of all people to drinking water. Much of the world's drinking water supply continues to depend on these community held and operated sources and institutions.

Where large cities have risen over the centuries, these informal mechanisms have typically given way to larger-scale, state dictated and managed systems of municipal water supply. The cities of the Indus Valley civilization in South Asia (more than 4,500 years ago!), for example, were characterized by carefully planned systems of wells and cisterns. The aqueducts of Rome are equally famous, as are the myriad canals of ancient Chinese cities. Throughout the history of civilization, therefore, drinking water has remained a common property resource, overseen and delivered by state authorities and engineers.

Early experiments in turning city water supplies over to private companies, especially in the United States during the late 1800s, were mostly abortive. Water provision for major cities like Los Angeles, for example, only briefly depended on private water companies, whose failure to provide sufficient and well-distributed supplies led to "municipalization" – the expectation and legal obligation of cities to provide water for citizens. Modern water grids and the sophisticated water treatment facilities that go with them are century-old products of city planning.

Much of the development and governance of cities over the twentieth century around the world, therefore, is fundamentally about getting water to flow to people's taps (as well as dealing with sewage, garbage, and roads). The global record on this, of course, has been mixed, with some cities and countries having distinct advantages in terms of money and taxes to develop these forms of infrastructure. Nevertheless, development worldwide has largely been synonymous with municipal and national governments providing safe drinking water to citizens.

## The rise of the water commodity

During the 1980s and 1990s, however, this expectation was challenged by new ways of thinking about water. Public utilities increasingly became "enterprise" units, who were expected to cover their costs by becoming more efficient and setting water fees/prices at appropriate levels. Some utilities were replaced altogether by private companies and water vendors. In cities like Buenos Aires, Argentina and Jakarta, Indonesia, municipal control over water sources and distribution have been surrendered entirely to private firms. In other cases, areas that were inconvenient or difficult to supply privately continued to be serviced by the government, while easier, more lucrative parts of cities were increasingly "cherry-picked" and turned over to private vendors. Where legal systems were poorly suited to these new approaches, regulations were changed or abandoned. Throughout all of these changes, the benefits of efficiency and market logic were widely touted, in keeping with theories derived from economics (Bakker 2004; Grand Rapids Press 2007; and see Chapter 3).

The rise of bottled water, while not an explicit part of the privatization of water services, coincides with this process. Bottled water, like privatization of city water, reflects an overall shift towards private responsibility and corporate profit in water provision. From the point of view of political economy, this trend is seen as largely detrimental and part of a larger set of discouraging trends. Specifically, from this vantage, a shift to bottled water (among other kinds of water privatization) turns water into a **commodity**, a thing marketed for profit. This process embodies two foundational concepts from political economy: **primitive accumulation** and **crisis of overproduction** (see Chapter 7 for further discussion).

**Commodity**  An object of economic value that is valued generically, rather than as a specific object (example: pork is a commodity, rather than a particular pig). In political economy (and Marxist) thought, an object made for exchange

**Primitive Accumulation**  In Marxist thought, the direct appropriation by capitalists of natural resources or goods from communities that historically tend to hold them collectively, as, for example, where the common lands of Britain were enclosed by wealthy elites and the state in the 1700s

**Overproduction**  In political economy (and Marxism), a condition in the economy where the capacity of industry to produce goods and services outpaces the needs and capacity to consume, causing economic slowdown and potential socioeconomic crisis

## Bottled accumulation: Selling back nature

In the first case, it is clear that water sources around the world, which have been historically freely available, are increasingly controlled privately. Even the history of spring waters in the United States reflects this long trend. At the time they were first "discovered" in the eighteenth century, spring water sources were controlled by native peoples and shared within their communities, where they would have been open and freely available (though not necessarily to other tribes or communities). Having been wrested from the control of native peoples during the conquest of North America, the rights to tap and market these water sources were given as concessions to private interests rather than making them public goods. Though in rare cases these water sources were established in public or state parks (consider Yellowstone National Park), generally they were privatized. In the language of political economy, this is the accumulation of capital through direct expropriation from communities and the public.

In the case of bottled water, similar concessions to private companies have occurred on a smaller scale. The Nestlé company, for example, pumps 218 gallons of groundwater per minute from an aquifer in Michigan, based on a 99-year lease from the state. This water otherwise feeds local lakes and streams. Nestlé, who essentially pumps the water for free and received significant tax breaks to locate the company at this site, bottles the water under the brand name "Ice Mountain." The company has faced down community opposition that insists that this water rightfully flows to nearby Osprey Lake and local wetlands (Marketing Week 2005). In essence, public water is made private and then sold back to consumers from whom it has been expropriated.

Throughout the developing world, the connection of bottled water to appropriated public sources is more complex, but no less evident. Many water vendors in India, for example, take control of publicly piped water either legally (through contracts) or illegally (through bribes), and then sell it to households or to marketers. The case of Tijuana, described earlier, is compelling evidence of the way publicly held resources, in this case municipal water, are bottled for free and sold at significant markups to people whose connection to municipal services is increasingly denied.

From a market-oriented perspective, it can certainly be argued that capitalists are merely filling a need here. The failure of Tijuana to make provision for its citizens provides an unfortunate gap into which innovative entrepreneurs enter to help poorer people and make a modest profit, from this angle. But given that the state actually provisions and treats the water at taxpayer expense, the seizure and exploitation of public goods as private commodities might also be viewed as a kind of basic theft. To the degree that the state sanctions or subsidizes that activity, moreover, political economy would suggest that this sort of private "initiative" is intimately tied to the state's willingness and ability to neglect its own civic responsibilities. The water is effectively a "free good." It is tapped at whatever rate firms see fit to set. It is sold at whatever price the market will bear. This primitive accumulation of water supplies provides businesses with free profits, drawing directly on natural water sources previously held by communities. These are sold back to the public in bottles.

## Bottled overproduction: Producing demand

On the other hand, in places like San Diego, with fairly complete and still publicly available water supplies, the problem is less supply (which apparently can be wrested from the public reasonably easily) but instead *demand*. In a culture with far more soda, beer, and juice choices than could possibly be consumed by the population, it is difficult to increase the demand for beverages. This presents a limit for capital accumulation for firms and for investors seeking profits by buying into major multinational companies like Pepsi, Coca-Cola, or myriad others. Beverages, a multi-billion-dollar industry, would appear to be reaching a maximum, raising the specter of stagnating markets.

The only solution to this problem is to *create* demand, that is, to make people want your product. In the case of bottled water, as we have seen, it is not complicated to secure public sources and put water in a bottle. It is a trickier, however, to produce demand. The industry is no longer concerned with securing a valuable product, but instead with transforming the public imagination of water, health, risk, safety, and good living. As a result, the lion's share of investment in bottled water goes into packaging and marketing.

The total number of separate, albeit effectively indistinguishable, bottled water brands underlines this point. The Nestlé corporation alone sells no fewer than *15* different brands of bottled water. This branding allows people to identify with a specific product, making the selection of a particular bottle an essential part of their lifestyle and identity. The expansion of this bottled water market is aimed at eating away at whatever non-commodity goods remain in people's lives (i.e., tap water). This represents the only real possible expansion of markets and economic growth, staving off the grim prospect of flat sales.

Revealingly, in more recent years, concern about the environmental effects from bottled water consumption (plastics, transport, energy costs, etc.) has threatened to reduce or flatten sales. As a result, the industry is attempting to expand and cement consumption habits among younger consumers and children, and establish long-term brand loyalty. As one marketing trade journal complained, people are becoming *too confident* in their tap water, and new, less well-informed consumers must be found:

> The increasing consumer confidence in tap water and in asking for it at restaurants and bars may well prove to be the key in effecting further rises in both penetration and volume consumption per head among adults. But the attitudes and behavior of children, carried into adult consumption patterns, may well counter this trend and prove to be more significant in the long run. (National Petroleum News 2007: 34)

And this is only the trade journal for the industry making the bottles! The global petroleum industry, which depends on packaging as one of its key growth areas for plastics production, also tracks the growth of bottled water consumption with tremendous interest.

In political economic terms, this effort to get people to pay money for something they already receive for free is a symptom of a larger problem: overproduction. In this view, the capacity of firms to produce consumer goods far outstrips the desire or capacity for them to be consumed, especially by working people who live off the limited wages provided by these same firms. This leads to a crisis of under-consumption, where the economy cannot digest goods (like beer and soft drinks) fast enough to maintain profits. Force-feeding such an economy demands that new needs be created, preferably essential rather than optional ones. Bottled water is a perfect fit. As long as consumers increasingly see bottled sources as superior to municipal ones, consumption of water as a commodity will continue. Given that water itself is necessary to survive (and as long as state-provided resources are overlooked, eschewed, or eliminated), life itself becomes a marketable good.

It is not difficult to see past the illusions surrounding bottled water, of course, especially if we know that bottled sources are not much different than tap water. But as noted previously, people's behaviors do not always flow directly from their knowledge. A bombardment of attractive packaging, coupled with appeals to "nature" (e.g., "Ice Mountain,"

"Polar Springs," "Mountain Valley," "Deer Park") and to "health and safety," maintains a steady consumer base for a manufactured need. In sum, from a political economy perspective, *scarcity can be produced*, and this occurs for reasons convenient to firms and inconvenient to citizens and working people.

Taken together then, from a political economy point of view, there are two linked trends in bottled water: 1) In poorer regions and countries, there is a trend towards expropriating community water supplies, and marketing that water back to residents in bottles. 2) Among wealthy consumers and in well-developed areas, there is a simultaneous effort to create demand for bottled water even where plenty of clean water is already available. Critically, these are both part of the same system.

## The Bottled Water Puzzle

In this chapter, we have learned that:

- Bottled water consumption has exploded around the world in the past two decades, coming to compete with and replace other sources of drinking water.
- The production, packaging, and transportation of water in a bottle means that it is not an environmentally benign way to get and consume water.
- Meeting human water needs is increasingly difficult in a rapidly growing and urbanizing world, but bottled water may only be viewed as a response to scarcity by acknowledging differing kinds of scarcity: hydrological, socioeconomic, and perceptual.
- One of the central drivers behind bottled water consumption is the perception that other water supplies are less safe or less healthy.
- Risk assessment shows no significant advantages of bottles over tap water, raising questions about human biases in risk perception.
- The simultaneous growth of markets for bottled water in underdeveloped countries as well as wealthier ones suggests that a common economic driver – commoditization of nature – may be behind water privatization.

This brings us back to our water consumers in Tijuana and San Diego. Does bottled water represent the most efficient and safe way to provision drinking water in the twenty-first century to rich and poor alike? Or does it instead suggest a surrender of the environment to private ownership? Can and should municipal water (or clean air, garbage collection, parks …) be made available to the world's poor through social provisioning instead of markets? Can better risk communication allow consumers to make better informed judgments about the risk and threats of water (or spinach or peanuts or toys …)? To what degree does allowing firms to control the sources and distribution of water (or public space, conservation areas, trees …) increase efficiencies in delivery, and to what degree does it inflate unnecessary consumption and lead to a loss of public control over the environment?

Bottled water is a puzzle that shares a lot, therefore, with an enormous range of natural goods and environmental services. The revolutionary and rapid growth of bottled water around the world is either the leading edge of a desirable consumer revolution or a warning sign about the decline of our responsibility to one another and the earth.

## Questions for Review

1. "Dasani" and "Aquafina" are two of the best-selling brands of bottled water in the United States. Who owns and markets these brands? Where do they get the water that goes into the bottles?
2. How (and why) has the recent and rapid urbanization in countries like Mexico, India, and Egypt affected the demand for bottled water?
3. Discuss the three discrete types of scarcity that are combining to drive global water consumption steadily upward.
4. In the United States, are municipal drinking water and bottled water held to the same regulatory standards? Explain.
5. What does it mean to say that many places worldwide have recently witnessed the "accumulation" of their freshwater supplies?

## Suggested Reading

Cech, T. V. (2003). *Principles of Water Resources: History, Development, Management, and Policy*. New York: John Wiley and Sons.

Kahrl, W. L. (1982). *Water and Power*. Berkeley, CA: University of California Press.

Opel, A. (1999). "Constructing purity: Bottled water and the commodification of nature." *The Journal of American Culture* 22(4): 67–76.

Royte, E. (2008). *Bottlemania: How Water Went on Sale and Why We Bought It*. New York: Bloomsbury USA.

Senior, D. A. G., and N. Dege (eds.) (2005). *Technology of Bottled Water*. New York: Wiley-Blackwell.

Wilk, R. (2006). "Bottled water: The pure commodity in the age of branding." *Journal of Consumer Culture* 6(3): 303–25.

## Exercise

In this chapter, we have reviewed how the puzzle of bottled water might be addressed in terms of population, risk, and political economy. Explain how you might understand this problem using a social construction framework (as described in Chapter 8). You may want to consider, in particular, the way bottle labels actually *look*. What images and text are utilized? What ideas of nature, geography, and history do they invoke? How does this matter to the way consumers think about themselves and the product? With what implications for the environment?

# 14

# French Fries

*Credit*: Joao Virissimo/Shutterstock.

## Keywords

- Anthropocentrism
- Columbian Exchange
- Conservation
- Ecocentrism
- Genome
- Globalization
- Monocrop
- Preservation
- Spatial fix

## Chapter Menu

### MMM-MMM Good

You are driving down the interstate, on your way home from school on vacation. You think you packed enough food for the trip. But, the more exits that whiz by, the more the roadside restaurants call to you. Golden arches lure you with the promise of hot, crispy French fries – just the thing to get you through the next couple of hours on the road (and more attractive than the sandwich you managed to pack in your post-exam haze). Try as you might to avoid it, you soon find yourself in the McDonald's drive-through, waiting with a dozen other cars. Unfortunately, you've managed to stop right at lunch hour.

Lucky for you, though, as is true of most stores in the mega-brand, this particular McDonald's has one person on fry duty for the whole lunch rush. For 90 minutes or more, your counterpart on the inside pours the required amount of perfectly cut, almost uniform, frozen fries into the fry basket and drops them in hot vegetable oil. After a couple of minutes, when the timer rings and the fries are just the right color, the basket is removed from the oil and the fries are dumped into the adjacent warming area, where they are sprinkled with a generous dose of salt. From there, the fries are scooped into a range of differently sized packets – medium, large, and the one you ordered, supersized. This process is repeated thousands of times during a typical lunch hour.

The line has moved efficiently and before long, you've paid the cashier and driven to the second window. Here, you are handed a white bag with the McDonald's logo on it containing your supersized order of fries, some packets of ketchup and salt, and a few napkins. As you drive away, you wrestle with a system for salting and putting ketchup on your fries one-handed, while keeping your eyes on the road and one hand on the wheel. When you are finally able to take the first bite, you find the fries to be just as you imagined them – hot, crispy, salty – yummm!

You, and the millions like you, who consume French fries every day have become part of the billions served. Not just in the United States, but around the world. Sure, the particulars change – in England you order chips and get salt and vinegar to eat them with, in Mexico, you can get a side of jalapeno salsa to put on your *papas fritas* – but the fries themselves are reliably, dependably, the same.

What does it take to produce this uniformity around the world? How does your super-sized bag of fries connect you, not only to millions of consumers across the globe, but also to the environment? These are the questions that will be answered in this chapter. First, you will learn the history of the French fry, from the lowly and much maligned potato to the mass-produced perfectly formed treat of today. Next, by applying what you learned in Chapters 6, 7, and 5 you will gain an understanding of how the puzzle of the modern fry might be unraveled using a risk approach, a political economy approach, or an ethics approach.

### A Short History of the Fry

To understand the French fry as an environmental puzzle one must first understand its most important raw material – the potato. Today the potato is the world's fourth largest crop, after corn, wheat, and rice (Jackson 1985). Most of this production is for food. While fresh potatoes are still consumed around the globe, much of the world's potato production is used for frozen potato products, a high percentage of which are the fries supplied to restaurants. About 10 million tons of frozen fries are consumed in the world each year.

Wild potatoes are inedible, bitter, and toxic – a far cry from the Russet Burbank used for French fries today. In this section, we will trace the history of the potato from the many varieties that were dietary staples of the indigenous people of South America, to the domesticated, **monocropped**, and industrial potato of today. Along the way, the journey of the potato highlights a history of colonization, the emergence of industrial agriculture, **globalization**, and an increasing distance between food production and food consumption.

> **Monocrop**   A single crop cultivated to the exclusion of any other potential harvest
>
> **Globalization**   An ongoing process by which regional economies, societies, and cultures have become integrated through a globe-spanning network of exchange
>
> **Columbian Exchange**   The movement of species across the Atlantic Ocean, from the New World to the Old World and vice versa, and the resulting ecological transformations

## What a long, strange trip it's been: Cultivation and use of *Solanum tuberosum*, from the New World, to the Old, and back again

The potato was first domesticated 7,000 years ago on the Andean altiplano where several varieties grew wild. There were blue potatoes, red potatoes, yellow and orange potatoes in myriad shapes, which ranged in flavor from bitter to buttery to sweet. Andean peoples developed a spud to suit each environment, encouraging variety and increasing the heartiness of the crop. In addition to purposeful cross-cultivation, domesticated potatoes continued to be hybridized with local weed potatoes.

When the Spanish arrived in South America, they acquired not only silver, gold, and dyes, but also many new foodstuffs that were then introduced to Europe, as part of what is called the **Columbian Exchange**, the movement of species from the New World to the Old World across the Atlantic and vice versa, with all the ecological transformations that this entailed. For some people, the potato is the most important of these, in part because of its significant role in the growth of European populations. Despite its importance, however, it is likely that the potato was an accidental tourist to Europe (Pollan 2001).

The potato's first significant foothold in Europe was Ireland, where it arrived in 1588. Though adapted to the climate, the potato carried with it a load of cultural baggage that discouraged its spread into other European countries for the next century. The new food was suspected of causing leprosy and encouraging immorality. The fact that it grew underground was also viewed as suspicious. These suspicions were heightened by the fact that the potato was not mentioned in the Bible, and that it had come from "backwards" uncultured tribes in the Americas. Racist assumptions associated with the experience of colonialism left traces on food. Moreover, the colonization of Ireland by England during the nineteenth century meant that, once again, the potato was associated with the history of a people constructed (see Chapter 8) as inferior to the English, French, and Germans. In fact, the very success of the potato on Irish soil, where it produced much more food than the cereals that had been grown there, was used by some English officials to argue against the widespread cultivation of potatoes in that country, as things Irish were viewed as immoral, uncivilized, and unhealthy. On the other hand, the potato also allowed the Irish to support themselves, rather than having to rely on the more powerful and economically dominant English. The potato was easy to grow and to maintain. It did not have to be grown in straight rows (like the cereal crops popular on the continent), and could grow in marginal environments, the kinds to which Irish cultivators were relegated by English colonialism.

All of the colonial suspicions about the potato, however, could not prevent its spread to continental Europe as cereal crops began to fail. Despite initial resistance to the crop among the elite and the peasants alike, European royalty began to see the potato as a way to keep their subjects fed. In Germany, Frederick the Great forced peasants to grow them, as did Catherine the Great in Russia. In France, Louis XVI took a more subtle approach, encouraging Marie Antoinette to wear potato flowers in her hair and planting a plot of potatoes that he had guarded until midnight. After midnight, peasants (thinking that anything the king had guarded must be pretty special) would sneak in and steal them. Indeed, by the 1800s, the English were the last to hold out against the potato. In 1794, the wheat crop across Northern Europe failed and the use of the potato as a fill-in was debated. This debate centered on the "Irish question." Did the potato really make the Irish more self-sufficient, thus proving its worth as a food crop? Or did it simply cause a population boom (Malthus found the potato to be an excuse to procreate (see Chapter 2)), that drove down wages and made them dependent on the plant? After a series of moralizing debates about the foodstuff, the practical need to keep people fed won out and the English did adopt the potato.

Meanwhile, Irish dependence on the potato, or more precisely, one cultivar, commonly known as the lumper, did prove to be fatal in the late summer of 1845, and again in 1847 and 1848, when the disease caused by *Phytophthora infestans* arrived on the island and devastated the potato crop. One million people died, and thousands more were ill over the three years. But, this disaster cannot be understood outside of the history of the English colonization of Ireland, which had pushed peasants off the little arable land and into marginal areas and, at the same time, limited access to other food sources through trade restrictions. Existing poor laws (see Chapter 2) worsened the situation for many people. In order to be eligible for full aid, one had to own less than a quarter acre of land. This meant that a lot of desperate people gave up their land in order to receive aid and many more fled to the United States.

The Irish brought some potatoes (back!) to the Americas in the century before the potato famine. It was not, though, a widely grown or eaten crop. Rather than being a popular foodstuff, the potato was first spread by missionaries hoping to pacify and settle Native American tribes. In 1836, for example, potatoes were brought to Idaho by the Presbyterian missionary Henry Harmon Spalding (1804–74), who established a mission in Lapwai. In an effort to bring Christianity to the Nez Perce Indians, Spalding first sought to teach them to practice formal, settled agriculture rather than using hunting and gathering. The potato was one of the crops Spalding used. Owing, however, to continued hostilities between Native Americans and the missionaries, as well as prejudice against the Irish, the potato did not really become a major foodstuff in the United States until after 1872. In that year, American horticulturalist Luther Burbank (1849–1926) developed what we think of as the standard Idaho spud, the Russet Burbank potato. Burbank, improving on the varieties that had been brought to the United States by Scotch Irish settlers, developed a potato with a russeted skin that was more resistant to disease. He later sold the rights to his variety for $150, which he then used to further develop the potato. In the 1900s, the Russet Burbank was a widespread crop in Idaho. The United States Department of Agriculture estimates were first made for the state of Idaho in 1882. The total value of the potato market in Idaho that first year was $250,000. Just over 20 years later, in 1904, the value of the potato crop in Idaho came in at over $1.3 million.

## The advent of the fry and the American century

Despite the slow spread of the potato in the United States, where it had long been viewed as a lowly and, possibly, poisonous food, methods of cooking the potato had been developed and refined throughout Europe. There is a long-standing debate over whether the French or the Belgians were the first to fry potatoes. It is rumored that Thomas Jefferson served French fries in Monticello in the early 1800s, but it was not until the early twentieth century, when American soldiers back from serving in Europe in World War I demanded them, that the fried potato became popular in the United States.

While the United States was not the originator of French fries, they have become emblematic of the cuisine and culture of the country. At the time that war veterans demanded them, frying potatoes was a task better undertaken in a restaurant with sufficient equipment to do it. It is at this point the history of the French fry becomes irrevocably entangled with the emergence of new types of dining experience: the drive-through and its companion form, the fast-food restaurant. The first drive-through restaurant opened in Dallas in 1921 (Plummer 2002). Quick on its heels came White Tower – one of the first to use standardized architecture and menus. The growing number of automobiles on the road fueled this trend, which really exploded after World War II.

In 1951, milkshake salesman Ray Kroc went to Richard and Maurice McDonald, owners of a fast-food emporium in San Bernadino, California, with a proposal. In 1955 the first McDonald's outlet opened in Des Plaines, Illinois on the outskirts of Chicago. It was the first to use the red and white building and the golden arches. A second and then a third outlet quickly followed. By 1965 there were 700 McDonald's restaurants across the United States.

Perhaps no fry is more emblematic than the McDonald's fry. At first, Ray Kroc insisted that all fries be hand-cut from fresh potatoes. By the 1960s, however, this became unwieldy. Seeking to lower costs and increase efficiency, in 1966 Kroc made a deal with J. R. Simplot whose company had been working on the proper technology for freezing potatoes. Simplot agreed to dedicate one factory to producing frozen fries. In the following decades frozen potato products became popular with the public in the United States. In 1960 consumers in the United States ate about four pounds of frozen French fries per year and about 81 pounds of fresh potatoes. Four decades later, in 2000, they ate about 30 pounds of frozen fries a year and 50 pounds of fresh potatoes (Schlosser 2001). McDonald's today is still the largest buyer of potatoes in the United States. This is not surprising given that trade sources estimate that 90 percent of the potatoes consumed in the United States today are sold by fast-food franchises. Nor is this a phenomenon particular to the United States. Worldwide exports of frozen potato products (of which 90 percent are fries) were valued at almost 2 billion dollars in 2000 (Yen et al. 2007).

## The demands of the Russet Burbank and contemporary frozen French fry production

The increasing popularity of fast food and the need for a constant supply of frozen potatoes have been a boon for the Russet Burbank. Though, as discussed above, the Russet Burbank is a relatively recent cultivar, only one of countless potato varieties domesticated in the Andes and bred around the globe in the centuries since, it is now the dominant potato throughout the world. In contrast, the Irish lumper is virtually extinct. The potato's dominance is due in part to its rapid spread in the United States in the late nineteenth and early twentieth centuries. It is also, however, an accident of history. The Russet Burbank was gaining popularity as a crop at the same time that the automobile and post-war prosperity made drive-throughs and fast-food restaurants so popular.

Despite its popularity, the Russet Burbank is not easy to grow. Or, at least, it is not easy to produce specimens appropriate to the frozen fry industry which requires uniformity in size and shape. Mass production of the Russet Burbank for the frozen fry industry requires multiple inputs, including water, chemicals, and energy (Table 14.1).

**Table 14.1**  Energy inputs and costs of potato production per hectare in the United States

| Inputs | Quantity | Kcal | Costs ($) |
|---|---|---|---|
| Labor | 35 hours | 1,964,000 | 350.00 |
| Machinery | 31 kg | 574,000 | 300.00 |
| Diesel | 152 L | 1,735,000 | 31.92 |
| Gasoline | 272 L | 2,750,000 | 78.88 |
| Nitrogen | 231 kg | 4,292,000 | 142.60 |
| Phosphorus | 220 kg | 911,000 | 121.00 |
| Potassium | 111 kg | 362,000 | 34.41 |
| Seeds | 2408 kg | 1,478,000 | 687.00 |
| Sulfuric acid | 64.8 kg | 0 | 73.00 |
| Herbicides | 1.5 kg | 150,000 | 13.50 |
| Insecticides | 3.6 kg | 360,000 | 14.40 |
| Fungicides | 4.5 kg | 450,000 | 180.00 |
| Electricity | 47 kWh | 135,000 | 3.29 |
| Transportation | 2779 kg | 2,307,000 | 833.70 |
| Total | | 17,470,000 | 2863.70 |
| Potato yield | 40,656 kg | 23,296,000 | Kcal input/output 1:1.33 |

*Source*: Pimentel et al. (2002)

Consider the steps one potato farmer listed in Michael Pollan's *Botany of Desire: A Plant's Eye View of the World* (2001). First, he applies a soil fumigant that kills all microbial life in the soil. Next, during planting a systemic insecticide is added. Third, when the plant is six inches tall, it gets an herbicide. After this, the plant receives weekly cycles of chemical fertilizers. This is followed by a fungicide when the rows begin to close. Often, the next step is to use crop dusting to protect against pests like aphids. During all of this, of course, the crop also requires from 500 to 700 mm of water (Food and Agriculture Organization). Though this is not as much water as many cereal crops need, it still averages about six gallons of water to produce one serving of French fries. Add to this the cost of irrigation and energy and the average cost per acre to the commercial potato farmer is $1,950. Fast-food companies will pay $2,000 for the 20 tons of potatoes that are produced on each acre. Given this very slim margin, it is hard for farmers to get off of the input treadmill. Even if farmers would like to switch to other varieties, they cannot because fast-food suppliers demand that they produce Russet Burbanks, so they are forced to do what they can to make this work. So, they have little choice but to keep adding more and more chemicals, water, and energy, even if it produces only marginally better results.

Even so, the environmental costs of producing frozen French fries do not end with the farm. After the potatoes are grown, harvested, and shipped to a plant, processing takes energy. The potatoes usually go through a dehydration process and are sliced, partially cooked, and frozen. The energy required to transport over nine million metric tons of bags and boxes of frozen fries each year is also significant. While freezing certainly allows us to transport food around the world, it takes a large amount of energy to keep those fries from thawing on their journey.

## The puzzle of French fries

French fries are ubiquitous. As such, they represent an object that certainly seems straight-forward. Take a potato, cut it, fry it, voilà! But, as we have seen in the above socio-natural history, there is much more to the fry than meets the eye (or the mouth!). Several major characteristics, then, of the French fry as an environmental puzzle must be taken into account:

- Not just any potato can become a fry. The mass production and consumption of the French fry demands a select breed of potato that can stand up to industrial production and that behaves the way that we expect when we slice and fry it. This leads to a virtual monoculture of a single variety of potatoes around the world.
- The Russet Burbank itself is an accident of history, beginning with European colonial-ism and ending with the subjugation of indigenous peoples in the United States.
- Its production is tied to increasing intensification of cultivation, with widespread envi-ronmental implications, including high inputs of water, fertilizer, and pesticides.
- The French fry is a cultural phenomenon, one whose immense popularity is wrapped up with the advent of the car and fast food, and the homogenizing processes of globalization.

- The expectation of such homogenous "French fries" leads to heavy processing of the potato in fast-food production.
- This puts the fry smack in the middle of contemporary debates about the types of foods we eat and their effects on both human bodies and the environment.

The environmental puzzle of the French fry is therefore that its universal success in diverse contexts (from Slovenia to Guam) has led to a system of production that is anything but diverse. This allows us to think about the connection between how we make what we eat and how we remake the environments in which we must live. In the following sections, we consider the implications of this puzzle, beginning with French fry consumption as a matter of individual food choices. Here, we can begin to understand how what we put in our body is often tied to a complex process of risk assessment. These individual choices, though, take place in a context of consumption and production on a societal and global scale. Therefore, the socio-environmental puzzle of French fries may also be approached through political economy frameworks. Finally, we consider how the need to produce more and more mass-marketed food products like French fries presents us with difficult questions about the ethics of food production.

## Risk Analysis: Eating What We Choose and Choosing What We Eat

In Chapter 6, we discussed how socio-environmental puzzles might be explained in a framework of hazards and risks. In this section, we consider how such a framework sheds light on efforts to mitigate the impacts of the French fry on the human body. McDonald's was sued in 2005 by the group Ban Trans Fats (BTF). They claimed that the fast-food chain had not followed up on promises it made to consumers in 2002, when the company promised to reduce the use of trans fats (the common name for a type of unsaturated fat containing trans-isomer fatty acids – also called partially hydrogenated vegetable oil) in its products. The company settled the suit and donated 1.9 million dollars to the American Heart Association (see the archives of the organization at www.bantransfats.com for a history of the movement).

Why was there so much pressure on this and other fast-food restaurants to switch from trans fats to more healthy fats to fry their food? Because, increasingly, the use of trans fats became an unacceptable risk to individual consumers of fast food, and at an aggregate level, a burden on people's health. Here, we will address the environmental problem of the French fry at the scale of the human body. While we are used to thinking about the environment as something removed from us, as we have discussed in preceding chapters on environmental ethics, political economy, and social construction, some scholars and environmental activists have argued that we must consider humans and nature indivisible. This means that human health and planetary health are intimately linked. Where is this link more evident than in the food we put into our bodies?

Dietary and nutrition sciences have long addressed the role of food in human health. Our understanding of the effects that food has on human bodies has changed throughout the centuries. It also varies across cultures. Remember that Europeans first rejected the potato because it looked funny, was not in the Bible, and was associated, first, with a "backward" tribe in the Americas, and later with the Irish? While these may seem like odd ways to think about what are and are not healthy foods, they reflect that period's ideas about health, diet, and morality. Clean foods that came from above ground, grew in straight rows, and had more regular shapes were assessed to be healthier than tubers that grew underground, in irregular patches. While many religions still ban the consumption of certain foods on moral or ethical grounds, few people would argue that these restrictions are linked to the nutritional value of the food, per se.

Nor should we imagine that such ideas are that far removed from the way we think about food in the United States and many European countries today. After all, we want French fries that have a predictable shape, that are not marred by black spots, and that come from a clean, well-lit, modern restaurant. But, our assessment of the health risks (the possible negative consequences) of consuming certain foods is increasingly shaped by different understandings of food and nutrition, based in *scientific* research. This different type of information and our access to it has influenced some consumers' willingness to eat certain foods.

In the case of trans fats, how has the level of risk involved in eating French fries been established by scientists and communicated to the public? How do people, given this type of information, make decisions about what they eat? And, if there are risks associated with food choices, whose responsibility is it to mitigate those risks?

## The science of good and bad fats

While trans fats may not be what we think of as a typical environmental hazard (like a tsunami, volcano, or drought, for example), it is, as we discussed in Chapter 6, a produced hazard that may threaten human health. Trans fats come from a process of hydrogenization. Hydrogenization is a process of using pressure to force hydrogen atoms into unsaturated fatty acids such as vegetable fats and oils. Partial hydrogenization is the process of adding hydrogen atoms until a food is at a desired consistency. The process was first developed in the 1890s by a French chemist, who was later awarded the Nobel Prize. The use of hydrogenization to produce stable cooking oils was developed and patented in 1902 by Wilhelm Normann. Soon after trans fats became the first *human-made* fats to enter the food supply. In 1911, Procter and Gamble began selling Crisco, the first widely available trans fat product in grocery stores. At first, Crisco and its competitors (e.g., margarine) were not widely used. It was not until the rationing of butter during World War II that margarine and other trans fat products were widely adopted.

In 1957, the American Heart Association declared that saturated fats, like those in butter and other animal products, were a factor in heart disease. This meant that, in terms of human health, trans fats were positioned as a less risky alternative. Acting on this information, in 1984 consumer advocacy groups were successful in getting fast-food restaurants, many of which used beef tallow in their fryers, to switch to trans fats instead. The use of such oils skyrocketed. Given the ever-increasing popularity of fast food, trans fats became a significant part of the typical American, and to some extent the European, diet.

But, even though many people argued that trans fats were the healthier alternative, several scientific studies in the early 1990s associated the use of trans fats with a rise in what we now call bad (LDL – low density lipoprotein) cholesterol. Increased levels of LDL were associated not only with heart disease, but also with diabetes and a host of other serious ailments. Before the decade was over, legislation was proposed in the United States to make food producers list the amount of trans fats on nutrition labels. This was not passed. In 2003, however, Denmark went a step further than requiring labeling. They actually put a cap on the amount of trans fats in certain foods. That same year, due to continued consumer activism, the US Food and Drug Administration began requiring that manufacturers list trans fats on nutrition labels. While many food manufacturers and restaurants, including fast-food chains, reduced their use of trans fats, not all did so to the satisfaction of lobbying groups like Ban Trans Fats. This resulted in law suits, like the one that opened this section.

## Better information, healthier choices?

As information about the health risks of dietary choices changed, individuals adjusted their behavior to account for these. One source of this information is the American Heart Association (AHA) who started a "Face the Fats" consumer education campaign in April of 2007. To assess the success of this and similar campaigns, the AHA commissioned a study to track people's understandings and perceptions of trans fats and the impact that these had on eating behaviors. The study, undertaken by a group of medical doctors, nutrition experts, and marketing researchers, concluded that awareness about trans fats among the American public increased between 2006 and 2007 (Eckel et al. 2009). The AHA had already recommended that people consume no more than 1 percent of their total daily energy intake (calories) in trans fats. In January of 2006, the US Food and Drug Administration instituted a requirement to list trans fat content as a separate item on nutrition labels on packaged foods. Researchers found that between 2006 and 2007, consumer concerns over both total fat content and the types of fat in foods increased (Figure 14.1). During the same period news coverage of trans fats doubled from 516 articles to 1,138.

Not only had more consumers heard about trans fats, but more were able to link them with particular health risks in 2007. The percentage of people who identified trans fats as a potential contributor to heart disease increased 10 percent over 2006, with 73 percent of people responding that this fat source increases one's risk of heart disease. In this case, increased information and awareness also changed respondents' behaviors. In 2007, significantly more people said that they reviewed trans fat information before making purchases (29 percent vs. 23 percent the previous year). They also looked for "zero trans fat" labels and substituted zero or low trans fat products at a higher rate. At the same time, more people knew which common foods contained trans fats. French fries, in this case, were most often identified as containing trans fats, with 53 percent of respondents in 2007 saying that they contain trans fats. This was an increase over 41 percent the previous year. People were beginning to think about trans fats, and in a big way.

How concerned, if at all, are you with the amount of fat that you consume in foods?

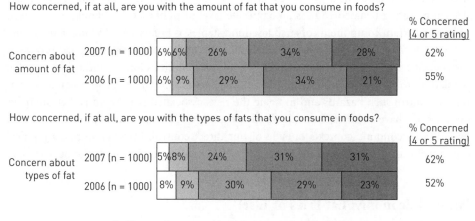

How concerned, if at all, are you with the types of fats that you consume in foods?

□1 - Not at all concerned    ■2    ■3    ■4    ■5 - Extremely concerned

**Figure 14.1**   Concern over amount and types of fat. *Source*: Eckel et al. (2009), p. 290.

## Risk, choice, and regulation

These findings about people's knowledge of trans fats and risky foods support other studies on fast food, risk, and perception, stressing that knowledge does not always lead to changes in behavior. For example, one study of midsized cities indicated that certain groups of women identified fast-food restaurants as hazards in their neighborhood (Colapinto et al. 2007). That study also found, though, that despite knowing the risks that fast-food consumption poses to their health, people still choose to eat in these restaurants. The long-term risks are not significant, or immediate, enough to outweigh the convenience of these foods for many people. Unlike a salmonella infestation, a melamine contamination, or a food allergy – all risks that typically produce immediate, noticeable behavioral changes in consumers – the negative consequences of a diet high in French fries and other fast foods are experienced long after that food is consumed. So, even though better information is available the risks posed by fry consumption do not always produce the immediate (gut reaction!) responses.

While it is easy to blame people for their eating behaviors, and more recently to vilify heavier people for everything from lower productivity to strains on the health care system (Guthman and DePuis 2006), the social, economic, and political contexts in which food choices are made represent a significant influence that cannot be ignored. In the study on trans fats mentioned above, for example, even though awareness of the health effects of such fats increased overall, neither this awareness nor associated behavioral change was evenly distributed across all groups that were sampled. They were more concentrated among women, 45 or older, who have at least some college education, incomes of $75,000 or higher, and who live in the Northeast or the South.

So, what is to be done? Given the proven risks of eating a diet high in French fries and other fast food and the public health concerns over the "epidemic" of obesity, how do we make decisions about who gets to eat what? Should individual choice and free markets determine eating behavior? Or, is an aggregate level of public health enough of a reason to ban important components of such fast-food staples as French fries? While many would argue the former, as of 2008 approximately 20 states and local jurisdictions had proposals for reducing trans fats in restaurants or schools and 11 cities and counties had adopted regulations to restrict trans fats in restaurants. One reason that many people turn to regulation to control such hazards and mitigate the risks associated with them is that often, we do not *really* have a choice about the foods we consume. Rather, our selections are driven by political-economic contexts outside of our direct control. In other words, a political economy of food, in this case the fry, is at work. We will examine this in the next section.

## Political Economy: Eat Fries or Else!

In Chapter 7, we discussed the insights that a political economy framework can provide in thinking about environmental puzzles. One key to this framework is to think about how socio-environmental objects are wrapped up in the drive for the creation of surplus value in a capitalist system. This process of surplus value production and circulation holds true as much for the food we consume as it does for the tennis shoes we buy. In this section we will see how, if we frame the environmental puzzle of the French fry as a problem of political economy, we could ask questions not only about the food choices that people make, but also about how those are related to broader socioeconomic conditions, including fast-food marketing, the concentration of fast-food restaurants in marginalized communities, the globalization of fast-food restaurants, and the food production system itself.

### You want what we say you want – marketing and food choices

Whenever you walk into a fast-food restaurant, chances are someone asks you "What would you like?" Some places even let you "have it your way." And, whether you answer immediately (fries and a shake, please!) or it takes you a while (um, well, I guess I'll have the Value Meal with chicken strips and a cola), the choice is yours. Darn the consequences – I'll have that apple pie! Or – wait, I'm going to the beach next week – better make that a diet cola! But, is it really so simple? Is it just that you've *weighed* your options and made the appropriate choice? Or, is something else going on? What makes us choose the food we do?

Of course, dieticians, nutritionists, medical doctors, weight loss counselors, gym trainers, psychologists, and beauty magazine editors all espouse theories accounting for people's behaviors. Psychological conditions (stress!), nutrient deficiency, lack of exercise, overabundance of certain hormones, or peer pressure, many sources argue, get in the way of making rational food choices. All individuals need to do is make more *rational* choices.

But, to what degree are such choices individual in the first place? To what degree are these social decisions, out of the hands of isolated people? It has long been accepted that food is a cultural phenomenon and that different cultures have their own way of eating. While this is sometimes tied to larger histories of colonialism and racial exploitation (remember how no one wanted to eat like the *Irish*, despite their remarkable ability to provide food for themselves in a hostile colonial context), food is often considered an artifact of culture, and many groups pride themselves on their culinary identity. Beyond cultural factors, however, may also be political and economic ones. How are your food choices shaped by marketing, including not just commercial advertisements, but also more subtle selling points, like sizing?

The development of fast food was an exercise in branding. Early White Tower restaurants were identified by the shape of the building, and this is still the case. When you see golden arches on a road sign, you know what to expect. This is true because with this branding comes standardization of products. Chances are you know the difference between a Quarter Pounder with Cheese, a Whopper, a Junior Deluxe, and a Double Double without giving it much thought. But, what makes you choose one over the other, or over different options? Not surprisingly, advertising exposure and acceptance are key determinants in who eats fast food, and where they eat it. Knowing this, fast-food restaurants spend a lot of money on advertising each year. They must do this, because as we know from Chapter 7 on political economy, competition among firms leads to lower surpluses, unless a business can somehow compensate for the potential loss of customers. In many cases, companies can maintain surpluses by cutting labor costs. But, branding allows firms to try to maintain a loyal customer group by differentiating their products from others that are basically the same.

Once the restaurant has lured you in, it may find other, sometimes more subtle ways to ensure profits. What, for example, ever happened to the small-sized fry? The much-discussed practice of supersizing is often used to increase profit margins. As this practice has become more widespread (even though it was condemned in the popular movie *Supersize Me*), our own notions of portion size change. There is some evidence that children exposed to fast-food advertising begin to view these super-portions as the norm (Schlosser 2001). Good, perhaps, for the bottom line, but not the waistline. Well, all's fair in business, you might say. Even if advertising influences our choices, we can still opt for healthier food if we really want or need to. Of course you can, *unless*, by dint of where you live and work, you have no healthier choices.

## Demand or supply: The geography of fast food

According to Eric Schlosser, author of *Fast Food Nation: The Dark Side of the All-American Meal*, Americans spent six billion dollars on fast food in 1970 and more than 110 billion in 2000. The French fry is the most widely sold foodservice item in the United States. Almost half of the money spent on food in the United States today is spent in restaurants, mostly fast-food restaurants. None of this is particularly surprising. What might be more surprising to some is that the availability of fast food (and especially healthier alternatives) varies greatly by where people live. Numerous studies have shown that not only does fast-food advertising get concentrated in certain places, but also that the actual restaurants themselves have patterns to their locations.

Many poor neighborhoods exhibit a high density of fast-food restaurants, meaning that the number of restaurants located in that area is higher than the number of such restaurants in other communities of the same size. In a review of the available literature on restaurant density, socioeconomic indicators and obesity rates, Larson et al. (2009) found that neighborhood residents who have better access to supermarkets and limited access to convenience stores tend to have healthier diets and lower levels of obesity. Conversely, there was evidence that residents with limited access to fast-food restaurants have healthier diets and lower levels of obesity. Further, residents of low-income, minority, and rural neighborhoods were most often affected by poor access to supermarkets and healthful food while the availability of fast-food restaurants and energy-dense foods has been found to be greater in lower-income and minority neighborhoods. From these data, the authors concluded that policy intervention is necessary to assure equitable access to healthy foods (Apple 1996; Larson et al. 2009). So, what may appear as individual choices about food consumption are directly related to political and economic factors.

## Globalization as McDonaldization

In the 1970s, farm activist Jim Hightower warned of the McDonaldization of America. By this he meant that we would soon see a food economy dominated by large corporations and homogenization. Some would go beyond just agreeing with this statement to argue that it does not go far enough, literally. What Hightower saw on the American horizon is now a global phenomenon. Fast-food restaurants are everywhere. Other people have picked up the term McDonaldization to describe the spread of a homogenous eating culture throughout the world. While a market approach might explain this spread as a response to global demand, a political economy approach would explain this as a **spatial fix** – the tendency of capitalism to temporarily solve its inevitable periodic crises by establishing new markets, new resources, and new sites of production in other places. While it is true that fast-food consumption has increased rapidly in the United States in the last several decades, competition between firms has eaten up some of the surplus that each company would like to receive. Rather than relying simply on finding more consumers in the United States (and it would be hard to find anyone that did not already eat at least some fast food), the companies looked for customers abroad. By extending consumption of fast food into new places, companies have fixed their problem of diminishing surpluses. In the early 1990s, for example, McDonald's had about 3,000 restaurants outside of the United States. Now that number is closer to 20,000 in 120 different countries. Of the approximately five restaurants that the company opens each day, four are located overseas. There is much evidence that the same kind of problems with weight and obesity that we see in the United States also spread with these restaurants to other industrialized, and then, industrializing countries (Schlosser 2001). Obviously, as worldwide consumption of fast food increases, we clearly need more resources to feed the corporate machine.

> **Spatial Fix**   The tendency of capitalism to temporarily solve its inevitable periodic crises by establishing new markets, new resources, and new sites of production in other places

## We need more fries!!!

Whether we believe that our individual decisions are influenced by psychology, marketing, or the spatial distribution of choice, what is clear is that what we put in our bodies has

**Box 14.1**  Slow Food

A typical meal in the United States may last only a few short minutes. Ironically, the length of time and distance over which the components of the food probably comes is likely astronomically larger. Indeed, the average lunch might be a heavily imported affair; consider that food imports to the United States include not only vegetables and fruits but also poultry, beef, seafood and dairy. That lunch is also probably a heavily processed event, with the food likely worked over with fillers, treated for preservation, frozen, and then reconstituted, heated, and served. The convenience of this kind of food is undeniable, but its costs are harder to measure, on health (preservatives are not always good for you), on the environment (every road mile food travels requires carbon emissions), and on your psyche (can you really fully enjoy a meal like that?).

Founded in the 1980s to counter the trend towards this kind of eating and living, the Slow Food movement was established based on the principle that a meal should be: "Good", in the sense of its quality as food and experience, "Clean", in the sense that it should have a light environmental footprint in terms of the way it was both grown and prepared, and "Fair", in the sense that it promotes socially just food economies.

In terms of the personal and experiential part of the meal, Slow Food should be patiently waited for, and consumed at a leisurely pace. It therefore represents a change in the pace of typical modern lifestyles, drawing the brakes on the reckless momentum of our daily lives. More radically, Slow Food proposes a totally new food economy, favoring the preservation of many disappearing local varieties of foods, from rare apples to special regional cheeses, and the defense of small-scale, local markets and craft farmers, rather than supermarkets and factory farms. It also stresses local preparation in small restaurants and kitchens. Though it began in Italy and thrives in Europe, the movement is making great inroads in the United Kingdom and the United States.

A possible criticism of the Slow Food movement, at least as it has been advanced so far, is that it can be elitist. United States Federal law does not actually require breaks for workers, though it does require that any such breaks be compensated. Lunch breaks for the average worker last 30 minutes and are uncompensated. Organic foods are often more expensive than more generally available varieties. Many poor people do not even live within a short drive of a grocery store. Who can afford the time and money for Slow Food, one might ask? A new food economy, therefore, will have to address fundamental problems of wealth and work time around the world as it tackles the way we eat.

Even so, Slow Food's growing popularity suggests that people are increasingly interested in reclaiming not only environmental sustainability but also simple joys in their daily lives, pointing to a possible happy convergence of new kinds of societies and environments in the future.

**Reference**

Petrini, C., and G. Padovani (2006). *Slow Food Revolution: A New Culture for Eating and Living*. New York: Rizzoli.

repercussions, not just for us, but for society and environment at a global scale. In an essay called "Education, Identity, and Cheap French Fries," Michael Apple (1996), an expert in critical pedagogy, tells the sad story of one such effect. He is visiting a friend and colleague in what he calls a "remote Asian country." They are driving through heat, humidity, and glaring sun, down a two-lane road. He is jarred from his internal musings over higher education by a familiar sign on the side of the road. The sign, for him, is too familiar. It is the iconic symbol of an American fast-food restaurant that he is careful not to name. Surprised and confused, he asks his friend, a local fellow, what the sign is doing there.

Surely there is no such restaurant around here. But, his friend is also surprised; he says to him "Michael, you don't know what these signs signify? … These signs signify exactly what is wrong with education in this nation" (p. 2).

His friend then describes how the country, in an effort to lure foreign investment, had offered large tracts of land to international agribusinesses. The land they were driving through had been bought up cheap by a potato grower and supplier of French fries for the restaurant whose signs they had passed. The company, of course, had jumped at the chance to grow potatoes cheaper. Some of the farm workers for the company's US-based suppliers had tried to unionize, and there was little chance of that in this Asian country. It was also the case that the available land was perfect for new production technology. There was only one problem – hundreds of people lived on that land. The government pushed the people, few of whom had formal deeds, despite having lived there for hundreds of years, off the land. Progress, in the form of intensive potato production, had come to the region. But, this progress had costs beyond the loss of subsistence farming on which hundreds of people had depended. Because the company was given tax breaks as incentives to locate potato production there, there was no money to attend to public services like utilities or schools. "Michael," his friend concluded, "there's no schools because folks like cheap French fries" (p. 4).

The need, on the part of fast-food companies, to continue to extract surplus value required the internationalization, not only of fry consumption, but also of its production. Internationally, more potatoes are now produced in the developing world than in the developed (FAO 2008) for the first time in several centuries (see Table 14.2). While some of this is for local consumption, many of the potatoes grown in these places are meant for the frozen fry market. As we see in Apple's example above, this can be a driving factor in landscape change.

As smallholders and subsistence growers are pushed off the land to make room for factory farming, the countryside switches from a landscape of diversity to one of large-scale monocropping. When you mechanize production, it is most efficient to grow only one crop in one place. You cannot spend time readjusting your equipment and other inputs every other row. As we have seen above, producing French fries for commercial use also requires numerous other inputs, including chemical fertilizers, pesticides, water, and energy. The amount of inputs required also increases when the crop is not indigenous to the area in which it is being grown. So, imagine the experience of the Idaho potato farmer on the input treadmill (described above) being replicated across the globe.

**Table 14.2**   World potato production, 1991–2007

| Countries | 1991 | 1993 | 1995 | 1997 | 1999 | 2001 | 2003 | 2005 | 2007 |
|---|---|---|---|---|---|---|---|---|---|
| | Million tonnes | | | | | | | | |
| Developed | 183.13 | 199.31 | 177.47 | 174.63 | 165.93 | 166.93 | 160.97 | 159.97 | 159.89 |
| Developing | 84.86 | 101.95 | 108.50 | 128.72 | 135.15 | 145.92 | 152.11 | 160.01 | 165.41 |
| Total | 267.99 | 301.26 | 285.97 | 303.35 | 301.08 | 312.85 | 313.08 | 319.98 | 325.30 |

*Source*: FAOSTAT (http://faostat.fao.org/). www.potato2008.org/en/world/index.html

In addition to pushing more people out of the market – even if they have access to land, not everyone can afford the costs of growing potatoes for the frozen food industry – increased inputs also mean increased environmental damage. It is important to note, moreover, that although the production of frozen French fries is global in scale, environmental costs are always felt more acutely at a locality (local soil, regional lakes, or municipal air). Of course, many of these often circulate. A river takes contamination from one site to another, for example. But, the fact remains that the costs of such production are far removed from the people who benefit. In a new global division of labor, those who consume the food grown under such conditions are often distanced (both spatially and socially) from the producers. While some might argue that this can be addressed by markets and contracts (Chapter 3), a political economist would see this as a problem of environmental justice. Is it fair that people in a small Asian country lose their livelihoods and find their soils and streams polluted so that other people can have cheap fries?

Such questions of justice, while the purview of political economic approaches, might easily be thought of in ethical terms, especially if they are extended to the condition of the non-human world. In this case, how has the French fry's transformation of agriculture led to changes in environments around the world, and what are human responsibilities, especially concerning global biodiversity, in the wake of the Russet Burbank revolution? When we feed the world, what do we owe the planet? In the next section, we will explore these issues by considering the socio-natural puzzle of the French fry through the lens of an ethics approach.

## Ethics: Protecting or Engineering Potato Heritage?

In Chapter 5, we discussed ethical approaches to environmental issues. This approach raised questions about environmental values and how we might measure and attempt to balance them against human demands. A large part of debates over these issues centers on the value of species and species diversity. In this section we will emphasize the ethical questions, especially regarding biodiversity, that may surround the French fry.

### Agrodiversity

An ethical approach to the puzzle of the French fry might begin by taking us back to the Andes and the many varieties of potatoes domesticated there. As the potato made its journey to Europe and then back to the Americas, new varieties were developed, but more were lost. The Irish lumper potato was once dominant across the globe. Now, it has virtually disappeared while today's most prevalent potato species, the Russet Burbank used in the frozen fry market, has reached a level of dominance that is virtually unprecedented (Pollan 2001). The rise, spread, and fall of food crops is nothing new, of course. From the time of the Columbian Exchange certain varieties of the potato were destined to disappear from the planet. Many food species in the Americas were pushed out by species that accompanied European "explorers" to the New World. Crops like wheat exploded across the continent, wiping out native species on the way (Crosby 1986). On the other hand, the

**Ecocentrism**  An environmental ethical stance that argues that ecological concerns should, over and above human priorities, be central to decisions about right and wrong action (compare to anthropocentrism)

**Anthropocentrism**  An ethical standpoint that views humans as the central factor in considerations of right and wrong action in and toward nature (compare to ecocentrism)

global reach of Europeans also ensured that species from the Americas found homes in other climes (United Nations 2008). Ironically, though, the most successful species, those that like the Russet Burbank have been picked up by the global agriculture industry, are the most vulnerable of all.

We discussed monocropping and the intense inputs it requires above. These inputs are necessary, not only to produce a consistent specimen, and thus to satisfy commercial demands, but also *to protect the species from its own success*. The more a species is allowed and encouraged to exist on its own, after all, without other species, the more susceptible it is to widespread disease and other forms of crop failure. By planting vast tracts of Russet Burbanks, we have not only aided in the elimination of other potato species and any critters, from the smallest bacteria to the largest mammal, that might be dependent on them (a violation of an **ecocentric** approach to environmental ethics), but we have also doomed the species to potential failure, endangering food supplies for human populations (a violation of an **anthropocentric** valuing of nature). Remember, after all, that the Irish lumper was not the only species that suffered from *Phytophthora infestans*; one million people died, and thousands more fell ill.

Whether we define our environmental ethics in ecological or anthropocentric terms, it is clear that we face a question of right action. How do we think about and remedy the decline in agrodiversity engendered by the global corporate agriculture of French fries? While there are a number of possible avenues, we will focus on two, both of which present further questions of their own. First, while international potato production is driven by the need for frozen fries and dominated by one species, groups in some areas are trying to recover species diversity, largely through local innovation and the search for alternative agricultural networks. Conversely, other interests are using controversial biotechnology to protect the Russet Burbank itself from disease, often trying to reduce the need for other inputs at the same time.

## Rescuing diversity: Back to the future?

In the Andean mountains of Peru, where our story of the potato began, we can today find four species of potato under extensive cultivation by farmers, three of which are only found in the immediate region. In 2008, the government created a national potato registry to store genetic material from local varieties. This was done in response to concern over the threat to many indigenous species and varieties from commercial growing for urban and international populations. In addition to this, other steps toward species conservation have been taken.

Lino Mamani, a member of the Sacara farming community, has come together with people from five nearby communities to create a 12,000-acre potato park. They were able to do this with assistance from the International Potato Center. According to Mamani, there are about 1,000 varieties of potatoes in the park, 400 of which had to be planted from material from the International Potato Center. These potatoes are dispersed throughout the park to allow them to grow in the altitude to which they most easily adapt. Much like the ancient Incan farming system, each variety is allowed its own niche, but the varieties are allowed to grow side by side with their wild relatives. This enhances the resilience of the potato park. On the United Nations page for the Year of the Potato, Mamani describes it this way: "Our native varieties live well with their wild relatives, which you will find all around here. They have a good relationship, like a family. But our potatoes don't live well with modern varieties. The potatoes you see here belong to us. They came to us from our ancestors and will go on with our children" (Sullins 2001). He goes on to describe the many inputs required by commercial potatoes and the havoc they wreak on the local environment.

Projects like this one address the concern over species and varietal diversity, by bringing native species back to their original home. And, many people would argue that this is a good thing. But, such projects have their limits. To reproduce a "native" environment, many would argue that you must pick a moment in time to "go back to." True **preservationists** would look for a pristine nature from which all human activity should then be excluded. Such a preservationist ethic would say that this park might be a step in the right direction, but that it does not go far enough because it allows human use of the land and harvest.

> **Preservation**  The management of a resource or environment for protection and preservation, typically for its own sake, as in wilderness preservation (compare to conservation)
>
> **Conservation**  The management of a resource or system to sustain its productivity over time, typically associated with scientific management of collective goods like fisheries or forests (compare to preservation)

On the other hand, an argument from a **conservationist** also might not clearly fit. Here, while human use of land might be encouraged, this kind of park does not set as its goal either the maximization of the land's utility, nor the assurance of its future productivity for human use.

As a middle road, the establishment of a potato park might reflect a *land ethic* like that proposed by Aldo Leopold (see discussion in Chapter 5). Leopold's land ethic is between conservation and preservation and supports ideas of what we might call today sustainable agriculture. In this way, a potato park might be seen as a responsible solution to the environmental problem of species loss. The trick comes when you consider Leopold's famous maxim: "A thing is right when it tends to preserve the integrity, stability, and beauty of the biotic community. It is wrong when it tends otherwise." Who, in this case, gets to decide what is right and what is wrong? How do we measure the effect of species reintroduction on the "integrity, stability, and beauty of the biotic community"?

Whether you believe in preservation, conservation, or something in between (or none of the above, for that matter), we face different ethical questions when we consider what right action might be when people can, not just protect, but actually create and alter species.

## Biotechnology

As a variety, it should be clear that Russet Burbank potatoes require a lot of inputs to produce those predictable uniform fries we all love. They are also susceptible to something called net *necrosis*, a potato disease, that threatens to wipe out whole crops. But, farmers are stuck producing them. Because so much of the potato market is now in frozen potatoes, destined for fast-food chains, individual potato producers have little choice but to keep growing the Russet Burbank. In an attempt to make this variety a little more farmer (and environmentally) friendly, companies have turned to biotechnology. The result is genetically modified (GM) potatoes.

One example of this is the New Leaf potato, a now defunct variety developed by Monsanto. The New Leaf is basically a Russet Burbank, genetically engineered to produce its own pesticide. This is intended to stave off the Colorado potato beetle, which is a threat to commercial potato crops. The New Leaf produces bacterial (*Bacillus thuringiensis* or BT) in each and every part of the plant, including the spud, which itself is actually registered as a pesticide with the Environmental Protection Agency (Pollan 2001).

The uneasiness provoked by the idea of consuming pesticides (which we unknowingly do all the time) is just one part of the controversy over such GM organisms (GMOs). Because these organisms have been developed and created by a person in a lab, companies patent them to limit their free spread. On the one hand this might be considered a good thing as the GMOs are not allowed to "contaminate" other species, but they often do get loose as was the case with GM corn in southern Mexico. Here, despite assurances to the contrary, GMOs actually present a threat to native, wild, or even other commercially grown species. Moreover, because these plants are patented, the farmer is not free to replant from his own stock year after year. He or she is legally required to continue buying new seeds and plants from the company, in this case Monsanto.

**Genome**   The complete set of genes of an organism, species, etc.

While humans have modified plants for a long time (native Andeans mixed strains of domesticated potatoes), new biotechnologies are now allowing us to do that at the level of the **genome**. From an ethical perspective, we can ask: Is this the right thing to do? We might be modifying potatoes and other crops to help us feed more people more efficiently, but how do we weigh this against our valuation of the natural world? Who gets to decide what new species get to exist, and for what purposes? How do these new socio-natural objects impact the rest of the physical world?

Although other BT species (like cotton) have met with more success, the New Leaf itself did not last long. McDonald's, one of the first promoters and large-scale buyers of this GM potato, was forced to abandon plans to use it in 2002, due to consumer concerns. Apparently, that's one line consumers are not yet willing to cross to satisfy their hunger for cheap fries.

## The French Fry Puzzle

In this chapter we have learned that:

*   Frozen French fries, just like other foods, have a history that spans centuries and continents.
*   Mass French fry production is intimately tied to the boom and spread of the fast-food industry from about the 1950s on.
*   Not just any potato can become a fry. Mass production of frozen French fries demands a species that can be widely planted and that responds well to the conditions of com-

mercial agriculture. When grown with significant inputs, including chemicals, water, and energy, the Russet Burbank is the perfect fry potato.

• Risk approaches to understanding the environmental puzzle of the French fry could focus first at the scale of the human body. What we put in our bodies is, in part, dependent on what, if any, negative consequences we might associate with it. Our assessment of risk is often influenced by larger societal contexts, including but not limited to the production of scientific knowledge.

• A political economy framework helps us to understand that fry consumption is not always as much as an individual choice as it seems. It is tied instead to broader political economic processes that influence the consumption and production of foods, often at a global scale.

• An ethics approach emphasizes the costs to biodiversity of the dominance of Russet Burbank potatoes over other varieties less suitable to mass commercial production. We might be able to protect ecological systems by reintroducing lost potato species. On the other hand, we may be able to make the Russet Burbank better through biotechnology, a controversial tool that allows us to change and create new species.

## Questions for Review

1. The potato arrived in Europe via the "Columbian Exchange." Explain.
2. Describe some of the environmental costs of our voracious consumption of French fries.
3. Trace, from 1957 to the present, the "rise and fall" of trans fats.
4. How does the growing "McDonaldization" of developing countries spell more bad news for their small farmers? (This, despite the fact that all the new restaurants will, of course, be buying plenty of agricultural products grown on farms.)
5. Why (perhaps) should it concern us that, every year, the total global potato production is concentrated into increasingly fewer varieties of potatoes? Discuss the actions being taken in Peru in the face of this trend.

## Suggested Reading

Crosby, A. W. (1986). *Ecological Imperialism: The Biological Expansion of Europe, 900–1900.* Cambridge: Cambridge University Press.

Goodman, D., and M. Watts (2002). *Globalising Food: Agrarian Questions and Global Restructuring.* New York: Routledge.

Guthman, J. (2004). *Agrarian Dreams: The Paradox of Organic Farming in California.* Berkeley, CA: University of California Press.

Kloppenburg, J. (1988). *First the Seed: The Political Economy of Plant Biotechnology.* Cambridge: Cambridge University Press.

Schlosser, E. (2002). *Fast Food Nation.* New York: Penguin.

**Exercise**

In this chapter, we have reviewed how the puzzle of French fries might be addressed in terms of risk, ethics, and political economy. Starting from the notion that markets can be harnessed for social and environmental good (as described in Chapter 3), how might we understand this problem differently? You may want to consider how the health and environmental risks associated with French fry production and consumption might be addressed through pricing and marketing, in particular. What economic opportunities are there for alternatives? To what degree is there unmet demand for other kinds of foods and food production systems? How might supply be unleashed?

# Glossary

**Acid rain**  Deposition of rain or snowfall with unusually high acidity, resulting from the emission of sulfur dioxide and nitrogen oxides into the air, typically from industrial emissions. This form of precipitation is harmful for plant life and aquatic ecosystems

**Affect**  Emotions and unconscious responses to the world that influence decision-making

**Animal liberation**  Named after Peter Singer's ground-breaking 1975 book, a radical social movement that aims to free all animals from use by humans, whether those uses are for food, medical testing, industry, personal adornment, entertainment or anything else

**Animal rights**  An ethical position and social movement that states that non-human animals, particularly intelligent mammals, should be granted rights as ethical subjects on par or at least similar to human beings

**Anthropocentrism**  An ethical standpoint that views humans as the central factor in considerations of right and wrong action in and toward nature (compare to ecocentrism)

**Apex predators**  Also known as "top carnivores," the animals in any ecosystem occupying the top trophic level; apex predators do not have any natural predators

**Background extinction rate**  Usually given in numbers of plant and animal species per year, the estimated average rate of extinction over long-term, geologic time, not counting mass extinction events

**Biodiversity**  The total variability and variety of life forms in a region, ecosystem, or around the world; typically used as a measure of the health of an environmental system

**Birth rate**  A measure of natural growth in a population, typically expressed as the number of births per thousand population per year

**Bycatch**  Non-targeted organisms incidentally caught by commercial fishing operations, including many fish species, but also a large number of birds, marine mammals, and sea turtles

**Cap and trade**  A market-based system to manage environmental pollutants where a total limit is placed on all emissions in a jurisdiction (state, country, worldwide, etc.), and individual people or firms possess transferable shares of that total, theoretically leading to the most efficient overall system to maintain and reduce pollution levels overall

**Capital accumulation**  The tendency in capitalism for profits, capital goods, savings, and value to flow towards, pool in, and/or accrue in specific places, leading to the centralization and concentration of both money and power

**Carbon cycle**  The system through which carbon circulates through the earth's geosphere, atmosphere, and biosphere, specifically including exchanges between carbon in the earth (e.g., as petroleum) and the atmosphere (as $CO_2$) through combustion and back again through sequestration

**Carbon sequestration**   The capture and storage of carbon from the atmosphere into the biosphere or the geosphere through either biological means, as in plant photosynthesis, or engineered means

**Carrying capacity**   The theoretical limit of population (animal, human, or otherwise) that a system can sustain

**Climax vegetation**   The theoretical assemblage of plants arising from succession over time, determined by climatic and soil conditions

**Coase theorem**   A thesis based in neoclassical economics, holding that externalities (e.g., pollution) can be most efficiently controlled through contracts and bargaining between parties, assuming the transaction costs of reaching a bargain are not excessive

**Collective action**   Cooperation and coordination between individuals to achieve common goals and outcomes

**Columbian Exchange**   The movement of species across the Atlantic Ocean, from the New World to the Old World and vice versa, and the resulting ecological transformations

**Command-and-control**   Forms of regulation that depend on government laws and agencies to enforce rules, including such things as regulated limits on pollution or fuel efficiency standards; contrasts with market-based or incentive-based approaches

**Commodification**   The transformation of an object or resource from something valued in and for itself, to something valued generically for exchange. In Marxist thought, the rise of the exchange value of a thing, over its use value

**Commodity**   An object of economic value that is valued generically, rather than as a specific object (example: pork is a commodity, rather than a particular pig). In political economy (and Marxist) thought, an object made for exchange

**Common property**   A good or resource (e.g., bandwidth, pasture, oceans) whose characteristics make it difficult to fully enclose and partition, making it possible for non-owners to enjoy resource benefits and owners to sustain costs from the actions of others, typically necessitating some form of creative institutional management

**Concept**   A single idea, usually captured in a word or a phrase

**Conditions of production**   In political economic (and Marxist) thought, the material or environmental conditions required for a specific economy to function, which may include things as varied as water for use in an industrial process to the health of workers to do the labor

**Conservation**   The management of a resource or system to sustain its productivity over time, typically associated with scientific management of collective goods like fisheries or forests (compare to preservation)

**Conservation biology**   A branch of scientific biology dedicated to exploring and maintaining biodiversity and plant and animal species

**Constructivist**   Emphasizing the significance of concepts, ideologies, and social practices to our understanding and making of (literally, *constructing*) the world

**Consumer boycott**   A method of protest that aims to pressure corporations into changing their practices by urging people to forgo purchasing products associated with the targeted corporations

**Co-production**   The inevitable and ongoing process whereby humans and non-humans produce and change one another through their interaction and interrelation

**Cultural theory**   A theoretical framework associated with anthropologist Mary Douglas that stresses the way individual perceptions (of risk, for example) are reinforced by group social dynamics, leading to a few paradigmatic, typical, and discrete ways of seeing and addressing problems

**Death rate**   A measure of mortality in a population, typically expressed as the number of deaths per thousand population per year

**Deep ecology**   A philosophy of environmental ethics that distances itself from "shallow" or mainstream environmentalism by arguing for a "deeper" and supposedly more truly ecologically-informed view of the world

**Demographic transition model**   A model of population change that predicts a decline in population death rates associated with modernization, followed by a decline in birth rates resulting from industrialization and urbanization; this creates a sigmoidal curve where population growth increases rapidly for a period, then levels off

**Desalinization**    A technology that removes salts and other minerals from water, especially sea water; prohibitively expensive in most contexts, current techniques are highly energy demanding

**Discourse**    At root, written and spoken communication; thicker deployments of the term acknowledge that statements and texts are not mere representations of a material world, but rather power-embedded constructions that (partially) make the world we live in

**Disturbance**    An event or shock that disrupts an ecological system, thereafter leading either to recovery of that system (e.g., through succession) or movement of the system into a new state

**Dolphin safe tuna**    Tuna caught without killing dolphins as bycatch

**Dominion thesis**    Arising from the Book of Genesis, the dominion thesis states that humans are the pinnacle of creation; as such, humans are granted ethical free rein to use nature in any way deemed beneficial

**Ecocentrism**    An environmental ethical stance that argues that ecological concerns should, over and above human priorities, be central to decisions about right and wrong action (compare to anthropocentrism)

**Eco-feminism**    Any of a number of theories critical of the role of patriarchal society for degrading both the natural environment and the social condition of women

**Ecological footprint**    The theoretical spatial extent of the earth's surface required to sustain an individual, group, system, organization; an index of environmental impact

**Ecology**    The scientific study of interactions amongst organisms and between organisms and the habitat or ecosystem in which they live

**Ecosystem services**    Benefits that an organic system creates through its function, including food resources, clean air or water, pollination, carbon sequestration, energy, and nutrient cycling, among many others

**Emissions trading**    A system for exchanging the right to emit/pollute limited amounts of determinant materials (like greenhouse gases). These rights or credits are exchangeable between emitters, but subject to a total regulatory limit

**Environmental justice**    A principle, as well as a body of thought and research, stressing the need for equitable distribution of environmental goods (parks, clean air, healthful working conditions) and environmental bads (pollution, hazards, waste) between people, no matter their race, ethnicity, or gender. Conversely, environmental injustice describes a condition where unhealthful or dangerous conditions are disproportionately proximate to minority communities

**Ethics/ethical**    The branch of philosophy dealing with morality, or, questions of right and wrong human action in the world

**Exchange value**    In political economy (and Marxist) thought, the quality of a commodity that determines the quantity of other goods for which the commodity might be traded at a given moment. Compare to use value

**Exclusive economic zones (EEZs)**    Usually extending 200 nautical miles off the coasts of sovereign states, EEZs are sea-zones within which states claim ownership over fishery and mineral resources

**Exponential growth**    A condition of growth where the rate is mathematically proportional to the current value, leading to continued, non-linear increase of the quantity; in population, this refers to a state of increasingly accelerated and compounded growth, with ecological implications for scarcity

**Externality**    The spillover of a cost or benefit, as where industrial activity at a plant leads to pollution off-site that must be paid for by someone else

**Extinction crisis**    The current era of anthropogenically induced plant and animal extinction, estimated to be between one thousand and ten thousand times the historical average, or background extinction rate

**Factory farms**    Intensive animal-raising agricultural operations; factory farms attempt to maximize production by raising as many animals in as little space as possible, often resulting in significant air and water pollution

**Fertility rate**    A measure describing the average number of children birthed by an average statistical woman during her reproductive lifetime

**First contradiction of capitalism**   In Marxist thought, this describes the tendency for capitalism to eventually undermine the economic conditions for its own perpetuation, through overproduction of commodities, reduction of wages for would-be consumers, etc., predicted to eventually lead to responses by workers to resist capitalism leading to a new form of economy. Compare to the second contradiction of capitalism

**Fordism**   Relations of production dominant in many industrialized countries in the first several decades of the twentieth century; marked by large, vertically integrated corporations, high wages and rates of consumption, and considerable state power

**Forest transition theory**   A model that predicts a period of deforestation in a region during development, when the forest is a resource or land is cleared for agriculture, followed by a return of forest when the economy changes and population outmigrates and/or becomes conservation-oriented

**Game theory**   A form of applied mathematics used to model and predict people's behavior in strategic situations where people's choices are predicated on predicting the behavior of others

**Genome**   The complete set of genes of an organism, species, etc.

**Globalization**   An ongoing process by which regional economies, societies, and cultures have become integrated through a globe-spanning network of exchange

**Green certification**   Programs to certify commodities for the purposes of assuring their ecological credentials, such as organically grown vegetables or sustainably harvested wood products

**Green consumption**   Purchasing of products that are purportedly environmentally friendlier or less harmful than their alternatives; a model of environmental protection that relies on consumer choices to change the behavior of firms or industries rather than regulation

**Green Revolution**   A suite of technological innovations, developed in universities and international research centers, which were applied to agriculture between the 1950s and 1980s and increased agricultural yields dramatically, but with a concomitant rise in chemical inputs (fertilizers and pesticides) as well as increased demands for water and machinery

**Greenhouse effect**   The characteristic of the earth's atmosphere, based on the presence of important gases including water vapor and carbon dioxide, to trap and retain heat, leading to temperatures that can sustain life

**Greenwashing**   The exaggerated or false marketing of a product, good, or service as environmentally friendly

**Hazard**   An object, condition, or process that threatens individuals and society in terms of production or reproduction

**Holism**   Any theory that holds that a whole system (e.g., an "ecosystem" or the earth) is more than the sum of its parts

**Ideologies**   Normative, value-laden, world views that spell out how the world is and how it ought to be

**Induced intensification**   A thesis predicting that where agricultural populations grow, demands for food lead to technological innovations resulting in increased food production on the same amount of available land

**Institutions**   Rules and norms governing collective action, especially referring to rules governing common-property environmental resources, like rivers, oceans, or the atmosphere

**Intrinsic value**   The value of a natural object (e.g., an owl or a stream) in and for itself, as an end rather than a means

**IPAT**   A theoretical formula holding that human Impact is a function of the total Population, its overall Affluence, and its Technology; this provides an alternative formulation to a simple assumption that population alone is proportional to impact

**Jevons' Paradox**   The somewhat counterintuitive observation, rooted in modern economic theory, that a technology that increases the efficiency of resource use actually increases, rather than decreases, the rate of consumption of that resource

**Kuznets curve (environmental)**   Based in the theory that income inequality will increase during economic development and decrease after reaching a state of overall affluence, this theory predicts that environmental impacts rise during development, only to fall after an economy matures

**Life cycle analysis**   The rigorous analysis of the environmental impacts of a product, service, or object from its point of manufacture all the way to its deposition as waste; also know as cradle-to-grave assessment

**Longliners**    An industrial fishing method deploying lines baited with hundreds or thousands of hooks; longlines are usually several miles long and often result in significant bycatch

**Market failure**    A situation or condition where the production or exchange of a good or service is NOT efficient; this refers to a range of perverse economic outcomes stemming from market problems like monopoly or uncontrolled externalities

**Market response model**    A model that predicts economic responses to scarcity of a resource will lead to increases in prices that will result either in decreased demand for that resource or increased supply, or both

**Masculinity**    The socially agreed upon characteristics of behavior associated with men in any society; these may vary significantly between cultures, locations, and periods of history

**Maximum sustainable yield**    The largest seasonal or annual amount of any particular natural resource (e.g., timber, fish) that can be harvested indefinitely

**Means of production**    In political economic (and Marxist) thought, the infrastructure, equipment, machinery, etc. required to make things, goods, and commodities

**Monocrop**    A single crop cultivated to the exclusion of any other potential harvest

**Monopoly**    A market condition where there is one seller for many buyers, leading to perverted and artificially inflated pricing of goods or services

**Monopsony**    A market condition where there is one buyer for many sellers, leading to perverted and artificially deflated pricing of goods or services

**Moral extensionism**    An ethical principle stating that humans should extend their sphere of moral concern beyond the human realm; most commonly, it is argued that intelligent or sentient animals are worthy ethical subjects

**Narrative**    A story with a beginning and end; environmental narratives such as "biological evolution" and "the tragedy of the commons" aid our comprehension and construction of the world

**Natural resource management**    Both the academic discipline and professional field dedicated to the management of environmental conditions, goods, or services for social goals, which may range between instrumental human utility to ecological sustainability

**Naturalistic fallacy**    A philosophically invalid derivation of an ethical "ought" from a natural "is"

**"Nature"**    The natural world, everything that exists that is not a product of human activity; often put in quotes to designate that it is difficult if not impossible to divvy up the entire world into discrete natural and human components

**Neo-Malthusians**    Present-day adherents to a position – established by Malthus in the nineteenth century – that population growth outstrips limited natural resources and presents the single greatest driver of environmental degradation and crisis

**NEPA**    The National Environmental Policy Act of 1970 commits the US government to protecting and improving the natural environment; after NEPA, the federal government is required to write environmental impact statements (EIS) for government actions that have significant environmental impact

**Niche**    In ecology, the location of an organism or species within a larger ecosystem, typically fulfilling an ecological function

**Overaccumulation**    In political economy (and Marxism), a condition in the economy where capital becomes concentrated in very few hands (e.g., wealthy individuals) or firms (e.g., banks), causing economic slowdown and potential socioeconomic crisis

**Overproduction**    In political economy (and Marxism), a condition in the economy where the capacity of industry to produce goods and services outpaces the needs and capacity to consume, causing economic slowdown and potential socioeconomic crisis

**Photosynthesis**    The process through which plants use the sun's energy to convert carbon dioxide into organic compounds, especially sugars that are used to build tissues

**Political ecology**    An approach to environmental issues that unites issues of ecology with a broadly defined political economy perspective

**Post-Fordism**   Arising in the last decades of the twentieth century, the current relations of production in most industrialized countries; marked by decentralized, specialized, and often subcontracted production, the prominence of transnational corporations, and diminished state power

**Power/knowledge**   A theoretical formulation associated with the philosopher Michel Foucault, which holds that what is known and held as true in a society is never separate from power, such that knowledge reinforces relationships of power but also that systems of power are associated with their own specific regimes of knowledge

**Pragmatism**   A branch of philosophy that arose in late-nineteenth-century North America, pragmatism considers real-world consequences and effects to be constituent components of truth and reality

**Preservation**   The management of a resource or environment for protection and preservation, typically for its own sake, as in wilderness preservation (compare to conservation)

**Primitive accumulation**   In Marxist thought, the direct appropriation by capitalists of natural resources or goods from communities that historically tend to hold them collectively, as, for example, where the common lands of Britain were enclosed by wealthy elites and the state in the 1700s

**Prisoner's Dilemma**   An allegorical description of a game-theoretical situation in which multiple individuals making decisions in pursuit of their own interests tend to create collective outcomes that are non-optimal for everyone

**Production of nature**   In political economic thought, the idea that the environment, if it ever did exist separate from people, is now a product of human industry or activity

**Purse-seine fishing**   An effective fishing method for species that school near the surface; a large net is encircled around the targeted catch, after which the bottom of the net is drawn tight like the strings of a purse, thus confining the catch in the net

**Race**   A set of imaginary categories distinguishing types of people, typically based on skin color or body morphology, which varies significantly between cultures, locations, and periods of history

**Reconciliation ecology**   A science of imagining, creating, and sustaining habitats, productive environments, and biodiversity in places used, traveled, and inhabited by human beings

**Relations of production**   In political economic (and Marxist) thought, the social relationships associated with, and necessary for, a specific economy, as serfs/knights are to feudalism and workers/owners are to modern capitalism

**Relativism**   Questioning the veracity of universal truth statements, relativism holds that all beliefs, truths, and facts are at root products of the particular set of social relations from which they arise

**Rewilding**   The restoration of natural ecological functioning and evolutionary processes to ecosystems; rewilding often requires the reintroduction or restoration of large predators to ecosystems

**Risk**   The known (or estimated) probability that a hazard-related decision will have a negative consequence

**Risk assessment**   The rigorous application of logic and information to determine the risk – possibility of an undesirable outcome – associated with particular decisions; used to so reach more optimal and rational outcomes

**Risk communication**   A field of study dedicated to understanding the optimal way to present and convey risk-related information to aid people in reaching optimal and rational outcomes

**Risk perception**   A phenomenon, and related field of study, describing the tendency of people to evaluate the hazardousness of a situation or decision in not-always-rational terms, depending on individual biases, culture, or human tendencies

**Scientism**   Usually deployed as a term of derision; refers to an uncritical reliance on the natural sciences as the basis for social decision-making and ethical judgments

**Second contradiction of capitalism**   In Marxist thought, this describes the tendency for capitalism to eventually undermine the environmental conditions for its own perpetuation, through degradation of natural resources or damage to the health of workers, etc., predicted to eventually lead to environmentalist and workers' movements to resist capitalism, leading to a new form of economy. Compare to the first contradiction of capitalism

**Secondary succession**  The regrowth of vegetation and return of species to an area cleared or reduced by disturbance, as where a forest recovers its "climax vegetation" cover after a fire

**Shifting cultivation**  A form of agriculture that clears and burns forest areas to release nutrients for cropping. Also known as "swidden," this method is highly extensive, typically rotating through areas of forest land for short periods of use, allowing previously used forest land to recover

**Signifying practices**  Modes and methods of representation; the techniques used to tell stories, introduce and define concepts, and communicate ideologies

**Social construction**  Any category, condition, or thing that exists or is understood to have certain characteristics because people socially agree that it does

**Social context**  The ensemble of social relations in a particular place at a particular time; includes belief systems, economic relations of production, and institutions of governance

**Social Darwinism**  The use of Darwinian evolutionary theory to explain social phenomena; as individuals are viewed as naturally and inherently competitive and selfish beings, social Darwinism typically rationalizes war, poverty, and hierarchically stratified social systems

**Social ecology**  A school of thought and set of social movements, associated with the thinker Murray Bookchin, asserting that environmental problems and crises are rooted in typical social structures and relationships, since these tend to be hierarchical, state-controlled, and predicated on domination of both people and nature

**Social reproduction**  That part of the economy, especially including household work, that depends on unremunerated labor, but without which the more formal cash economy would suffer and collapse

**Spatial fix**  The tendency of capitalism to temporarily solve its inevitable periodic crises by establishing new markets, new resources, and new sites of production in other places

**Stakeholders**  Individuals or groups with a vested interest in the outcome of disputed actions

**Stewardship**  Taking responsibility for the property or fate of others; stewardship of land and natural resources is often used in a religious context, such as "caring for creation"

**Succession**  Ecologically, the idealized tendency for disturbed forest areas to recover through stages of species invasion and growth, progressing from grassland, to shrubs, and eventually back to tree cover

**Superfund**  The environmental program established to address abandoned hazardous waste sites in the United States

**Surplus value**  In political economic (and Marxist) thought, the value produced by underpaying labor or over-extracting from the environment, which is accumulated by owners and investors

**Sustainable/sustainability**  The conservation of land and resources so as to secure their availability to future generations

**Transaction costs**  In economics, the cost associated with making an exchange, including, for example, drawing a contract, traveling to market, or negotiating a price; while most economic models assume low transaction costs, in reality these costs can be quite high, especially for systems with high externalities

**Transnational corporations (TNC)**  Corporations operating facilities in more than one country; also commonly called multinational corporations (MNCs)

**Trophic levels**  Parallel levels of energy assimilation and transfer within ecological food webs; in terrestrial ecosystems, photosynthetic plants form the base trophic level, followed "up" the web by herbivores and successive levels of carnivores

**Uncertainty**  The degree to which the outcomes of a decision or situation are unknown

**Uneven development**  The geographic tendency within capitalism to produce highly disparate economic conditions (wealth/poverty) and economic activity (production/consumption) in different places

**Use value**  In political economy (and Marxist) thought, the quality of a commodity derived from its actual usefulness and importance for individuals, since it fills a need or purpose. Compare to "exchange value"

**Utilitarian**  An ethical theory that posits that the value of a good should be judged solely (or at least primarily) by its usefulness to society; following the eighteenth–nineteenth-century philosopher Jeremy Bentham, usefulness is equated with maximizing pleasure or happiness and minimizing pain and suffering

**Utopia/utopian**  Imaginary, idealized social conditions arising from socio-political systems that facilitate cooperation over competition

**Wilderness**  A wild parcel of land, more or less unaffected by human forces; increasingly, wilderness is viewed as a social construction

**Zero population growth**  A condition in a population where the number of births matches the number of deaths and therefore there is no net increase; an idealized condition for those concerned about overpopulation

# References

Anadu, E. C., and A. K. Harding (2000). "Risk perception and bottled water use." *Journal of the American Water Works Association* 92(11): 82–92.

Andrews, R. N. L. (1999). *Managing the Environment, Managing Ourselves: A History of American Environmental Policy*. New Haven, CT: Yale University Press.

Apple, M. W. (1996). *Education, Identity and Cheap French Fries*. New York: Teachers College Press.

Arvai, J. (2003). "Testing alternative decision approaches for identifying cleanup priorities at contaminated sites." *Environmental Science and Technology* 37(8): 1469–76.

Back, W., E. R. Landa, et al. (1995). "Bottled water, spas, and early years of water chemistry." *Ground Water* 33(4): 605–14.

Bakker, K. (2004). *An Uncooperative Commodity: Privatizing Water in England and Wales*. Oxford: Oxford University Press.

Barclay, P. D. (2002). "A 'curious and grim testimony to a persistent human blindness': Wolf bounties in North America, 1630–1752." *Ethics, Place and Environment* 5(1): 25–34.

Barlow, C. (1999). "Rewilding for Evolution." *Wild Earth* 9(1): 53–6.

Barnes, T. J., and J. S. Duncan (1992). "Introduction: Writing worlds." In T. J. Barnes and J. S. Duncan (eds.) *Writing Worlds: Discourse, Text, and Metaphor in the Representation of Landscape*. New York: Routledge, pp. 1–17.

Beck, U. (1999). *World Risk Society*. Oxford: Blackwell Publishers.

Beder, S. (1996). "Charging the earth: The promotion of price-based measures for pollution control." *Ecological Economics* 16(1): 51–63.

Bender, K. (2006). "Men chained to tree to protest UC." Oakland, CA: *Oakland Tribune*.

Beverage Marketing Corporation (2009). *Beverage Marketing's 2007 Market Report Findings*. International Bottled Water Association 2009 [cited April 3, 2009]. Retrieved April 3, 2009, from www.bottledwater.org/public/statistics_main.htm.

Bloomberg Newswire (2008, 16 June). "Pacific Nations Ban Tuna Boats To Stop Stock Collapse." Retrieved April 11, 2009, from www.bloomberg.com/apps/news?pid=20601101&sid=aqOdnFHygH1k.

Blum, J. (2005). "Exxon Mobil profit soars 75%." *Washington Post*, October 28, p. D1.

Bonanno, A., and D. Constance (1996). *Caught in the Net: The Global Tuna Industry, Environmentalism and the State*. Lawrence, KS: University of Kansas.

Boserup, E. (1965). *Conditions of Agricultural Growth: The Economics of Agrarian Change under Population Pressure*. Chicago, IL: Aldine.

Brower, K. (1989). "The destruction of dolphins." *Atlantic Monthly* July: 35–58.

Bullard, R. D. (1990). *Dumping in Dixie: Race, Class, and Environmental Quality*. Boulder, CO: Westview Press.

Castree, N., and B. Braun (1998). "The Construction of Nature and the Nature of Construction: Analytical and political tools for building survivable futures." In

N. Castree and B. Braun (eds.) *Remaking Reality: Nature at the Millennium*. London: Routledge, pp. 3–42.

Chambers, N., C. Simmons, et al. (2002). *Sharing Nature's Interest*. London: Earthscan.

Ciriacy-Wantrup, S. V., and R. C. Bishop (1975). "Common Property as a Concept in Natural Resources Policy." *Natural Resources Journal* 15: 713–27.

Coase, R. H. (1960). "The Problem of Social Cost." *Journal of Law and Economics* 3(October): 1–44.

Cohen, S. E. (2004). *Planting Nature: Trees and the Manipulation of Environmental Stewardship in America*. Berkeley, CA: University of California Press.

Colapinto, C. K., A. Fitzgerald, et al. (2007). "Children's preference for large portions: Prevalence, determinants, and consequences." *Journal of the American Dietetic Association* 107(7): 1183–90.

Commoner, B. (1988). "The Environment." In P. Borelli (ed.) *Crossroads: Environmental Priorities for the Future*. Washington, DC: Island Press, pp. 121–69.

Commons, J. R. (1934). *Institutional Economics*. New York: Macmillan.

Constance, D., and A. Bonanno (1999). "Contested terrain of the global fisheries: "Dolphin-safe" tuna, the Panama Declaration, and the Marine Stewardship Council." *Rural Sociology* 64(4): 597–623.

Cronon, W. (1995). "The trouble with wilderness; or, getting back to the wrong nature." In W. Cronon (ed.) *Uncommon Ground: Toward Reinventing Nature*. New York: W.W. Norton, pp. 69–90.

Crosby, A. W. (1986). *Ecological Imperialism: The Biological Expansion of Europe, 900–1900*. New York: Cambridge University Press.

Cruz, R. V., H. Harasawa, et al. (2007). "Asia." In M. L. Parry, O. F. Canziani, J. P. Palutikof, et al. (eds.) *Climate Change 2007: Impacts, Adaptation and Vulnerability. Contribution of Working Group II to the Fourth Assessment Report of the Intergovernmental Panel on Climate Change*. Cambridge: Cambridge University Press, pp. 469–506.

Davis, D. K. (2007). *Resurrecting the Granary of Rome: Environmental History and French Colonial Expansion in North Africa*. Athens, OH: Ohio University Press.

Delcourt, H. R. (2002). *Forests in Peril: Tracking Deciduous Trees from Ice-Age Refuges into the Greenhouse World*. Blacksburg, VA: McDonald and Woodward Publishing.

Demeny, P. (1990). "Population." In B. L. Turner, W. C. Clark, R. Kates, et al. (eds.) *The Earth as Transformed by Human Action*. Cambridge: Cambridge University Press.

Demeritt, D. (2001). "Being constructive about nature." In N. Castree and B. Braun (eds.) *Social Nature: Theory, Practice, and Politics*. Malden, MA: Blackwell.

Doria, M. F. (2006). "Bottled water versus tap water: Understanding consumers' preferences." *Journal of Water and Health* 4(2): 271–6.

Douglas, M., and A. Wildavsky (1983). *Risk and Culture: An Essay on the Selection of Technological and Environmental Dangers*. Berkeley, CA: University of California Press.

Eckel, R. H., P. Kris-Etherton, et al. (2009). "Americans' awareness, knowledge, and behaviors regarding fats: 2006–2007." *Journal of the American Dietetic Association* 109(2): 288–96.

Ehrlich, P. R. (1968). *The Population Bomb*. New York: Ballantine Books.

Ehrlich, P. R., and J. Holdren (1974). "Impact of population growth." *Science* 171(3977): 1212–17.

Ellis, R. (2008). *Tuna: A Love Story*. New York: Knopf.

Emel, J. (1998). "Are you man enough, big and bad enough? Wolf eradication in the US." In J. Wolch and J. Emel (eds.) *Animal Geographies*. New York: Verso, pp. 91–116.

Field, B. C. (2005). *Natural Resource Economics: An Introduction*. Long Grove, IL: Waveland Press.

Fischer, H. (1995). *Wolf Wars : The Remarkable Inside Story of the Restoration of Wolves to Yellowstone*. Helena, MT: Falcon Press.

Fischhoff, B., P. Slovic, et al. (1978). "How safe is safe enough? A psychometric study of attitudes towards technological risks and benefits." *Policy Sciences* 9(2): 127–52.

Food and Agriculture Organization (2008). "International Year of the Potato." Retrieved April 2009, from www.potato2008.org/en/perspectives/mamani.html.

Foreman, D. (2004). *Rewilding North America*. Washington, DC: Island Press.

Foster, J. B. (2005). "The treadmill of accumulation: Schnaiberg's environment and Marxian political economy." *Organization and Environment* 18: 7–18.

Foucault, M. (1980). *Power/Knowledge*. New York: Pantheon.

Gosliner, M. L. (1999). "The tuna–dolphin controversy." In J. R. Twiss Jr. and R. R. Reeves (eds.) *Conservation and Management of Marine Mammals*. Washington, DC: Smithsonian Institution, pp. 120–55.

Gottlieb, R. (1995). *Forcing the Spring: The Transformation of the American Environmental Movement*. Washington, DC: Island Press.

Grainger, A. (2008). "Difficulties in tracking the long-term global trend in tropical forest area." *Proceedings of the National Academy of Sciences of the United States of America* 105(2): 818–23.

Grand Rapids Press (2007). "Ruling dampens challenge to water rights; Group can't sue over Nestlé's private land, court says." *Grand Rapids Press*, July 26, p. A1.

Guthman, J., and M. DuPuis (2006). "Embodying neoliberalism: Economy, culture, and the politics of fat." *Environment and Planning D: Society and Space* 24(3): 427–48.

Hacking, I. (1999). *The Social Construction of What?* Cambridge, MA: Harvard University Press.

Haraway, D. (2003). *The Companion Species Manifesto: Dogs, People, and Significant Otherness*. Chicago, IL: Prickly Paradigm Press.

Haraway, D. J. (1989). *Primate Visions*. New York: Routledge.

Hardin, G. (1968). "The Tragedy of the Commons." *Science* 162: 1243–8.

Hartmann, B. (1995). *Reproductive Rights and Wrongs: The Global Politics of Population Control*. Boston, MA: South End Press.

Harvey, D. (1996). *Justice, Nature, and the Geography of Difference*. Cambridge, MA: Blackwell.

Harvey, D. (1999). *The Limits to Capital*. London: Verso.

Hornaday, W. (1904). *The American Natural History*. New York: Scribner's Sons.

Hunt, C. E. (2004). *Thirsty Planet*. New York: Zed Books.

Intergovernmental Panel on Climate Change (2007). *Climate Change 2007: The Physical Basis. Contribution of Working Group I to the Fourth Assessment Report of the Intergovernmental Panel on Climate Change*. S. Solomon, D. Qin, M. Manning, et al. Cambridge: Cambridge University Press.

Jackson, K. (1985). *The Crabgrass Frontier: The Suburbanization of the United States*. New York: Oxford University Press.

Johnson, B. B. (2002). "Comparing bottled water and tap water: Experiments in risk communication." *Risk: Health, Safety and Environment* 13: 69–94.

Johnson, G. P., R. R. J. Holmes, et al. (2004). *The Great Flood of 1993 on the Upper Mississippi River: 10 Years Later*. US Geological Survey Fact Sheet 2004–3024. Washington, DC: Department of the Interior.

Joseph, J. (1994). "The tuna–dolphin controversy in the Eastern Tropical Pacific: Biological, economic and political impacts." *Ocean Development and International Law* 25: 1–30.

Katz, C. (2001). "Vagabond capitalism and the necessity of social reproduction." *Antipode* 33(4): 709–28.

Katz, E. (1997). *Nature as Subject: Human Obligation and Natural Community*. Lanham, MD: Rowman and Littlefield.

Kitman, J. L. (2000). "The secret history of lead." *The Nation* March 20: 11–30.

Klein, N. (2007). *The Shock Doctrine: The Rise of Disaster Capitalism*. New York: Metropolitan.

Kropotkin, P. (1888). *Mutual Aid: A Factor in Evolution*. Boston, MA: Porter Sargent.

Landi, H. (2008). "Bottled water report." *Beverage World* 127(4): S12–14.

Larson, N. I., M. T. Story, et al. (2009). "Neighborhood environments: Disparities in access to healthy foods in the US." *American Journal of Preventive Medicine* 36(1): 74–81.

Leopold, A. (1949). *A Sand County Almanac*. New York: Oxford University Press.

Leopold, A. (1987). *A Sand County Almanac, and Sketches Here and There*. New York: Oxford University Press.

Lohmann, L. (2006). "Carbon trading: A critical conversation on climate change, privatization, and power." *Development Dialogue* 48.

Malthus, T. R. (1992). *An Essay on the Principle of Population (selected and introduced by D. Winch)*. Cambridge: Cambridge University Press.

Marcus, E. (2005). *Meat Market: Animals, Ethics, and Money*. Boston, MA: Brio Press.

Margulis, L., and M. F. Dolan (2002). *Early Life: Evolution on the PreCambrian Earth* (2nd edn). Boston, MA: Jones and Bartlett.

Marketing Week (2005). "Profits flow in from bottled-water market." *Marketing Week* 24.

Marx, K. (1990). *Capital: A Critique of Political Economy, Volume I*. New York: Penguin.

McKibben, B. (1990). *The End of Nature*. New York: Random House.

McKinley, J. (2008). "Berkeley tree protesters climb down." *The New York Times*, September 10.

Mech, D. (1970). *The Wolf*. Garden City, NY: Natural History.

Mech, L. D., and L. Boitani (eds.) (2003). *Wolves: Behavior, Ecology, and Conservation*. Chicago, IL: University of Chicago Press.

Naess, A. (1973). "The shallow and the deep, long-range ecology movement: A summary." *Inquiry* 16: 95–100.

Napier, G. L., and C. M. Kodner (2008). "Health risks and benefits of bottled water." *Primary Care* 35(4): 789–802.

National Petroleum News (2007). "Convenience consumer: Bottled water." *National Petroleum News* 99(1): 16.

Newbold, K. B. (2007). *Six Billion Plus: World Population in the Twenty-First Century*. Oxford: Rowman and Littlefield.

Newsweek (1990). "Swim with the dolphins." *Newsweek* 115(17): 76.

Nie, M. A. (2003). *Beyond Wolves: The Politics of Wolf Recovery and Management*. Minneapolis, MN: University of Minnesota Press.

O'Connor, J. (1988). "Capitalism, nature, socialism: A theoretical introduction." *Capitalism, Nature, Socialism* 1: 11–38.

Ostrom, E. (1990). *Governing the Commons: The Evolution of Institutions for Collective Action*. Cambridge: Cambridge University Press.

Ostrom, E. (1992). *Crafting Institutions for Self-Governing Irrigation Systems*. San Francisco, CA: Institute for Contemporary Studies.

Ostrom, E. (ed.) (2002). *The Drama of the Commons*. Washington, DC: National Academy Press.

Owusu, J. H. (1998). "Current convenience, desperate deforestation: Ghana's adjustment program and the forestry sector." *Professional Geographer* 50(4): 418–36.

Parayil, G. (ed.) (2000). *Kerala: The Development Experience*. London: Zed Books.

Päster, P. (2009). "Exotic bottled water." Retrieved April 3, 2009, from www.triplepundit.com/pages/askpablo-exotic-1.php.

Pearce, F. (2007). *With Speed and Violence: Why Scientists Fear Tipping Points in Climate Change*. Boston, MA: Beacon Press.

Perz, S. G. (2007). "Grand theory and context-specificity in the study of forest dynamics: Forest transition theory and other directions." *Professional Geographer* 59(1): 105–14.

Pimentel, D., Doughty, R., Carothers, C., Lamberson, S., Bora, N., and Lee, K. (2002). "Energy inputs in crop production in developing and developed countries." In R. Lal, D. O. Hansen, N. Uphoff, and S. A. Slack (eds.) *Food Security and Environmental Quality in the Developing World*. Boca Raton, FL: CRC Press.

Plummer, C. (2002). "French fries driving globalization of frozen potato industry." *Frozen Food Digest* 18(2): 12–15.

Pollan, M. (2001). *The Botany of Desire: A Plant's Eye View of the World*. New York: Random House.

Population Reference Bureau (2008). *World Population Data Sheet*. Retrieved October 7, 2009, from www.prb.org/Publications/Datasheets/2008/2008wpds.aspx.

Poundstone, W. (1992). *Prisoner's Dilemma*. New York: Anchor Books.

Proctor, J. (1998). "The social construction of nature: Relativist accusations, pragmatic and realist responses." *Annals of the Association of American Geographers* 88(3): 353–76.

Ranganathan, J., R. J. R. Daniels, M. D. S. Chandran, P. R. Ehrlich, and G. C. Daily (2008). "Sustaining biodiversity in ancient tropical countryside." *Proceedings of the National Academy of Sciences of the United States of America* 105(46): 17852–4.

Rees, J. (1990). *Natural Resources: Allocation, Economics, and Policy*. New York: Routledge.

Richards, J. F. (1990). "Land transformation." In B. L. T. Turner, W. C. Clark, R. W. Kates, et al. (eds.) *The Earth as Transformed by Human Action*. Cambridge: Cambridge University Press, pp. 163–78.

Robbins, P. (2004). *Political Ecology: A Critical Introduction*. New York: Blackwell.

Robbins, P. (2007). *Lawn People: How Grasses, Weeds, and Chemicals Make Us Who We Are*. Philadelphia, PA: Temple University Press.

Robertson, M. M. (2006). "The nature that capital can see: Science, state, and market in the commodification of ecosystem services." *Environment and Planning D: Society and Space* 24(3): 367–87.

Rogers, P. (1996). *America's Water*. Boston, MA: MIT Press.

Rorty, R. (1979). *Philosophy and the Mirror of Nature*. Princeton, NJ: Princeton University Press.

Rosenzweig, M. L. (2003). *Win–Win Ecology: How the Earth's Species Can Survive in the Midst of Human Enterprise*. Oxford: Oxford University Press.

Safina, C. (2001). "Tuna conservation." In B. A. Block and E. D. Stevens (eds.) *Tuna: Physiology, Ecology and Evolution*. San Diego, CA: Academic Press, pp. 413–59.

Schlosser, E. (2001). *Fast Food Nation: The Dark Side of the All-American Meal*. Boston, MA: Houghton Mifflin.

Seager, J. (1996). "'Hysterical housewives' and other mad women: Grassroots environmental organizing in the

United States." In D. Rocheleau, B. Thomas-Slayter, and E. Wangari (eds.) *Feminist Political Ecology: Global Issues and Local Experiences*. New York: Routledge, pp. 271–83.

Sessions, G. (2001). Ecocentrism, wilderness, and global ecosystem processes. In M. E. Zimmerman, J. B. Callicott, G. Sessions, et al. (eds.) *Environmental Philosophy: From Animal Rights to Radical Ecology*. Upper Saddle River, NJ: Prentice Hall, pp. 236–52.

Shrader-Frechette, K. S. (1993). *Burying Uncertainty: Risk and the Case against Geological Disposal of Nuclear Waste*. Berkeley, CA: University of California Press.

Simon, J. L. (1980). "Resources, population, environment: An oversupply of false bad news." *Science* 208(4451): 1431–7.

Singer, P. (1975). *Animal Liberation: A New Ethics for our Treatment of Animals*. New York: New York Review (distributed by Random House).

Slovic, P. (2000). *The Perception of Risk*. London: Earthscan.

Sluyter, A. (1999). "The making of the myth in postcolonial development: Material-conceptual landscape transformation in sixteenth century Veracruz." *Annals of the Association of American Geographers* 89(3): 377–401.

Smith, D. W., and G. Ferguson (2005). *Decade of the Wolf: Returning the Wild to Yellowstone*. Guilford, CT: Lyons.

Smith, N. (1984). *Uneven Development: Nature, Capital, and the Production of Space*. New York: Blackwell.

Smith, N. (1990). *Uneven Development: Nature, Capital, and the Production of Space* (2nd edn). Oxford: Blackwell.

Smith, N. (1996). "The production of nature." In G. Robertson, M. Mash, L. Tickner, et al. (eds.) *FutureNatural: Nature/Science/Culture*. New York: Routledge, pp. 35–54.

Stone, C. D. (1974). *Should Trees Have Standing? Towards Legal Rights for Natural Objects*. Los Altos, CA: William Kaufmann.

Sullins, T. (2001). *ESA: Endangered Species Act*. Chicago, IL: American Bar Association.

Terborgh, J., J. A. Estes, et al. (1999). "The role of top carnivores in regulating terrestrial ecosystems." In J. Terborgh and M. E. Soulé (eds.) *Continental Conservation: Scientific Foundations for Regional Reserve Networks*. Washington, DC: Island Press, pp. 39–64.

TerraChoice Environmental Marketing Inc. (2007). *The Six Sins of Greenwashing: A Study of Environmental Claims in North American Consumer Markets*. Reading, PA: Author.

Thompson, A. (2006). "Management under anarchy: The international politics of climate change." *Climatic Change* 78(1): 7–29.

Tierney, J. (1990). "Betting on the planet." *The New York Times*, December 2, pp. 52–3, 76–81.

Tiffin, M., M. Motimore, et al. (1994). *More People, Less Erosion*. New York: John Wiley and Sons.

Turner, B. L. II, and S. B. Brush (eds.) (1987). *Comparative Farming Systems*. New York: Guilford Press.

Twain, M. (1981). *Life on the Mississippi*. New York: Bantam Books.

United States Fish and Wildlife Service (2008). *Factsheet: American Alligator: Alligator mississippiensis*. Arlington, VA: Author.

van Mantgem, P. J., N. L. Stephenson, et al. (2009). "Widespread increase of tree mortality rates in the western United States." *Science* 323(5913): 521–4.

Vandermeer, J., and I. Perfecto (2005). *Breakfast of Biodiversity: The Truth about Rainforest Destruction* (2nd edn). Oakland, CA: Food First.

Watts, M. J. (1983). "On the Poverty of Theory: Natural Hazards Research in Context." In K. Hewitt (ed.) *Interpretations of Calamity*. Boston, MA: Allen and Unwin, pp. 231–62.

White, G. F. (1945). *Human Adjustments to Floods: A Geographical Approach to the Flood Problem in the United States*. Chicago, IL: University of Chicago, Dept. of Geography, Research paper no. 29.

White, Jr., L. (1968). "The historical roots of our ecologic crisis." In L. T. White, *Machina Ex Deo: Essays in the Dynamism of Western Culture*. Cambridge, MA: MIT, pp. 57–74.

Whynott, D. (1995). *Giant Bluefin*. New York: Farrar Straus Giroux.

Williams, R. (1976). *Keywords: A Vocabulary of Culture and Society*. New York: Oxford University Press.

Williams, T. (2000). "Living with wolves." *Audubon* 102(6): 50–7.

Wilson, E. O. (1993). *The Diversity of Life*. New York: W.W. Norton.

Wilson, E. O. (2002). *The Future of Life* (1st edn). New York: Alfred A. Knopf.

Wilson, R. S., and J. L. Arvai (2006). "When less is more: How affect influences preferences when comparing low and high-risk options." *Journal of Risk Research* 9(2): 165–78.

World Resources Institute (2007). *Earth Trends Data.* Retrieved October 7, 2009, from http://earthtrends.wri.org/.

Yen, I. H., T. Scherzer, et al. (2007). "Women's perceptions of neighborhood resources and hazards related to diet, physical activity, and smoking: Focus group results from economically distinct neighborhoods in a mid-sized US city." *American Journal of Health Promotion* 22(2): 98–106.

Young, S. P., and E. A. Goldman (1944). *The Wolves of North America, Part I.* London: Constable.

# Index

Figures in *italic*, tables in **bold**